MAPS AND DIAGRAMS

THEIR COMPILATION AND CONSTRUCTION

MAPS AND DIAGRAMS

THEIR COMPILATION AND CONSTRUCTION

F. J. MONKHOUSE

and

H. R. WILKINSON

LONDON

METHUEN & CO LTD

First published 30 October 1952
Second edition, revised and enlarged, 1963

First published as a University Paperback in 1963
Reprinted five times
Third edition, revised and enlarged, 1971
Reprinted 1972, 1973, 1974 and 1976

© *1963 and 1971 F. J. Monkhouse and H. R. Wilkinson*

Printed in Great Britain by
Richard Clay (The Chaucer Press), Ltd
Bungay, Suffolk

ISBN 0 416 07450 2

Distributed in the USA by
HARPER & ROW PUBLISHERS, INC.
BARNES & NOBLE IMPORT DIVISION

Contents

Preface to the First Edition

'Maps are drawn by men and not turned out automatically by machines,' wrote the eminent American geographer, J. K. Wright, in his delightful essay 'Map Makers are Human'.[1] But proficiency in the art of drawing maps is not attained without some systematic training. The discipline necessary to attain this proficiency involves three things: (1) the handling of raw material and data; (2) a critical knowledge of cartographical principles and techniques; and (3) actual drawing practice to attain some degree of manual dexterity and skill in the execution of maps.

Problems concerned with the methodology of the compilation of data and with experiment in the technique of presentation constitute vital aspects of cartography; emphasis on these has perhaps figured less prominently in text-books than, for example, on the history of cartography, on practical surveying, and on map projections. The scope of cartography is so vast that no apology is needed for a selective approach, which omits consideration of these other aspects.

The map is the traditional medium of the geographer, and it is with the training of geographers that this book is primarily concerned. It must not be overlooked, however, that its contents should prove useful to the historian, to the economist and in fact to all who may have reason to handle or produce maps.

The maps and diagrams which a geographer is required to produce in the course of his training and development may be classified into three groups. The first group comprises those maps drawn as a series of formal cartographical exercises, to provide a course complete in itself, but naturally closely linked with the content of a particular geography syllabus, whether at school, training college or university. At Liverpool, for example, all first-year students work through a systematic series of twenty-six exercises, with the object of replacing the extremely varied standards of school cartography by some general level of attainment.

[1] J. K. Wright, *Geographical Review*, vol. 32, p. 527 (New York, 1942).

The second group involves those maps drawn to illustrate the more or less elaborate dissertations and theses which form so important a part of more advanced work in training colleges and universities. Frequently the dissertation comprises some type of regional survey, and the maps produced to illustrate it are of great importance. In fact, in some universities the dissertation presented for the first degree consists basically of a set of original maps, with a brief explanatory text.

The third group comprises maps which are drawn to be reproduced by line-blocks for publication. Many hundreds of books and periodicals containing maps are published every year. All too often the maps they contain are poorly drawn, with inadequate lettering, either over-reduced so that legibility has suffered or under-reduced so that they are crude and empty in appearance, and frequently with heavy obliterating stipples. As V. C. Finch wrote, 'A survey of geographic publications shows all too clearly how ill-adapted many of the maps and diagrammatic devices employed are to the purposes for which they were intended, and how often only a few minor changes would have improved their effectiveness had the author been more familiar with the range of devices and techniques at his disposal.'[1]

The book has been divided into six chapters. The first is devoted to a preliminary discussion of materials and techniques, while the remaining five chapters deal with specific maps and diagrams which fall within the purview and scope of the geographer's interest.

The phrase 'maps and diagrams' has been interpreted in the broadest possible sense, for geographical data lend themselves to many possibilities and varieties of cartographical and diagrammatic treatment. In addition to the accepted concept of a map – a conventionalized depiction of spatial distributions viewed vertically – a wide range of diagrams has been discussed, including graphs, block-diagrams, profiles and landscape sketches. After all, a gradual transition can be traced from a straightforward topographical map, through various degrees of conventionalization and selective emphasis, to a simple diagram.

An attempt has been made to increase the value of this book by the inclusion of references to source material and data for map compilation, and to original articles, both those discussing the principles of some particular cartographical method, and also those illustrating the successful application of the various techniques. Bibliographical references are quoted fully in footnotes or in captions to maps and diagrams.

[1] V. C. Finch, 'Training for Research in Economic Geography', *Annals of the Association of American Geographers*, vol. 34, p. 207 (Lancaster, Pa., 1944).

ACKNOWLEDGEMENTS

This book substantially embodies the cartography courses which have developed during the last five years in the Department of Geography in the University of Liverpool. Many of the exercises discussed have, in fact, been carried out in the departmental studios, and a large proportion of the illustrations in this book is derived from such exercises. In addition, invaluable experience has been gained in the drawing of the numerous maps contained in the several volumes of the Liverpool Studies in Geography, either already published by the University Press, or in the course of preparation.

We therefore owe more than we can adequately express, first to Professor H. C. Darby, and then to Professor Wilfred Smith for constant encouragement, stimulus and criticism, while our other colleagues have throughout contributed suggestions and advice, as well as a number of original maps.

Professor S. H. Beaver was kind enough to read through the proofs, and to suggest several amendments which we were glad to incorporate.

The most valuable part of a book of this nature is the maps and diagrams which illustrate it; these have been drawn by Mr A. G. Hodgkiss, with the assistance of Mr D. H. Birch.

Liverpool   F. J. M.
June, 1951   H. R. W.

Preface to the Second Edition

During the decade that has elapsed since the original publication of *Maps and Diagrams*, while the main principles and tenets have held good the science and art of cartography have progressed, and certain modifications and enlargements to the original text have now become necessary. The authors have received a great deal of helpful correspondence, suggesting various amendments, criticisms and elaborations, many of which have been incorporated in this revised text. They would like to take this opportunity of thanking collectively these many correspondents, whom it would be invidious to list by name in case any of the more informal, yet none the less useful, contacts should pass unacknowledged. The sources of additional maps included in this edition have been duly acknowledged in the captions. For the drawing of these maps, and for helpful advice generally, the authors would like to thank Mr A. Carson Clark, Senior Cartographer in the Department of Geography in the University of Southampton, and Mr R. R. Dean, Senior Technician in the Department of Geography in the University of Hull. Miss G. A. Evans and Miss J. Bailey have shared the laborious task of typing the new material.

During recent years statistical methods and procedures have assumed an important role in the practice of geography. Although these aspects were by no means neglected in the first edition, their implications have attained such proportions that the authors decided it was desirable to include a separate Appendix devoted to a comprehensive review of statistical methods in geography. This has been kindly undertaken by Mr R. G. Barry, Lecturer in Geography in the University of Southampton; the draft was critically reviewed by Dr S. Gregory of the University of Liverpool. Other new or substantially revised sections in the text include those devoted to morphometry in general (with particular reference to slope development), morphological mapping, maps of agricultural distributions, maps of the incidence of disease, the application of linear analyses to central place theory, the study of the distribution of settlement and population, and urban population mapping. In Chapter 1 details of certain new materials and instruments now widely used in drawing have been included. Of necessity the number of footnotes has had to be appreciably increased, in order to afford full reference both to the many new methods and to the successful application of the well-tried ones, most of which have appeared in recent periodical publications.

F. J. M.
H. R. W.

December 1962

Preface to the Third Edition

It has once again become necessary to produce a revised edition of *Maps and Diagrams*, as a result of changes and developments in cartography, which in fact seem to have accelerated during the last decade. Symptomatic of this has been the formation of the International Cartographic Association in 1959 (which now has a number of commissions and working groups, including Automation in Cartography and Thematic Cartography); of the British Cartographic Society in 1964 (with its attractively produced Journal, a forum of dissemination and discussion); and of the Society of University Cartographers in 1964.

In his kindly review of the second edition of *Maps and Diagrams*, I. A. G. Kinniburgh said: 'One would like to see, in a future rewrite of this book, the scope widened and the nettle of modern trends, in the advanced theoretical and practical aspects of the subject, firmly grasped.' The book was, of course, originally intended to cover the requirements of University cartography, to enable students to produce maps and diagrams to assist their investigations both at undergraduate and postgraduate level, and to help researchers to illustrate their articles and books. This must remain its major objective, but at the same time the authors appreciate that an increasing number of students hope to become professional cartographer-geographers, and that maps are widely used in the planning departments of Local Authorities, in government departments and in the offices of architects and civil engineers. They seek, therefore, in the limited additional space available, to give students some idea of the novel and exciting developments in tools, materials, techniques and methods (notably in the realms of quantification, mechanization and automation), which on the one hand immeasurably speed up the production of maps, and on the other hand enable them to be based on vast masses of information which could hardly be dealt with manually. The growth, amounting to an explosion, in data of all kinds available for map-making emphasizes the increasing need for a discerning use of statistical techniques. Inevitably the dependence on the computer for ordering and sifting data must grow, as must the degree of sophistication in the techniques employed.

But we must still firmly bear in mind the quotation by the late J. K. Wright which appears at the head of the Preface to the First Edition.

ACKNOWLEDGEMENTS

Ideas for the cover design are based on an illustration in the introduction to J. R. Passonneau and R. S. Durman, *Urban Atlas: Twenty American Cities* (M.I.T., Mass., 1966).

F. J. M.
H. R. W.

December 1969

Illustrations

I

Materials and Techniques

It is essential for a student who desires to acquire a reasonable cartographical technique to provide himself with an efficient set of drawing instruments, and to learn their use by systematic practice.[1]

Pencils

Much preliminary work is necessarily carried out in pencil. A hexagonal pencil is better than a circular one, as it prevents rolling or slipping. Leads are made of graphite compounds of varying degrees of hardness; thus 8H or 9H leads are of metallic hardness and will scratch or cut the surface of the paper, while a 6B lead is extremely soft. For most drafting work, either an HB or an H pencil is adequate. A 6B or a solid carbon pencil may occasionally be used for such operations as hill-shading, but such a map should be protected from smudging with an overlay of tracing- or tissue-paper. Mention should be made of the extremely useful 'Clutch' pencil, which is available with a variety of readily interchangeable leads, of different colours and degrees of hardness, and with its own self-sharpener.[2]

A pencil point must be kept really sharp. A mechanical sharpener is useless, and the lead should be sharpened with a keen knife and brought to a needle-point by rubbing it on a sand-paper block. For ruling lines against a straight-edge, sharpen the pencil to a chisel edge.

[1] Much useful information is contained in *Graphic Arts Technicians' Handbook* (numerous reprints, published by Hunter-Penrose-Littlejohn Ltd., 7 Spa Road, Bermondsey, London, E.16.

[2] Made by the Eagle Pencil Company, whose parent establishment is in Danbury, Connecticut, U.S.A., U.K. establishment Ashley Road, London, N.17.

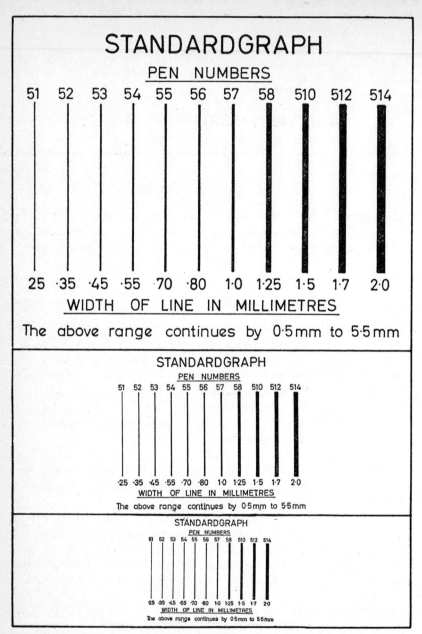

Figure I. STANDARD LINE THICKNESSES

Eleven standard thicknesses of line obtained by using eleven *Standardgraph* grades, are shown: *top:* full size; *centre:* reduced to one-half; *bottom:* reduced to one-third.

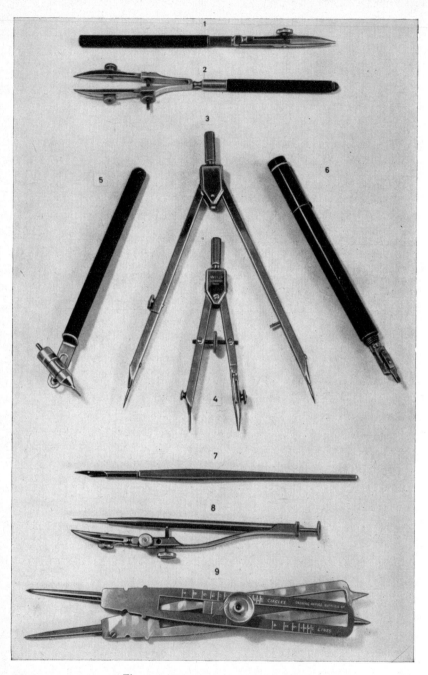

Figure 2. DRAWING INSTRUMENTS

Pens

Many types of pen of specialized design for free-hand techniques are on the market; most of these are chiefly of value to professional draughtsmen. A broad medium-sized stiff steel nib will be useful to the student. A mapping-pen can be used for fine line-work, but it is essential that perfectly even pressure should be applied on up-and-down strokes, for one of the main difficulties in its use is the varying line thicknesses produced. It can be employed for rivers, where a tapering line is required.

The main problem of the ordinary cartographer is to obtain a uniform thickness of line. One solution is to use a pen such as a *Standardgraph*. This consists of a reservoir, hollow tubular nibs with a wide variety of diameters, and a spring-loaded wire-plunger; the latter controls the flow of ink, so ensuring a line of uniform thickness and density. The pen fits a plastic or metal holder, as illustrated on Fig. 2 (no. 5); another type is provided with a rest to the pen in a level position when not in use. The various line thicknesses which can be drawn, together with their appearance upon half and third reductions, are shown in Fig. 1. The nib must be kept absolutely clean, and every time the ink in the reservoir is exhausted it should be dipped in water, the plunger carefully pulled up and down, and the nib wiped off with a piece of fluffless material. Other types of pen are available, mostly originating from America or Germany, notably the *Graphos* (no. 6), *Leroy*, *Speedball*, *Barch-Payzant*, *Rapidograph* and *Wrico*. These pens are available with various nib-sizes, thicknesses and shapes, including fine ones for free-hand work, different grades of ruling nib, tubular nibs for stencil lettering, slant nibs for square-ended lines, and chisel points. These instruments will enable draughtsmen to carry out virtually every kind of line-work and lettering with no more trouble than changing the nib (though admittedly this is apt to be rather messy). Some types are equipped with cartridge reservoirs of ink in black and other colours, which give a smooth flow and avoid the irritating and time-consuming troubles resulting from ink drying or coagulating on the nib or in the feed-duct. Mention should be made of the Barch–Payzant pen, which has a beak-like nib. With its adjustable ink-flow it is valuable for drawing lines of constant thickness.

Ruling-pens. The ruling pen consists basically of a bone or plastic handle, with two high-grade steel blades which taper to a point;

these blades are adjustable, in order to draw lines of various thicknesses, by means of a screw set through their shoulder (Fig. 2). These blades must be kept clean and sharp on an oilstone, otherwise really clean line-work cannot be attained. Avoid touching the inner sides of the blades. Any ink on the outside of the blades will result in disastrous blurring; the ink is inserted between them by a quill, or by a brush, or from a tube (see p. 8). The student should experiment with various settings of the screw and the resulting line-thicknesses, and it is always advisable to rule a few trial lines before starting on the map itself. After use, the surplus ink should be cleaned from the pen by passing a very thin wedge of soft linen or muslin between its points; this should never be forced, otherwise splaying of the blades may result, and the drawing of accurate lines will be found impossible.

Ruling-pens which rotate on their handles, and so can be used to draw curving lines, are available, but are not of any great value to the cartographer. Double ruling-pens, which consist of two ruling-pens fastened in a single holder, adjustable so that double lines can be ruled of varying thickness and distance apart, are sometimes of convenience for inserting railway-lines, roads and canals (Fig. 3).

Ruling-pens need considerable practice. They should be held lightly in a vertical position against a straight-edge, preferably of steel or plastic and bevelled to prevent the ink-lines from running; this edge must not be so high that the line wavers. The pen must not be so full that the line is blobbed, nor so dry that fine double lines are produced instead of a single line of correct thickness. However, when the technique is mastered, it enables clean-cut lines to be ruled, and it is of great value for line-shading (see p. 50). Where a thick marginal line is required, it is better to rule two parallel lines and then fill in the intervening space with a fine brush.

Quill-pens may be used, with practice, for lettering. 'No one who has learned to cut a quill will ever again be content with a steel pen. . . .'[1], though admittedly this applies to lettering as an art-form on diplomas, certificates and the like, rather than to the more utilitarian work on maps and diagrams. Either a goose or turkey quill is used; the former are more supple and wear longer, while the latter are much stiffer. For very fine work a duck or crow quill may be used, while for large work, such as wall-maps, a reed or cane is employed. The quill has to be cut cleanly with a very sharp knife to the desired shape of nib;

[1] G. Hewitt, *Lettering*, p. 251 (London, n.d.).

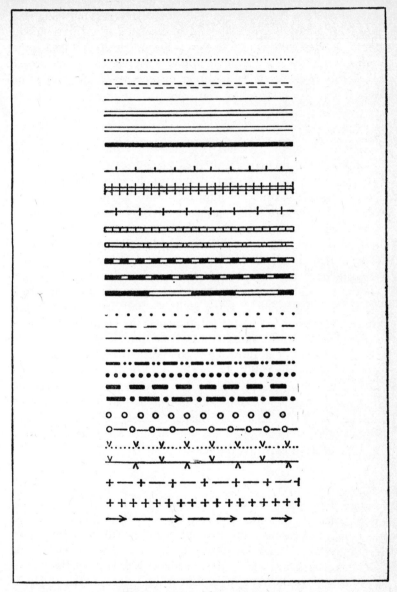

Figure 3. VARIETIES OF LINE

Various distinctive lines are used for categories of footpaths, roads, railways, waterways, political boundaries, electricity grid cables and aqueducts.

much practice and experiment is necessary to do this successfully and to find a shape which suits the individual draughtsman. Ink should be put into the quill with a filler. The great advantage of quill-pens is that lettering is speedy; only a single stroke is needed, for gradations of thickness are easily produced by different pressures, and 'building-up' is unnecessary (see p. 59).

Other Instruments

A variety of other drawing and geometrical instruments will at times be found useful. Minimum requirements comprise a pair of compasses (preferably fitted with both pencil and ruling-pen), dividers, a large celluloid set-square with bevel-edges for ink-work, a protractor, a hard-wood T-square for laying out margins, a steel straight-edge, hard-wood inch and centimetre scales, and parallel rulers (or more simply two set-squares) for setting out parallel lines or for line-shading. Various types of compasses are available. For large circles a model with a hinged arm or a lengthening bar, or even a 'beam compass' (a horizontal arm with a point at one end, a ruling-pen at the other) may be used. A 'spring-bow compass' (Fig. 2, no. 4) is required for small circles, and a 'pump compass' (Fig. 2, no. 8) for very minute ones; the needle is inserted on the centre point and by pressing down the 'pump' the arm with the ink-pen is pivoted. The drawing of curved lines requires not only a steady hand, but generally the use of some guide. 'French curves' consist of plastic templates, each with a number of curves of varying radii; a large number of templates are needed to meet all possible requirements. The 'Cemtex' variable curve consists of a bendable lead core between steel ribbons, with an outer plastic cover; the strip can be bent to the desired curve on the paper.

Scribers

In many large cartographic establishments, both of government agencies and private firms, scribing has virtually replaced drawing with a pen; increasingly the method is being taught in university cartography laboratories, especially in America. The student should know of its principles and practice, particularly if he is contemplating a career as a professional cartographer. A base-sheet of glass or plastic (such as *Vinylite*, *Mylar*, *Astrafoil*, *Astralon*) is smoothly coated with an opaque medium (for example, *Astrascribe*, *Scribecoat*), which is usually yellow or green in colour in order to reduce eye-strain. Into this coating is cut the detail to be shown on the map, using a 'scriber' or 'graver'; various types of these are available. Some are free-hand, pen-type scribers for fine lines; others are attached to tripod ball-castor carriages which hold the point perpendicular to the work, either rigidly or on swivels. Special gravers can produce evenly spaced dots and dashes; another type has a chisel-edge for broad lines, another is a double-line scriber for roads and even a triple-line one for dual carriageways. One type of 'turret-graver' has a number of needles of different sizes and points, each of which can be locked into position as required. Some types have wide-field magnifiers attached for viewing the work area. The actual points of the scribers may be of tungsten carbide or of sapphire; while the latter are more expensive, it is claimed that they never require to be resharpened. Scribing is done either as a negative, so that when laid on a sensitized metal plate a positive image is produced by exposure to light (which by offset-litho (see p. 72) will produce correct-reading copies), or as a positive. The operator follows a guide-image which has been photographically transferred on to the surface of the medium with which the base-sheet is coated. Should he make a mistake, the area can be painted out with a quick-drying opaque liquid, which when dry can be rescribed. Names and symbols can be stuck on to the plate in 'windows'.

The Drawing-table

Drawing-paper can be rested upon a hard-surfaced drawing-board, tilted at a slight angle. More convenient, indeed essential for good line-work and for copying, is a drawing-table, the top of which consists of heavy glass, and can be illuminated from below. This will enable maps to be copied easily on tracing-paper, and it is possible, given a sufficiently brilliant illumination, to trace on to apparently opaque drawing-paper. Lead weights, covered with baize, are better than drawing-pins to keep the tracing from slipping over the original or first draft, or the tracing may be stuck down by strips of drafting tape over each corner.

INKS AND COLOURS

Indian Ink

Indian ink consists of fine lamp-black suspended in a liquid medium. It is deep black, waterproof, and photographs well, but it dries very rapidly and the cartographer must keep his line-work moving. All cartographic work, whether for the exercise book, the dissertation, or for line-blocks, should be finished off in Indian ink, because of its clean, clear-cut effect. When exposed to the air the ink in a bottle coagulates and dries out rapidly, and unless it is corked when not in use, the ink soon deteriorates; it is economical to fill a small bottle periodically from a larger one. It is preferable for *Standardgraph* and ruling-pens to use ink supplied in plastic tubes, which on being squeezed at the base will exude from the nozzle the desired amount of ink of correct fluidity. It should be noted that plastic drafting film usually requires a special ink, such as *Pelikan K* or *TN*.

Water-colours

Water-colours can help, if used with discretion, to clarify maps for the cartography notebook and for the dissertation. Ordinary water-colours, made in tubes and cakes, or aniline powder dyes, may be employed. Several sable brushes, broad ones for washes and a few finer ones for detailed work, are needed. One brush should be kept for painting with Indian ink, where considerable areas have to be blackened, or double lines filled in. Another may be reserved for applying white paint to obliterate unnecessary lines on drawings for the block-maker, and a third for tinting with coloured waterproof inks. It is necessary to emphasize that brushes and paints should be kept scrupulously clean by washing them thoroughly in warm water after use.

Polymer Colours[1]

A new colour medium has been developed in recent years, which claims to combine the advantages of oils and water-colours. It is, in fact, an emulsion-based plastic, which can be used undiluted to give solid brilliant colours, but it can be water-diluted for the application of washes. It will adhere to almost any working surface, does not flake or crack, and is available in thirty different colours.

Other Colours

Crayons and coloured pencils are occasionally useful to colour exercises, but the results are generally cruder than if water-colours are used. They are exceedingly useful in colouring different land-use categories when working in the field with a base-map. For example, for the new

[1] Produced by Reeves and Sons, Ltd., Enfield, Middlesex, England.

1 : 25,000 Land Use Survey (see p. 267), a series of 'Derwent' crayons is recommended.[1] Coloured waterproof inks are helpful for line-work (as, for example, blue for rivers and red for railway-lines). Where the areas concerned are small these inks may be applied with a brush, but it is difficult to obtain a smooth wash with ink over a large surface.

Mechanical Colour Application

Where a considerable area of colour-tinting is to be applied, it may be convenient to use commercially available colour-sheets. For example, *Drycolour* and *Letrafilm* are produced by the Letraset Company.[2] The former comprises a range of forty different shades of translucent dry-coated ink, which can be readily applied to a drawing by the dry-transfer process (i.e. by simple pressure). The latter is a high-quality adhesive colour film, available in a range of seventy-seven colours in matt finish.

DRAWING MEDIA

Notebooks and Folders

Where a student is working through a systematic series of exercises, it is obviously desirable that these should be arranged and presented in some permanent form. There are three alternatives: a bound note-book, a loose-leaf file or a large envelope or folder. Each has its advantages.

A *notebook* should have stiff covers, well guarded at the binding to allow drawings to be pasted in, and interleaved with graph- and writing-paper. Written exercises, calculations, notes on methods, and various graphs and diagrams can be written or drawn straight in. Maps on drawing-paper may be mounted by means of *cow gum* and with or without one or more guarded folds. A carefully compiled notebook, containing a series of exercises executed week by week, will constitute a most useful and attractive reference book of carto-graphical methods.

A *loose-leaf file* should have good stout covers and some efficient binding system; it allows the convenient storage of work of different dimensions on varying media, and is flexible in arrangement. Where the course consists of a few major exercises, a large stout envelope is

---

[1] These colour pencils (available in seventy-two shades) are made by the Cumberland Pencil Co. of Keswick (London office, 134 Old Street, London, E.C.1).

[2] Letraset Ltd., St George's House, 195/203 Waterloo Road, London, S.E.1.

probably most useful. Alternatively, the maps may be kept in a folder consisting of two sheets of cardboard hinged together by a strip of adhesive linen, with the open ends secured by tapes.

Drawing- and Tracing-papers[1]

Drawing-paper should be used for all finished map-work, except where it is to be reproduced (when tracing-paper is adequate). Ordinary cartridge-paper, or a matt-surfaced machine-pressed paper, or thin card such as Bristol board, may be used for particular jobs. The cheaper media will not stand much erasure, nor will they take smooth colour-washes.

Tracing-paper is made in a wide range of qualities; a good paper is tough, of a smooth matt surface with no gloss, and highly translucent. An ideal tracing-paper for most map-work is an all rag-base paper such as *Gateway*. The grade is denoted by its weight; thus a 50-gram paper is light and thin, while a 150-gram paper is heavy, almost of a parchment quality. All weights of this particular paper are highly translucent. Maps for the block-maker can be drawn directly on to tracing-paper in Indian ink; copying by tracing, and any alterations and obliterations, can be more easily done than on card. Tracing-papers, however, expand very appreciably, particularly on damp days, and especial care must be taken with the exact register where drawings are intended for two-colour blocks. Tests show that a change of 40 per cent relative humidity may produce a distortion exceeding 2 per cent in either direction, which with a sheet a foot square may amount to a quarter of an inch. These papers tear easily, and tracings should have their edges guarded by adhesive tape of some kind; a small machine may be used to bind the edges with special linen tape.

Tracing-paper is also useful for overlays. For example, a tracing of geological outcrops may be superimposed on a topographical map to emphasize some significant relationship (see p. 187).

Various other media, such as vellum, cellophane, tracing-cloth, and drawing-paper mounted on linen or muslin, are used by professional draughtsmen for certain specialized work. One superb medium is *Syntosil*,[2] a material made from synthetic fibres which looks and handles like a high-quality paper, but has a high dry- and wet-tearing strength, is water-resistant, and has excellent qualities of ink adherence. Another

[1] For those requiring a detailed exposition, see F. H. Norris, *The Nature of Paper and Board* (London, 1966).
[2] Manufactured by the Zürich Paper Mill, Sihl, Switzerland.

is *Silbond*, made from polyester fibre;[1] it too is water-resistant, dirt repellent, washable, non-frayable, and takes smoothly inks of all kinds.

Plastic Media. Various plastic materials are now available for carto-graphical work. They have a high degree of dimensional stability, and are especially useful for colour overlays to ensure a high fidelity of register. Such a medium is *Permatrace*, made from the same raw materials as the plastic film known as *Melinex*, an I.C.I product. This is highly translucent, and has great stability over a wide range of temper-ature. Tests have revealed that a sheet of Permatrace 40 inches in length, subjected to a sudden change of temperature of 20° F, would suffer a dimensional distortion of less than the thickness of a fine pencil line; the makers, in fact, claim a coefficient of linear expansion of only 0·00027 mm. per 1°C. change. This material has a fine matt surface, and is equally suitable for pencil, ink and colour. It is particu-larly helpful where large areas of ink have to be blocked in, since it will remain completely flat with no suggestion of cockling, which invariably happens with tracing-paper. It can easily be cleaned, using a damp piece of muslin or similar material, which will wipe off either pencil or ink quite easily. Several varieties of plastic sheet are produced in America, such as *Vinylite, Copyrite, Cronaflex* and *Dyrite*. One variety consists of a sheet of plastic laminated with paper on either side. A very pleasant material to use is *Ilford Supermattex*, a drafting film with a polyester base; it is untearable, is sufficiently translucent for tracing, has a fine matt finish, takes both pencil and ink strongly and cleanly, and lines drawn on it can be easily erased without affecting the surface.

Graph-papers

It has been recommended that the cartography notebook be inter-leaved with arithmetic graph-paper, preferably ruled in inches and tenths of inches. This will enable a wide range of graphs to be drawn straight into the notebook. Sometimes a larger sheet may have to be mounted in the notebook with one, two or more folds.

Semi-logarithmic graph-paper,[2] which combines a horizontal arith-metic scale-ruling with a vertical logarithmic scale-ruling, may be needed where it is desired to plot rates of change (see pp. 297, 378).

[1] Manufactured in West Germany by *Fraserprodukte Gmbh*.
[2] D. W. Griffin, 'Semi-logarithmic Graphs in Geography', *Professional Geographer*, vol. 15, pp. 19–23 (Lawrence, Kansas, 1963); and T. Burke, 'Semi-logarithmic Graphs in Geography: a Pertinent Addendum', *Professional Geographer*, vol. 16, pp. 19–21 (Lawrence, Kansas, 1964).

This paper is available in a series of cycles, and can be scaled as desired in powers of 10. Thus with five-cycle paper, the lowest cycle may be 1–10, the second 10–100, the third 100–1,000, the fourth 1,000–10,000 and the fifth 10,000–100,000. Should larger numbers be needed, the lowest cycle might begin at 10,000 or 100,000. Sheets of two-way logarithmic paper (i.e. with both horizontal and vertical logarithmic scales) may occasionally be required for frequency graphs.

Circular graph-paper may prove very useful in drawing some projections, for wind-roses, and for diagrams showing seasonal distribution of climate and of human activity. *Percentage circular graph-paper*, in which the circle is divided into one hundred segments, facilitates accurate and speedy plotting of divided circles (see pp. 303, 304). *Triangular graph-paper* may be used when three variables have to be plotted, as for example, three related aspects of climate or relief (see p. 163 and Fig. 52). Another type of graph-paper, *isometric*, enables three-dimensional figures, such as block-diagrams (Fig. 60), to be constructed without recourse to angular measurement.

Probability graph-paper is invaluable for plotting graphs of normal probable distribution (Fig. 230), and can be used for any series of frequencies, such as slope data in geomorphology, temperature and rainfall statistics (see p. 236), and population growth. Arithmetic probability graph-paper has a vertical (ordinate) arithmetic scale, and as a horizontal (abscissa) scale the function, in terms of percentage values, of the 'normal' (Gaussian) distribution, from 0·01 to 99·99. The data being used are converted into percentages, and tabulated in cumulative form. The adjustment of spacing on the ordinate of the graph paper will cause any normal probability distribution to yield a straight sloping line of points. After plotting these values, a straight continuous line can be drawn through them. This will assist, for example, the forecasting of future growth of population, and it can also reveal significant deviations from normality. On occasions, should *rates* of change be involved, logarithmic probability graph-paper may be required, using a log-scale for the ordinate instead of arithmetic.[1]

[1] The following are useful: (i) F. E. Croxton and D. J. Cowden, *Applied General Statistics*, pp. 458–61 (New York and London, 1939, several editions), dealing specifically with population statistics; (ii) H. Landsberg, *Physical Climatology*, pp. 76–80 (Dubois, Penn., 1960), dealing with climatological statistics; and (iii) A. N. Strahler, 'Statistical Analysis in Geomorphic Research', *Journal of Geology*, vol. 62, pp. 7–9 (Chicago, 1954). A variety of graph-papers is illustrated in *The Chartwell List of Graph Data Sheets*, published by W. Heffer and Sons (Cambridge).

MAP COMPILATION[1]

Map Design

It would be quite wrong to suggest that map-makers in the past were indifferent to the design of their products; indeed, some of the Elizabethan county maps were among the most aesthetically attractive. But in recent years increasing attention has been paid to principles of map design, balance and layout, particularly in terms of their visual effectiveness; to 'consumer demand', i.e. what the potential user of the map desires; and to the use of sophisticated psychological investigations to determine a particular design for a specific type of user.[2] A map can be regarded essentially as an integrated assemblage or synthesis of four categories of information: points, lines, areas and names, which are presented in terms of different shapes, characters, patterns, symbols, sizes, thicknesses, forms and hues. But these have to be considered not only as individuals but also in terms of their inter-relationships, so as to give the maximum overall clarity, legibility and visual impact. One important concept of design involves the presentation of information as a series of 'visual planes', what has been called 'the depth-cue approach', so as to create for the map-user a clear separation of the various distributions to be shown on the map. This is important, because a whole map is visible at once, a fact which can make successful map-interpretation so difficult. An example of different visual planes is the tinting of the sea in blue as a background plane, against which the land and its detail are clearly distinguishable.

General Features

Whether a map or diagram is drawn by a student as an exercise, or by an author to illustrate his monograph, it is essential in most cases to compile a preliminary draft. The cartographer has to consider the aim or purpose of his map, the source material from which it is to be constructed, and the most striking and effective method to be used. He

[1] See A. G. Hodgkiss, *Maps for Books and Theses* (Newton Abbot, 1970).

[2] Among the many references may be cited: (i) A. H. Robinson, *The Look of Maps. An Examination of Cartographic Design* (Madison, Wisc., 1952), which has been described as . . . 'a worthy landmark in this difficult quest for good design'; (ii) A. H. Robinson and R. D. Sale, 'Cartographic Design', chapter 11, *Elements of Cartography* (3rd edition, New York, 1969); M. Wood, 'Visual Perception and Map Design', *Cartographic Journal*, vol. 5, no. 1, pp. 54–64 (1968), an extremely lucid analysis of this kind of investigation, with a bibliography of forty-three items; and R. J. Ferens, 'The Design of Page-size Maps and Illustrations', *Surveying and Mapping*, vol. 28, pp. 447–55 (Washington, D.C., 1968).

selects carefully his base-outline, the scale, and the size of the finished map or of the desired reduction, for on this depends the degree of simplification and generalization of the data presented. Indeed, this problem of generalization is one of the most difficult features of map compilation.[1] It often happens that a map, clear and attractive in its original form, is disappointing upon reduction; it may be under-reduced, so that it is crude and empty-looking, or (more frequently) over-reduced so that legibility has suffered. As V. C. Finch wrote, 'The editorial staffs of the geographical periodicals know to their sorrow how many good-looking manuscript maps and diagrams present almost insoluble problems in printing owing to failure of the authors to grasp the elements of scale and proportion in drawing for reproduction.'[2] While it is possible to standardize the size of lettering, line-thickness and density of shading suitable for any degree of reduction, it is often necessary to make an actual trial. A photographic reduction can be made, or when a large series of maps is contemplated it may even be well worth having a trial block made. It should be noted that the amount of reduction is expressed in linear terms, not areal, i.e. an original map of 10 by 5 inches will, for a half reduction, appear as 5 by 2·5 inches. The compiler must be careful, when concerned with an awkwardly shaped map, to mark the correct controlling dimension; thus an original map, 16 by 8·2 inches, marked for a reduction width of 4·1 inches, would be too long for a page which measures 7 by 4·1 inches. In this case, the length should be marked as 7 inches, even though the resulting block will be slightly narrow for the page. In general practice, a block should be made less than the actual page-size, in order to afford room for title and caption.[3]

The correct balance of the map is sometimes difficult to attain, especially when an awkwardly shaped or irregular area is involved. The most convenient positions are assigned to the key-panel, the scale-line and the cartouche (see p. 23). The compiler chooses in the light of his experience the correct line-thicknesses, stipples or line-shadings and symbols, and the character and size of the lettering. Care has to be

[1] G. F. Jenks, 'Generalization in Statistical Mapping', *Annals of the Association of American Geographers*, vol. 53, pp. 15–26 (Lancaster, Pa., 1963); and O. M. Miller and R. J. Voskuil, 'Thematic Map Generalization', *Geographical Review*, vol. 54, pp. 13–19 (New York, 1964).

[2] V. C. Finch, 'Training for Research in Economic Geography', *Annals of the Association of American Geographers*, vol. 34, p. 207 (Lancaster, Pa., 1944).

[3] An interesting general article on this subject by A. B. Clough, 'The Preparation of Maps and Illustrations for Geographical Articles and Theses', *Scottish Geographical Magazine*, vol. 50, pp. 77–82 (Edinburgh, 1934).

taken not to overload the map with detail which might obscure the main theme of the map.

As a rule, a first pencil draft should be carefully prepared, either to be inked in or to serve as a precise copy from which a finished tracing may be made. When a pencil draft is to be inked-in, rub the map lightly with a very soft 'soapy' eraser (such as '*Artgum*'), or with a powdered erasing medium, so that the pencil lines, while still visible, become as faint as possible, and dust the paper carefully. Indian ink should never be applied over heavy pencil lines, as the stability and permanence of the inked lines is thereby impaired. It may be mentioned that when the map is being traced from a direct source, there is no gain in preparing a preliminary pencil tracing. This not only doubles the time taken, but increases the possibility of minor errors. Instead, using a lighted tracing-table, draw the map straight away in ink; the only pencil lines needed will be guide-lines for the lettering.

Base-maps

A base-map is an outline map used for plotting information. It may consist of the coastline and frontier, or major or minor administrative divisions, contours, field-boundaries or natural drainage patterns. These outlines may be extracted from topographical maps or from atlases, and occasionally they are provided by national Institutes of Geography and other organizations.[1]

Duplicated outlines can be issued to a class as a basis for a specific exercise, and they can be used by a research worker who has to plot a number of distributions on the same base.

Map-Checking

All maps should be minutely checked upon completion. The compiler should where possible secure the help of one or more critics to whom the map is novel, for all too often he may miss the most obvious errors because of his familiarity with the map. The checking should be systematic; spelling, line-work, shading, key, scale-line, all must be checked in turn. The caption should be checked at the same time for correct title, source or acknowledgement, and explanatory comments.

[1] For example, the *Institut National de la Statistique et des Etudes Economiques* (*Ministère de l'Economie National, Paris*), published a base-map on a scale of 1 : 600,000, entitled *Régions Géographiques d'Economie National*. The map was specially prepared for use as a base-map for statistical plotting.

Key-maps

When a series of maps has been completed, either for a dissertation or a published monograph, the maps should be carefully numbered in order of first reference in the text, and a list of their titles included in the table of contents. A key-map, placed at either the beginning or end of the text, is often useful to locate the various maps. Thus a series of a dozen large-scale coast maps may be located on a small-scale map of the whole coast, each map being shown by a rectangle drawn to scale and numbered with the consecutive figure number. Similarly, a series of scattered village plans can be located on a map of the whole county. This is particularly helpful to the general reader when the monograph is of a detailed nature, and widely dispersed regional examples are used.

SCALES AND SCALE-LINES

Definitions

Through usage the word 'scale' has come to be employed in two distinct senses. In the first place, it denotes the relationship which the distance between any two points on the map bears to the corresponding distance on the ground, expressed either in words, as 'one inch to one mile', or as a representative fraction (R.F.), as 1/63,360 (see p. 18).

Used in the second sense, the word scale denotes a line-scale which enables distances on the map to be directly measured and read off in terms of distances on the ground. All maps should bear a line-scale.

When compiling a map, the first essential is to determine the scale to be used. It is obviously necessary to take into account the extent of ground to be depicted and the available size of the paper. The amount of detail which can be included is clearly a function of the scale, in that a large-scale map will show a smaller area in greater detail than a small-scale map. Thus the lower map on Fig. 5 is an enlargement of one portion of the upper Fig. 4, in an effort to clarify the complexity

---

Figures 4, 5. ENLARGEMENT OF SCALE

Based on a contemporary map, *The German Confederated States* (London, 1839).

    The complex portion of the upper map demarcated by a pecked line is shown on the lower map on a scale four times as great. This enlargement enables some state-names to be inserted for reference.

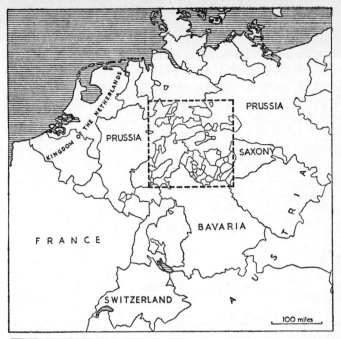

of the intricate pattern of state boundaries; the scale of the lower map is four times greater than that of the upper.

A further complication is introduced if the original drawing is to be reduced, either photographically for a dissertation, or to make a line-block. In either case, the scale should be indicated on the face of the map as a line-scale, and not expressed in words or as an R.F.; the last two will obviously be invalid upon reduction. The line-scale is of course automatically reduced in the same proportion as the map.

Useful R.F. Conversions

R.F. 1/to	Miles to 1 inch	Inches to 1 mile	Km to 1 cm	Cm to 1 km
Million	15·782	0·0634	10·0	0·1
633,600	10·0	0·1	6·336	0·1578
500,000	7·891	0·127	5·0	0·2
253,440	4·0	0·25	2·534	0·395
250,000	3·945	0·245	2·5	0·4
126,720	2·0	0·5	1·267	0·789
100,000	1·578	0·6336	1·0	1·0
63,360	1·0	1·0	0·6336	1·578
50,000	0·789	1·267	0·5	2·0
25,000	0·395	2·534	0·25	4·0
10,560	0·167	6·0	0·1056	9·468
10,000	0·1578	6·336	0·1	10·0
2,500	0·0395	25·34	0·025	40·0
1,250	0·01978	50·69	0·0125	80·0

There are two types of line-scale. The first is the long accurately divided line, used for large maps, which enables direct measurements with dividers to be made. The second is the short, inconspicuous line drawn on a small map which gives merely a general idea of actual distances involved, but is unlikely to be used for more than a casual measurement by eye.

Long Line-scales

The calculation and construction of a long line-scale, which should as a rule be about 6 inches in length, forms a useful cartographical exercise, and helps to drive home the principles involved. From experience, it is incredible how confused many students become when faced with an R.F. which they are requested to convert into a line-scale.

Suppose it is necessary to draw a line-scale, indicating miles, for a continental map of R.F. 1/50,000.

On the continental map, 1 *mile is represented by* $\dfrac{63,360}{50,000} = 1 \cdot 2672$ *inches*

∴ 5 *miles is represented by a line of length* $1 \cdot 2672 \times 5 = 6 \cdot 34$ *inches.*

Draw this line accurately, then divide it into five, by setting-off similar triangles. This is done by drawing any line five units in length, at a reasonable angle from the left end of the line-scale. Join the end of this line to the right end of the line-scale and rule lines parallel to this from the unit intersections on the upper line. This will give the divisions on the line-scale itself. It has been calculated that 1 mile on the ground

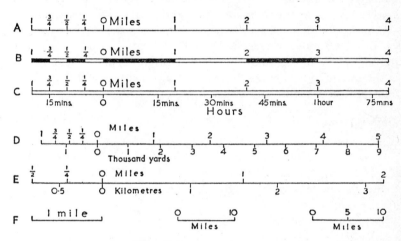

Figure 6. LINE-SCALES

A. Open divided line-scale; **B.** filled line-scale; **C.** time and progress line-scale (the time-scale represents a uniform speed of 3 miles per hour); **D., E.** double linear-unit line-scales; **F.** short line-scales.

would be represented by $1 \cdot 2672$ inches, and this distance can be stepped off five times with dividers. It is clear, however, that when a single line is drawn there is one possible error of measurement, which by division into five is distributed; when the five small lengths are stepped off, the errors of measurement are cumulative and may be quite appreciable. The division on the extreme left of the line can be further divided into four or eight to show quarter-miles or furlongs; this is known as an open divided line-scale. If the whole line is divided, it is known as a fully divided line-scale, but this is not usually necessary. When numbering the scale, zero should fall one interval along from the left, so that the left-hand end of the line will be numbered 1, and the divisions to the right of zero will be numbered 1, 2, 3 and 4. When measuring a length with dividers, place one leg on the nearest whole number on the right of zero, so that the fraction can be read off directly to the left of zero (Fig. 6, example A).

The line-scale should be drawn as a single clean line, marked with neat ticks of uniform height. Double lines, with alternate spaces filled black, are commonly used for effect, but should be avoided if exact measurements are desired, since the black sections are actually longer than the white by the thickness of the bounding line at either end (example B). A second scale can be placed on the underside of the line if necessary; it might be a uniform time and progress scale (example C), or a scale of different linear units, such as kilometres or thousands of yards below a mile scale (examples D, E). With the introduction of *S.I.* (metric) scales of measurement, scale-lines should, of course, be given in the metric form, though during the transition period it is desirable to include both metric and Imperial.

Diagonal Scales

It is occasionally necessary, particularly on large-scale plans, to be able to subdivide the first unit length with considerable precision, and a diagonal scale may then be used (Fig. 7). Suppose that it is necessary to draw a line-scale of R.F. 1/633,600 (i.e. 10 miles to 1 inch) to show

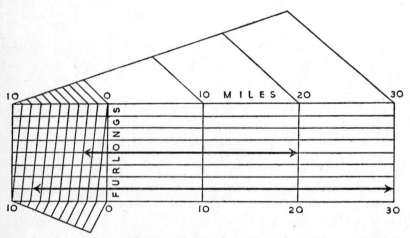

Figure 7. A DIAGONAL SCALE

This shows miles and furlongs. The upper arrow indicates a length of 22 miles, 4 furlongs, the lower of 37 miles, 7 furlongs.

miles and furlongs; direct division to represent a furlong on such a small scale would be out of the question, for it would be only 0·0125 inches in length.

Draw a line 4 inches long, dividing it into four portions of 10 miles each by setting-off proportional triangles. The left-hand unit can be divided into ten parts to indicate mile lengths, but these will only be 0·1 inches long, and are obviously indivisible into eight further portions to represent furlongs. To obtain these furlong divisions, construct a diagonal scale, as shown in Fig. 7; here the distances represented by arrows marked are 22 miles, 4 furlongs and 37 miles, 7 furlongs respectively.

Short Line-scales

These lines are used commonly on maps intended for reduction as line-blocks, and are indicative, rather than for use in precise measurement. They should be reasonably small; thus a line 1 inch in length is adequate for a map to be 4 inches wide on reduction, but if there is a large blank area in one corner it may look better to have a longer line. Care must be taken to ensure that the vertical dividing ticks are sufficiently large to appear when reduced. Choose carefully where to place the line-scale; this is, of course, part of the general lay-out and compilation of the draft. Avoid if possible a corner with crowded detail; it may be necessary to cut out a narrow panel and inset the line, or to put it inside the key panel or below the title. Each map must be viewed on its own merits, and the scale placed where it interferes least with the detail. Various alternative forms of line may be used (Fig 6, example F), but uniformity throughout a single series must be maintained. Consider carefully the style and size of the lettering and figures; O's especially are apt to look extremely ponderous.

Other Scales

On small-scale maps of extensive areas, a scale may occasionally be omitted if a graticule is drawn, or if meridian and parallel intervals are indicated in the margin; in either case, the projection should be stated. Otherwise a *variable scale* (Fig. 8) should be used, or the average scale for the central portion of the map stated.

The easiest way of giving the scale of a block-diagram is to state in the caption the distance between two clearly identifiable points on the diagram. A *perspective scale*, which decreases from the foreground to a vanishing point on the horizon, can be used if required for a landscape drawing, but again a statement of the distance between two points is usually adequate. The *horizontal* and *vertical scales* are marked on the base and left edge of sections and profiles, while the vertical exaggeration should also be stated (see p. 120).

Using the word 'scale' in its broadest sense, it can also include diagrams which enable quantitative measurements to be read off; thus

Figure 8. A VARIABLE SCALE

This scale, constructed for a Conical Orthomorphic Projection, with two standard parallels at 27° and 63° North latitude, enables distances to be correctly measured along parallels between 20° and 40° N.

there are horizontal and vertical scales on line-graphs and columnar diagrams, scales which enable proportional squares, circles and spheres to be drawn or evaluated, and scales of slopes. Each is discussed below in its relevant place.

FRAMES AND PANELS

Margins

A map must have a margin or frame; detail, except on simply drawn sketch-maps, must not be left 'hanging in mid-air', but must be continued to the margin. Graphs and diagrams may or may not have a frame, depending on their appearance. The margin, together with a panel for the key and, if needed, another for the title, should be ruled in as a preliminary to the inking-in of the pencil draft. The position of these panels depends on the layout of the detail on the map, and should be carefully considered during the preliminary compilation; usually they will be placed in empty corners, or centred on the bottom margin. Sometimes they may be conveniently placed where the detail is irrelevant to the main purpose of the map (Figs. 116–19).

Margins or frames should be as simple as possible. For a line-block, a single bounding line, cleanly ruled with sharp corners, is quite adequate. For larger manuscript maps, it is permissible to have an inner margin, with an outer one of a heavier line, or with a double line, either open or filled black and white in alternate spaces, possibly to show degrees or linear units. But elaborate scrolls, designs and ornamental corners should be strictly avoided.

Cartouches

This avoidance of elaboration applies also to the cartouche, a panel in which the title is placed if one is needed on the map. Elaborately decorated cartouches were the delight of the Dutch Renaissance cartographers and the Elizabethan county map makers, but they are usually out of place in a modern map (see Fig. 186). As a rule, the title need not appear on the face of a map intended for a line-block, but it can be set up below in type. On a large manuscript map the carefully lettered title is, however, an integral feature.

Key-panels

The key-panel should be carefully compiled so as to include a reference to all symbols, shading, stipples and special lines used on the map. Normally a frontier or coastline need not be keyed, but if different lines are selected for international, provincial and communal boundaries, they must be explained. All shading should be placed in small detached rectangles, which are clearer than divided columns. For relief map keys, put the darkest tint (highest land) at the top of the series; in geological maps put the stipple representing the youngest formation at the top. The key panel must be carefully laid out to give a balanced appearance, economizing in space, maintaining a strict rectilinearity of arrangement and using succinct definitions. A time- and space-saving idea is to number the items and list the numbers, with their definitions, in the legend below (Fig. 137); much fuller explanations are of course possible than could be lettered on the map, and moreover this saves valuable drawing time. Sometimes it is preferable, where a key is particularly involved, and when it is essential for it to appear on the face of the map, to have the definitions set up in a carefully chosen fount of type. These are cut out and pasted on to the drawing in their correct position, and then a line-block is made in the usual way. Where there is a series of maps with a common shading system, another economy is to use a single key on the first map in the series, to which reference may be made in successive maps.

Legends

The legend or caption can be conveniently mentioned at this point. A map, with its key and legend, should be complete in itself and self-explanatory, although to maintain the strict essential liaison between text and maps, the relevant places in the text should carry a reference, as '(Fig. *X*)'. The legend should be carefully compiled. It must state the source of information, unless the map is wholly original, or give the exact reference to the statistical data on which the map is based. Any expansion of key definitions and any significant comments will follow. On a manuscript map, this caption may be typed on white paper and then stuck centrally below the map. A caption for a line-block in a printed book should be set up in a smaller fount of type than the body of the text.

The North Point

As a rule, a north point is not necessary when a map is drawn conventionally with the north at the top margin. But a point should be inserted on port-plans, town-plans and the like, where frequent references to directions are needed, and certainly on architect's plans and similar large-scale work. The old map-makers lavished great skill on highly ornamental north points and compass-roses (Fig. 186), but a simple arrow, with a clean barb and a short cross-stroke mid-way across the vertical shaft, discreetly placed near a side margin, is quite adequate. It is necessary to distinguish between 'True North', 'Magnetic North' and 'Grid North'. The last is especially important in the United Kingdom, since the grid-lines of the National Grid System correspond to the sheet lines of the current Ordance Survey series. Thus, for example, the 10-km grid on the One-inch map is an index of the 1 : 25,000 series.

Lines of Latitude and Longitude

Many a good map has been spoiled by the superimposition of a heavy graticule or network of lines of latitude and longitude. For ordinary distribution maps, in which towns and coastlines are clearly marked, lines of latitude and longitude are unnecessary, although sometimes, as in the case of ocean-route maps, or of maps where an unusual projection has been used, the graticule must be drawn over the face of the map in some detail. Occasionally it is sufficient to mark off the degrees of latitude and longitude on the frame of the map.

Grid Lines

In the case of maps which are based on a topographical series with a local or a national grid, reference lines marked on the map are preferable to lines of latitude and longitude. Here again, grid lines need not be drawn across the map, but grid references may be marked off along the frame, or, in the case of small areas, at the four corners.

POINT-SYMBOLS

An immense variety of point-symbols can be used to specify distributions at particular points; Fig. 9 illustrates a number of these, but many others can be devised. They may consist of solid or outline geometrical figures, placed as near as possible to the centre of the place each is intended to indicate. They may be conventional,[1] or they may be chosen arbitrarily by the compiler to denote some specific feature. Pictorial symbols may occasionally be used effectively, though with discretion, particularly on the more graphic type of map;[2] thus bales, sacks, sheaves or barrels may indicate with some relevance the distribution of certain commodities. Literal symbols (initial letters), possibly of a size more or less roughly proportional to the quantities represented, may be used, but these tend to give a somewhat confused appearance to the map, particularly when other lettering has to be inserted. All symbols must of course be carefully keyed.[3]

Quantitative Dot Maps

The simplest form of symbol is the dot, and a very useful form of distribution map is one on which quantities or values are represented by dots of uniform size, each dot having a specific value.[4] This form of distribution map is especially useful when the values are distributed unevenly and sporadically, a fact which might well be concealed by some form of average density map. The dots are inserted within the particular administrative units for which statistics are available; obviously the smaller the statistical unit, the more accurate is the map.

[1] See, for example, the U.S. Geological Survey Publication, *Topographic Map Symbols* (New York, 1950); and M. Couzinet, 'Le Problème de l'Uniformisation des Signes Conventionnels sur les Cartes Topographiques', *Comptes rendus du Congrès International de Géographie, Lisbonne, 1949, Tome I. Actes du Congrès. Travaux de la Section 1*, pp. 190–4 (Lisbon, 1950).

[2] D. L. Mumby, *Industry and Planning in Stepney* (Oxford, etc., 1951), uses pictorial symbols in a series of maps to denote industrial distributions.

[3] A standard work of reference is R. L. Williams, *Statistical Symbols for Maps: Their Design and Relative Value* (New Haven, Conn., 1956).

[4] H. St. J. L. Winterbotham, 'Dots and Distributions', *Geography*, vol. 19, pp. 211–13 (London, 1934). An indication of the sophisticated approach to dot maps is given by R. E. Dahlberg, 'Towards the Improvement of the Dot Map', *International Yearbook of Cartography*, vol. 7, pp. 157–66 (London, 1967).

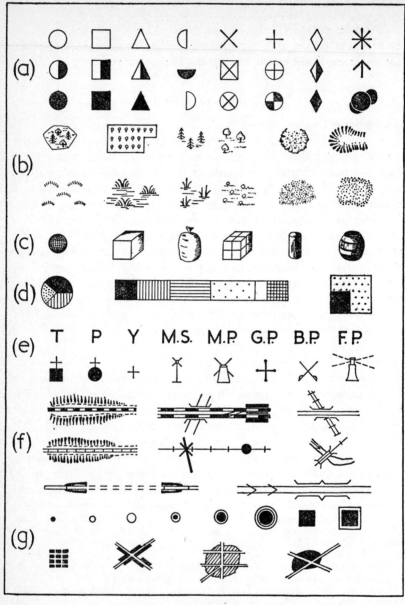

Figure 9. VARIETIES OF SYMBOL

(a) Solid and outline geometrical figures; (b) physiographical symbols; (c) pictorial symbols; (d) divided symbols; (e) and (f) conventional point and route symbols adapted from the Ordnance Survey One-inch and 1 : 25,000 series, (g) town symbols.

The first step is to examine the range of quantities involved, and then to select a value to be represented by each dot. The success of the map depends largely on the choice of this value; it should not be so low that there will be difficulty in inserting all the dots in units where the quantities are high, nor so high that there will be units with very few dots. However carefully the value is chosen, it frequently happens that there will be problems attached to the placing of the dots in the units with very high or very low figures. In the former case, it may be found that the dots will be so numerous as to coalesce. Some cartographers, in fact, claim that this result should be aimed at; the contrast between the white (dotless) and solid black areas, with a gradual transition between them as the dots increase in frequency, is regarded as an ideal form of visual presentation. Other cartographers insist that the dots should be countable and therefore must not be allowed to merge. In practice, dots are rarely counted, for obviously the person in search of more precise information must return to the statistical source upon which the map is based; this, of course, should always be stated under the dot map. However, if the high values do produce merging, it may be preferable to fill the unit concerned in solid black, and then insert the actual value represented in figures in a white panel. In the case of units with very low figures, on the other hand, even when the value attributed to the dot has been chosen with discretion, a single dot may have to represent the total value within the whole of a unit. The problem is where to put the dot; obviously the map will here cease to have any precise locational value.

If dots are to be placed evenly and uniformly within each unit, as in the absence of more precise information they must be, the boundaries of the units should, strictly speaking, be included on the map; in practice, these are often removed after the dots have been inserted. But straight rows of dots should be avoided, and so should lines of dots parallel to the boundary of each unit, which will produce patterns and whorls in the finished map. The dots should be placed evenly by eye, and some continuity attained with the dots in adjacent units.

The dots may be placed more precisely when detailed information is available concerning the exact location of the values, derived from field-work, such as land utilization survey for a crop map, or from larger scale maps, giving, for example, the actual positions of hamlets and farms for a rural population map. But more generalized information, such as is provided by relief or climatic maps, or even by vague geographical preconceptions, must be avoided; 'it is susceptible to wishful thinking, with some loss of objectivity. Sheep are not always found on uplands, nor barley in the drier parts.'[1] Occasionally an assumption is warrantable; 'in Arizona, alfalfa will be shown in the irrigated valleys and not in the arid uplands'.[2] But such an assumption must be used with very great discretion (Fig. 121).

The actual size of the dots to be used presents a further problem. Their size must depend on the scale of the base-map and on the number of dots to be inserted, but the dots must be neither so big that a coarse generalized effect is produced, nor so small that a blur is produced in areas of dense value.[3] A ball-pointed nib should be used to make dots of uniform size. Care must be taken when the drawn map is to be reproduced by means of a line-block that the dots are drawn sufficiently large to reduce, and the dots should be spaced rather more widely than might seem necessary on the manuscript map.

Instead of a pen, some form of stamp or die, using printer's ink, can be employed,[4] and if the scale is large and the dots can have a diameter of one-tenth of an inch or more, small rotating compasses can be used to draw outlines which can be filled in with a pen or a fine brush. Yet another method is to stick black dots directly on to the base-map.

Percentage Dot-maps. Instead of using a located dot with a precise absolute value, it may be desirable to let it represent 1·0 or 0·1 per cent of the total value involved in the particular distribution; in other words, such a map will bear 100 or 1,000 dots respectively. These are known as percentage dot-maps[5] and mille-maps.[6] They facilitate

[1] W. G. V. Balchin and W. V. Lewis, 'The Construction of Distribution Maps', *Geography*, vol. 30, p. 91 (London, 1945).

[2] E. Raisz, *General Cartography*, p. 249 (1st edition, New York, 1938).

[3] J. Ross Mackay, 'Dotting the Dot Map: An Analysis of Dot Size, Number and Visual Tone Density', *Surveying and Mapping*, vol. 9 (Washington, 1949).

[4] It is extremely difficult to maintain exact regularity in size and shape of dots, since even the difference between a full and nearly empty nib, for example, can make a big difference. W. G. Byron, 'Dotting the Dot Map with Steel Drill Stamps', *Professional Geographer*, vol. 8, no. 4, pp. 5–7 (New York, 1956), suggests a steel drill of the exact diameter, with the butt end ground flat, and an oil-based block printing-ink (which does not flow) squeezed on to a glass plate. He claims that 15–20 dots per minute can be located, of uniform diameter, varying according to the drill-size from 0·0135 to 0·228 inches, thus affording uniformity, sharpness and precision.

[5] J. Ross Mackay, 'Percentage Dot Maps', *Economic Geography*, vol. 29, pp. 263–6 (Worcester, Mass., 1953).

[6] L. D. Stamp, *The Land of Britain, its Use and Misuse*, pp. 102–7 (London, 1948); this contains twelve effective mille-maps. See also K. Buchanan and N. Hurwitz, 'The "Coloured" Community in the Union of South Africa', *Geographical Review*, vol. 40, p. 401 (New York, 1950).

arithmetical comparisons, and they afford ready information concerning fractional distributions and proportions. Their disadvantage is
that a dot will rarely represent a round number.

Colour-dots. If several types of distribution require to be shown on one
map, dots of differing colour and significance may be used. Colours
should be carefully chosen to provide contrasts, yet with some psychological relationship, if possible, to the commodities they represent
(see p. 290). The advantage of colour is that the proportions of the
colours vary with changes in distributions, producing similarly changing
patterns, and emphasizing transitional zones. Such a map combines the
advantage of accurate detail viewed closely, and broad areal patterns
viewed more generally.[1] The dots may be put on in coloured ink or
stuck in position.

Proportional Symbols

Some symbols can provide quantitative as well as locational information if they are drawn to scale. Circles, spheres, squares, rectangles,
columns, triangles and cubes[2] may be employed. The areas of two-
dimensional figures such as squares or circles, or the volumes of three-
dimensional figures such as cubes or spheres, are made proportional to
the quantities they represent. A wide variety of these symbols may be
used either as diagrams in themselves, or located on maps to illustrate
population and economic distributions; it will be seen that many of the
methods discussed especially in Chapters 5 and 6 rely on these devices.
In addition to showing total quantities, they may be divided proportionally to give further information.

The Calculation of Proportional Symbols. If a quantity, say 10,000, is
to be represented by a circle, the square root of the number must first
be found, i.e. 100. The area of a circle is πr^2, but π is a constant and

---

[1] G. F. Jenks, ' "Pointillism" as a Cartographic Technique', *Professional Geographer*,
vol. 5, no. 5, pp. 4–6 (New York, 1953). His work in connection with agricultural maps
is discussed on p. 290.

[2] J. Ross Mackay has made a careful study of cubic symbols, in 'A New Projection
for Cubic Symbols on Economic Maps', *Economic Geography*, vol. 29, pp. 60–2 (Worcester, Mass., 1962). He classes them as (i) *isometric*, with the front and sides of equal
length; (ii) *cabinet*, with sides half the length of the front, inclined at 30–40 degrees from
the horizontal; and (iii) *clinographic*, used only as a rule for precise drawings in crystallo-
graphy. He tried out tests with students, comparing the resemblance of a drawn cube
to the solid form, examining the space occupied, and deciding whether they looked
'thick' or 'thin'. His suggested cube has sides three-fifths the length of the front,
inclined from 30 to 50 degrees to the horizontal, preferably with a right-sided view.

A. Circles

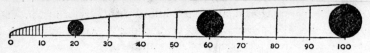

B. Circles

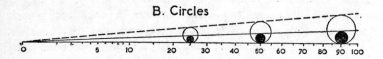

C. Spheres

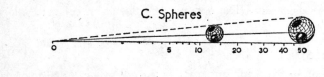

D. Squares

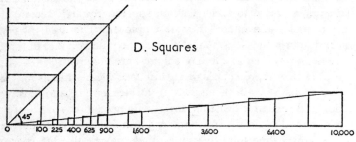

E. Cubes

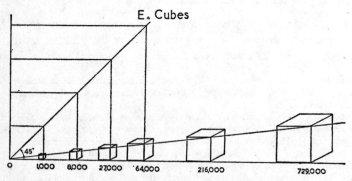

Figure 10. THE CALCULATION OF PROPORTIONAL SYMBOLS

may be ignored. A circle can then be drawn with a radius some given length equivalent to 100 (e.g. 0·1 inch). If a second quantity, say 50,000, has to be represented, its square root is 223·0, and so on the same scale a radius of 0·223 inch must be used. This will obviously be a lengthy business if many quantities are involved, although a slide-rule will ease matters.

The most convenient method is to employ some form of scale from which radii of circles, or spheres, or sides of squares, triangles and cubes, can be measured directly or stepped off with dividers.

One method is to rule an evenly graduated linear scale, as in Fig. 10 (example A). Calculate the square root of several numbers (in this example 5, 10, 40 and 100 were used), and draw from each of these points on the linear scale a perpendicular proportional in length to these square roots. The ends of each perpendicular are joined by a smooth curve. The diameter of a circle representing any intermediate value can be measured immediately.

Number Represented	Circles		Spheres	
	Square Root	Distance in cm from zero	Cube Root	Distance in cm from zero
5	2·24	4·48	1·71	3·42
10	3·16	6·32	2·15	4·30
20	4·47	8·94	2·71	5·42
30	5·48	10·96	3·11	6·22
40	6·33	12·66	3·42	6·84
50	7·07	14·14	3·68	7·36
60	7·75	15·50	3·92	7·84
70	8·37	16·74	4·12	8·24
80	8·94	17·88	4·31	8·62
90	9·49	18·98	4·48	8·96
100	10·00	20·00	4·64	9·28

An alternative method is to rule a line-scale, the divisions of which, measured from the zero on the left, are spaced at distances proportional to the square roots of the values (example B). This line may be of any convenient length; obviously a longer line will give a finer gradation between the values. Suppose that the maximum number to be represented is 100, then the length of the line is made proportional to the square root, i.e. 10. Intermediate square roots will be placed proportionally along the line, as shown in the table on p. 31, where the scale-line representing the value 100 is made 20 cm in length. Draw an oblique line at any convenient angle through the zero, the angle depending on how big the maximum symbol is to be (example B). A vertical line drawn from the horizontal scale at any point to meet the oblique line will give the radius of a circle, the area of which is proportional to the number whose square root is represented at that point. Thus, briefly, using a table of square roots, put one leg of a pair of compasses on the zero, the other the correct distance along the scale-line (e.g. if the number was 35, distance on this scale = 11·84 cm), then swing the first leg vertically to the oblique line. This distance will be the radius of a circle proportional in area to the number it represents.

The same procedure may be followed to find the radii of proportional spheres (example C), but cube-root values have to be plotted on the base-line instead of square-root values, as shown in the table above. This type of scale may easily be adapted to plot proportional squares and cubes (examples D, E).

The Drawing of Symbols

Symbols must be carefully selected and drawn so that they are readily distinguishable, and, particularly when intended for reduction, each must be sufficiently large to reproduce clearly. Sometimes minute black squares, discs and diamonds are indistinguishable on printed maps because of over-reduction. They must first be carefully plotted in pencil, their outlines inked in with a ruling-pen or compasses, and then blacked in with a fine brush or broad pen, unless outline symbols are required. It is preferable to ink in the symbols before other line-work and lettering is completed, so that they may be left clear to avoid blurring upon reduction. Fig. 113 exemplifies the use of varied symbols to illustrate particular distributional patterns.

It may be well worth the expense of having a celluloid stencil cut to order with the outlines of a number of the more common geometrical symbols, in several sizes. A compiler should, of course, maintain consistency of symbols throughout a particular dissertation or monograph, and a stencil is a great aid to this end.

Adhesive Symbols

A wide range of symbols is available, printed on transparent self-adhering plastic sheets, which can be applied directly to a map.[1] They

---

[1] *Artype* sheets of symbols, etc., are made by Artype Inc., Chicago, Ill., U.S.A.; the British agents are Hunter-Penrose-Littlejohn Ltd., 7 Spa Road, Bermondsey, London, E.16.

include almost every type of point or linear symbol which it is possible to visualize, in a wide range of sizes, ornamental frames, arrows, spheres, lines, rules, borders, in black and in colour. A symbol is carefully placed in its exact position on the map, and a square cut round it with a stylo; pressure is applied to make permanent adherence. Special mention may be made of the wide range of *Letraset* (dry adhesion by pressure) and *Sasco* (adhesive) symbols, which may be applied to paper, plastic and glass.

The Visual Evaluation of Symbols

An important fact, sometimes not fully appreciated or even overlooked by geographers who are plotting spatial distributions of quantitative facts by means of symbols, is that '. . . accurately drawn maps are not always accurately read', in the words of J. I. Clarke,[1] who has carried out some interesting experiments in visual perception. It is clear that the degree and amount of optical illusion in the evaluation of proportional symbols varies considerably. Clarke tested this concept by showing cards with various symbols to a group of thirty-three students, and then analysed the results, which revealed a remarkable diversity of attainment. He concluded that efforts should be made to improve this kind of map by including only one type of symbol per map, by using standard sizes shown in a carefully constructed key, by avoiding overlapping symbols, and unless absolutely necessary including no place-names, which may destroy the pattern.

GRAPHS AND DIAGRAMS

In its broadest sense the term 'graph' is used to denote divided circles ('pie-graphs' or 'wheel-graphs'), 'unit-graphs', 'bar-graphs' (i.e. columnar diagrams), 'star-graphs' or 'clock-graphs' and dispersion diagrams, as well as ordinary line-graphs, such as frequency graphs, probability curves and so on. In this book a distinction is made for convenience between *line-graphs*, in which a series of points is plotted by means of co-ordinates and then joined by a line, and *diagrams*, in which various graphical devices such as columns, rays and sectors are employed.

[1] J. I. Clarke, 'Statistical Map-Reading', *Geography*, vol. 44, pp. 96–104 (Sheffield 1959).

Line-graphs

The conventional Cartesian type of line-graph is that in which a series of points is plotted by means of rectangular co-ordinates. The abscissae are measured on a horizontal and the ordinates on a vertical scale. One scale may represent a series of equal time divisions (independent variable), the other a series of quantitative or percentage values (dependent variable). Less conventional, but with a special application to certain geographical problems, are polar, or circular, graphs in which the points are plotted by two co-ordinates, one an angle or bearing (the vectorial angle) the other a distance from the point of origin (the radius vector). A further type of line-graph is that in which oblique co-ordinates are used for plotting points. Where three oblique co-ordinates are employed, the graph is usually referred to as triangular.

In each of these line-graphs, the line joining the points is referred to as a curve. But this curve is frequently only a visual aid to overall changes in value and is not coincident with the true locus. The practice of joining points by smooth curves should therefore be avoided as far as possible. Curves can be smoothed by the employment of statistical methods, such as curve-fitting, but unless such methods are employed points should be joined by a series of short straight lines. Where continuous readings such as weather records yielded by thermographs and barographs are available, the resultant curves will, of course, be true line-graphs.

A *simple line-graph* shows only a single series of values, connected by one line. A *polygraph* or *multiple line-graph* (Fig 91) includes several sets of values connected by distinctive lines, usually involving some direct comparison. A *compound line-graph* (Fig. 125) (known also as a *band-graph* or an *aggregate line-graph*) shows trends of value in both the total and its constituents by a series of lines on the same frame; the area between two successive lines may be shaded distinctively.

A *histogram* is a type of graph which presents frequency distributions. The term was originally applied to rainfall (see p. 234), whereby the amounts are plotted as abscissae and the scale of frequencies as ordinates. The term now denotes a diagram of frequencies of any values, such as heights of land, akin to a columnar diagram or bar-graph which depict absolute (not percentage) quantities.

The Construction of Line-graphs. The series of figures to be graphed is examined, and the maximum value noted, for on this and on the available size of paper depends the vertical scale to be adopted. The top of the vertical scale should just exceed the maximum value. The whole vertical range of values must be represented when zero is significant, even at the cost of a considerable empty space at the bottom of the graph, since a false impression of relative variability in the trends of the lines may be given unless the whole range can be appreciated. A vertical line is ruled on either side of the horizontal scale-line, and the equal intervals representing quantities are marked by neat ticks.

The plotting of these various graphs must be carried out on accurately ruled graph-paper. It is preferable, however, on completing the plotting, to trace off the graph-lines, the frame and the scales, using tracing-paper and a ruling-pen. This gives a cleaner finish and combines the accuracy of plotting which graph-paper allows with the clarity of white paper unobscured by the close and unnecessary grid of the graph-paper. Moreover, graph-paper is rarely of sufficient quality to take ink well, nor will it reduce clearly, particularly if it is ruled in faint coloured lines. A more open grid of perhaps $\frac{1}{2}$-inch squares may occasionally be added to help the eye to follow the lines across to estimate the values involved.

The frame and grid of the graph must be ruled as lightly as possible, while the graph-lines will be made considerably heavier. Several different varieties of line (Fig. 3) can be used in drawing polygraphs. Care must be taken in scaling the values on the graph not to overload the figures; strings of noughts should be avoided on the vertical scale by, for example, lettering 'Thousand tons' at the top of the scale and numbering the units 1, 2, 3 and so on.

Diagrams

Columnar diagrams, sometimes known as bar-graphs, consist of a series of columns or bars proportional in length to the quantities they represent. They may be *simple*, when each bar shows a total value, or *compound*, when each bar is divided to show constituents as well as the total value. The bars may be placed either vertically (Fig. 128) or horizontally (Fig. 165), or in pyramidal form (Fig. 166); the first is usually most satisfactory when a time-scale is involved. After the scale of values has been determined, the exact length of each bar is computed and parallel lines are drawn; care should be taken not to make the bar too wide. It is preferable to leave a space slightly less wide than the bar itself between each, although in some cases, particularly for rainfall diagrams, there need not be any intervening space; these are often descriptively known as 'battleship-diagrams' from their profiles (Fig. 76). When the outline of each bar has been drawn with a ruling-pen, it can be filled, either in solid black or with diagonal shading.[1] In a compound bar-graph, several shadings and stipples may be used (Fig. 167).

Divided rectangles are akin to bar-graphs in that their lengths are directly proportional to the values they represent, but they are not plotted in series. When the scale is linear they are often deliberately widened in order to allow names and figures to be placed within the constituent divisions. The scale may, on the other hand, be areal, in which case the area both of the rectangle as a whole and of its divisions are directly proportional to the quantities they represent.[2] Either absolute values can be shown, or, for comparative purposes, the rectangles may be of uniform size and divided on a percentage basis.

Divided circles, sometimes termed 'wheel-graphs' or 'pie-graphs', have much the same advantages as divided rectangles, in that they can provide striking proportional effects and can carry considerable information (Fig. 132). A circle is divided into sectors, each of which is proportional to the value it represents. It is convenient to work with percentage circular graph-paper (see p. 12), on which 100 angles of 3·6 degrees each are set off from the centre of the circle. If only two or three major divisions are required, proportions of 360 degrees may be easily calculated. Care must be taken with residual quantities; it is tempting to set off angles representing major divisions first, with the result that the cumulative error is felt most in the smallest section. Unless there is a composite category of 'Others', the small sectors must be measured and demarcated first, so that the cumulative error will be absorbed into the large sectors where its effect will be negligible. Care should be taken in lettering a divided circle, so that it is not necessary to turn the page around in order to read the names.

Star-diagrams, sometimes called 'clock-diagrams', 'roses' or 'vector diagrams', are a form of graph in which values are plotted as radii from a point of origin. This is especially useful where vector values are involved as in wind-roses (see p. 240 and Fig. 97) Instead of using

---

[1] In point of fact, G. M. Schultz sounds a warning in 'Beware of Diagonal Lines in Bar Graphs', *Professional Geographer*, vol. 13, no. 4, pp. 28–9 (New York, 1961); she demonstrates the optical illusions of tilting and tipping induced by this shading, and recommends instead black, white and dots.

[2] E. Raisz, 'The Statistical Rectangular Cartogram', *Geographical Review*, vol. 24, pp. 292–6 (New York, 1934).

rays of proportional length, they may be drawn of proportional thickness, radiating in the correct direction, possibly to the actual destination, in respect, for example, of movement of commodities or of population.

Dispersion diagrams are of value where it is required to analyse the 'spread' or 'scatter' of a series of values (see Appendix, p. 485). The basic principle is that a single dot to represent one value is plotted

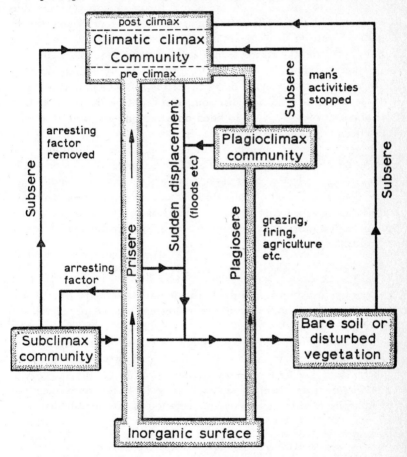

Figure 11. A DIAGRAM TO ILLUSTRATE PATTERNS OF SERAL DEVELOPMENT (after D. Watts)

This is based on an original diagram by S. R. Eyre. It inter-relates various stages in the development of plant communities in response to human interference and other factors.

alongside a vertical scale. The dispersion diagram has a special application to the analysis of rainfall (see p. 243 and Fig. 99). In addition, it
can be employed as an aid in determining critical values of density
ranges for choropleth and isopleth maps.

Flow, Linkage or Cycle-diagrams and Charts

This very revealing and informative type of diagram can be used when
it is required to present a sequence of interlinked topics, events or items
to illustrate the development or evolution of some theme, objective or
product. Figure 11 shows the relationship between different vegetation
communities and the effects of human intervention. Such a chart may
be prepared to show, for example, the progress of production of a map
from raw field-data, through the drawing and reproduction stages. This
enables adherence to specification, cost control, quality control,
assessment of progress, etc., to be rigorously maintained at all stages,
thus effecting economies in time and cost.

   Linkage analysis has become a highly sophisticated technique, including the reduction of a mass of data into a number of interconnected
classes, using a selected series of factors. For example, this method can
be used for the reduction of a number of regions into homogeneous
groups, inevitably losing detail at each stage of the grouping. This
technique has only really become practicable with the availability of
computers.

CHOROCHROMATIC MAPS

It is possible to compile and draw a wide variety of maps which illustrate spatial distributions. One group does not involve any consideration of quantities and values; thus the United States' cotton-belt can be
outlined and tinted, and British coalfields can be delineated in solid
black. Land utilization, soil and geological maps form a big proportion
of these maps. The term *chorochromatic* can be applied to this category of
map; in America it is often known as a 'color-patch' map, the term
colour being used in its widest interpretation of 'tint' or 'shade'.

CHOROGRAPHIC COMPAGE MAPS

This name was devised by J. F. Hart and E. Mather[1] to denote a
type of map on which is located a number of specific places or small

   [1] J. F. Hart and E. Mather, 'The Chorographic Compage Map', *Surveying and Mapping*, vol. 13, pp. 333–7 (Washington, D.C., 1953).

units for which detailed individual information is available. The data are presented in the form of graphs, divided circles, columns, etc., in boxes around the border of the map. This affords a clear presentation of comparative data without overcrowding the face of the map itself, yet enabling an instant appreciation of spatial distributions.

ISOPLETHS AND CHOROPLETHS

Definitions

There are two main groups of maps which involve some indication of quantity as well as of spatial distribution, known as 'quantitative areal maps'. The first group includes maps where quantities are indicated by lines of equal value, such as contours and isotherms. The second group consists of maps which depict average values per unit of area over some administrative region for which statistics are available, such as density of population per square mile, the percentage of land under cultivation, and the yield per acre of arable land.

It must be admitted that there is a considerable amount of confusion concerning the multiplicity of terms which have been coined to denote variants of these cartographical terms.[1] For example, to cover all lines representing constant values on maps, the terms *isopleth*, *isarithm*,[2] *isoline*,[3] *isobase*, *isogram*, *isontic line* and *isometric line* have been used at various times. J. K. Wright proposed that *isogram* be used for all lines of quantity, with two subdivisions.[4] The first group he called *isometric lines* (Greek, *metron*=measurement), a line such as a contour,

[1] Mention may be made here of a useful list of geographical terms, with their English–French equivalents, by A. Bargilliat, *Vocabulaire pratique anglais-français et français-anglais des termes techniques concernant le cartographie (géodésie, topographie, dessin, photomécanique, impression)* (Paris, 1944).

[2] The term isarithm is sometimes used as a generic name for all lines of constant value (from Gr. *arithmos*, number), as by D. I. Blumenstock, 'The Reliability Factor in the Drawing of Isarithms', *Annals of the Association of American Geographers*, vol. 43, pp. 289–304 (Lancaster, Pa., 1953); and by A. H. Robinson, *Elements of Cartography*, p. 139 (2nd edition, New York, 1960). It is used in a more limited sense by A. G. Ogilvie, *Report of the Commission for the Study of Population Problems*, p. 7 (Washington, 1952), to refer to isopleths of population density; A. H. Robinson, 'The Cartographic Representation of the Statistical Surface', *International Yearbook of Cartography*, vol. 1, pp. 53–63 (London, 1961), which discusses and illustrates isarithmic surfaces derived from the use of different intervals applied to the same data.

[3] E. Imhof, 'Isolinienkarten', *International Yearbook of Cartography*, vol. 1, pp. 64–98 (London, 1961); a very full summary of the varieties of isoline (here used as an all-embracing word), their use, treatment and limitations.

[4] J. K. Wright, 'The Terminology of Certain Map Symbols', *Geographical Review*, vol. 34 (New York, 1944).

an isotherm or an isobar 'representing a constant value or intensity pertaining to every point through which it passes'. The second group he called *isopleths*[1] (Greek, *plethos*=a multitude or crowd), ' a line that represents a quantity or enumeration assumed to be constant, pertaining to certain areas through which it passes', such as lines of equal density of population.

Wright further suggested that the term *chorogram* be used to define all quantitative areal symbols, with two main categories, *choropleths* and *chorisograms* (Greek, *choros*=area, space). He used the term choropleth to denote an areal symbol applied to an administrative subdivision, such as an indication of the density of population in a parish or county,[2] while he defined a chorisogram as a system of shading or colour applied for distinctiveness between two successive isograms. A further complication was introduced by a distinction between two other terms, *chorisometers*, where graded shadings and colours are put between two isometric lines, and *chorisopleths*, where these are put between two isopleths. One also comes across other formidable compound words, such as *chorisometrograms* and *chorisochores*.

This terminology is not yet recognized as standard usage. In this book, the term *isopleth* is used in its broadest sense to embrace all lines of quantity, and *isopleth maps* will denote all maps on which isopleths appear, with or without tinting. The category of *choropleth map* will be used for all quantitative areal maps, calculated on a basis of average numbers per unit of area.

Isopleth Maps

The contour-map is perhaps the most common example of an isopleth map, using the term as defined in its broadest sense. In addition, a wide range of maps can be drawn, the basis of each of which is the plotting of values for as many stations as possible, and the interpolation of iso-

[1] See also 'Isopleth as a Generic Term', an unsigned note in *Geographical Review*, vol. 20, p. 341 (New York, 1930). J. Ross Mackay, 'Some Problems and Techniques in Isopleth Mapping', *Economic Geography*, vol. 27, pp. 1–9 (Worcester, Mass., 1951), also uses the term isopleth in the restricted sense suggested by Wright as a line passing through 'ratios for areas', thus showing ratios and not absolute quantities. He defines the 'quantity of an isopleth' as that of the areal strip through which it passes.

[2] E. Raisz, *General Cartography*, p. 246 (1st edition, New York, 1938), emphasizes that the term choropleth need not be limited to administrative divisions, but that a map divided into squares for which a density could be calculated and then tinted or shaded would also be a choropleth map. He suggested that a more restricted term of *demopleth* be used for those maps based on administrative divisions.

(*Note.* This term was omitted from the 2nd edition (1948)).

pleths for specific values.[1] These lines are either drawn through stations with identical values, or are interpolated proportionally between them[2] (Figs. 12–15). Thus the 50° F isotherm may be assumed to pass midway between two stations with average temperatures of 45° F and 55° F. These isopleth maps can be used to depict climatic distributions (isotherms, isobars, isohyets and isonephs), salinity of the sea (isohalines), and in fact any feature where figures are available and can be plotted for a series of particular points. The term *isonoetic line*[3] has even been coined to indicate the distribution of intelligence percentages (Greek, *noetikos*=intellectual), and *isophore* (*phora*, a Greek term for a charge for carrying loads) to indicate a line connecting points of equal freight-rate from some point.[4] An *isobase* (Greek, *basein*=a rise) is a line drawn through a series of points which have been equally elevated (or, in minus quantities, depressed) by relative movements of land and sea.[5]

When an isopleth is used for a density or ratio map, such as the density of population or the percentage of arable land under a certain crop, the problem is rather more difficult. The four main variables determining the manner in which such isopleths may be drawn (and their resultant patterns) are: (i) the selected value intervals; (ii) the shapes and sizes of the units for which statistics are available; (iii) the situation of the plotting-points (which may or may not be central) within each unit; and (iv) the actual method of interpolation.[6] The

[1] A count by P. W. Porter, of the University of Minnesota, reveals that 440 isopleth maps were published in three premier American geographical periodicals (*Annals of the Association of American Geographers*, *Economic Geography* and the *Geographical Review*) between 1900 and 1957, 20 per cent of them in the peak period 1938–40.

[2] For a careful note of some of the statistical implications of possible unreliability, with methods of correction for bias, sampling and observational errors (with particular respect to temperature maps), see D. I. Blumenstock, 'The Reliability Factor in the Drawing of Isarithms', *Annals of the Association of American Geographers*, vol. 53, pp. 289–304 (Lancaster, Pa., 1953). See also J. Ross Mackay, 'Some Problems and Techniques in Isopleth Mapping', *Surveying and Mapping*, vol, 12, pp. 32–8 (Washington, 1952); this deals with control points, interpolation, intervals and reliability of ratios.

[3] P. Scott, 'An Isonoetic Map of Tasmania', *Geographical Review*, vol. 47, pp. 311–29 (New York, 1957).

[4] J. W. Alexander, S. E. Brown and R. E. Dahlberg, 'Freight Rates: Selected Aspects of Uniform and Nodal Regions', *Economic Geography*, vol. 34, pp. 7–18 (Worcester, Mass., 1958).

[5] H. Valentin, 'Present Vertical Movements of the British Isles', *Geographical Journal*, vol. 99, p. 303 (London, 1953).

[6] Valuable appraisals of these points and their significance are afforded by (i) J. Ross Mackay, 'The Alternative Choice in Isopleth Interpolation', *Professional Geographer*, vol. 5, no. 4, pp. 2–4 (New York, 1953); and (ii) P. W. Porter, 'Putting the

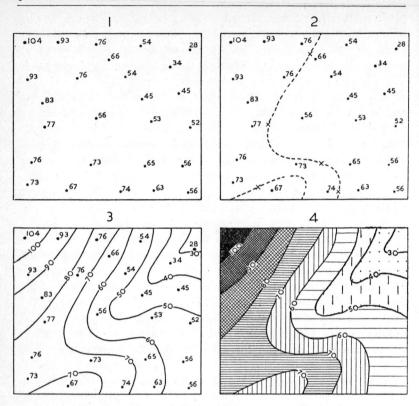

Figures 12–15. THE INTERPOLATION OF ISOPLETHS

1. The point-values are located; **2.** the critical isopleth of value 70 is interpolated, with the aid of crosses placed between pairs of values at a distance proportional to the value of each; **3.** other isopleths are similarly interpolated; **4.** a system of density shading is applied for clarity between the isopleths.

value intervals should be carefully selected,[1] based on the overall range of quantities to be mapped and on a trial scatter of distribution (see p. 478), or by means of a frequency graph; they may be isarithmic, that

---

Isopleth in its Place', *Proceedings of the Minnesota Academy of Science*, vol. 25, pp. 372–84 (St Paul, Minn., 1958). An elaborate mathematical analysis of a method for estimating the quantitative reliability of isoline maps is given by F. Stearns, 'A Method for Estimating the Quantitative Reliability of Isoline Maps', *Annals of the Association of American Geographers*, vol. 58, no. 3, pp. 590–600 (Lawrence, Kansas, 1968).

[1] J. Ross Mackay, 'An Analysis of Isopleth and Choropleth Class Intervals', *Economic Geography*, vol. 31, pp. 71–81 (Worcester, Mass., 1955); and 'Isopleth Class Intervals: a Consideration in their Selection', *Canadian Geographer*, vol. 7, pp. 42–5 (Ottawa

is, on an arithmetic interval basis (e.g. 10, 20, 30), geometric (2, 4, 8, 16), or they may be based on natural breaks in a frequency distribution (see p. 357). The size of the statistical units (whether communes, parishes, counties or provinces) will determine the number and density of evaluated points from which each isopleth can be interpolated, and therefore its precision. The location of the central point of a statistical unit is a difficult matter, particularly when its shape is irregular or elongated. Interpolation is usually done by assuming a uniform increase of value between two points and so placing the isopleth proportionally (as in Figs. 12–15); in this case, and in interpolating contours on a uniform slope, this is quite permissible and a fair degree of accuracy can be attained. But for agricultural and population densities this may be far from the case. Various techniques are suggested on pp. 276–9 and 357–8 to overcome some of these difficulties. The superposition of a grid over data already distributed in the form of dots enables these to be counted, and thus a value is obtained for the centre-point of each grid-cell. The grid may be of squares, or, as used by J. Ross Mackay and P. W. Porter (op. cit.), hexagonal,[1] involving isometric graph-paper (see p. 12). The hexagon is the closest approximation to a circle, wherein the border is a uniform distance from the central point, which will allow an area to be covered completely with these non-overlapping units. Moreover, the central points of adjacent hexagons form equilateral triangles, avoiding problems of interpolation inherent in square grids where identical values are involved.

Choropleth Maps

A wide variety of statistics may be used for the calculation and compilation of areal density maps. The actual administrative unit chosen as a basis depends upon the scale and upon the desired detail of the map.

1963). See also G. F. Jenks and M. R. C. Coulsen, 'Class Intervals for Statistical Maps', *International Yearbook of Cartography*, vol. 3, pp. 119–34 (London, 1963). The last is a most valuable analysis of the need for classing mappable data, a discussion of techniques for the selection of class intervals, the determination of class limits, test procedures and adjustments for readability. He uses plotted curves to decide the 'best fit' for the particular set of data.

[1] The value of the hexagonal grid, with illustrations of isopleth interpolation, is stressed by A. H. Robinson, J. B. Lindberg and L. W. Brinkman, 'A Correlation and Regression Analysis applied to Rural Farm Population Densities in the Great Plains', *Annals of the Association of American Geographers*, vol. 51, p. 214 (Lawrence, Kansas, 1961). Readers will be interested in the statistical methods here used to transform county population data into values for the control point in each hexagon.

Thus a small-scale map of Belgium might be based on the average figures for the nine provinces, but a detailed large-scale map must use the figures for the 2,670 communes. Where the shapes of the administrative units are irregular, however, a misleading pattern may result. The narrow parishes of many parts of England,[1] the long Canadian counties thrusting northward into the Laurentian Shield, the minute communes of the Brussels agglomeration, and the large heathland communes of north-eastern Belgium, all may result in curiously artificial patterns on density maps. Conversely, the remarkably uniform shapes and areas of the French *départements* produce quite effective maps.

It must be realized that these density maps, related as they are to the administrative units, reveal only average distributions over sometimes considerable areas. That being so, the misconception commonly arises that the grouping of a considerable unit under one average value implies distributional uniformity. This is in fact far from being the case, for the broad average may mask a vast range of local variations; obviously, the more extensive the administrative unit used, the more sweeping is the generalization presented in map form. Only occasionally can the compiler depart from administrative boundaries, when, by using detailed ground information, more accurate density boundaries can be computed (see p. 337).

The first step in the construction of a choropleth map is the calculation of the average density for each administrative unit. Density calculation is a laborious process, but it can be speeded up, with the help of a slide-rule, a desk calculator, or by inspection from a graph. If three or more variables are involved, then a *nomograph* may be employed.[2] This is a graphical method of solving the functions of three or more variables (Fig. 16). The graph consists of three or more related scales, and values are read off by laying a straight-edge across them. Thus, if population densities are being calculated in terms of persons per square mile for any administrative unit, and only totals of population and acreages are available, then a nomograph can be constructed. This requires the use of semi-logarithmic graph-paper, otherwise the graph will become too unwieldy. The acreages are plotted on the left-hand

[1] J. T. Coppock, 'The Parish as a Geographical–Statistical Unit', *Tijdschrift voor Economische en Sociale Geografie*, vol. 51, pp. 317–26 (Rotterdam, 1960).

[2] F. T. Mavis, *The Construction of Nomographic Charts* (Scranton, Pa., 1948); and S. A. Emery, 'Nomography applied to Land Surveying', *Surveying and Mapping*, vol. 14, pp. 59–63 (Washington, D.C., 1954).

side of one-cycle logarithmic graph-paper, with values from top to bottom. The populations are plotted similarly on the right-hand side of a second sheet, with values from bottom to top. On a third sheet of two-cycle semi-logarithmic graph-paper, a scale of persons per square mile is plotted from bottom to top; this acts as a 'dial' which has to be set relative to the other two scales. In this particular case, the dial is moved until the 10 on it is aligned between 6,400 acres on the left

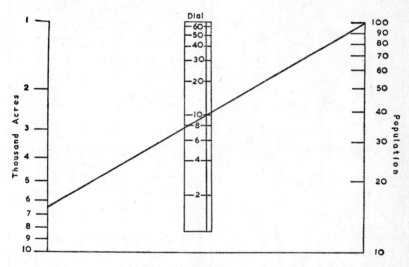

Figure 16. A NOMOGRAPH

Based on J. W. Alexander and G. A. Zahorchak, 'Population-Density Maps of the United States: Techniques and Patterns', *Geographical Review*, vol. 33, p. 458 (New York, 1943).

scale and 100 persons on the right scale, because 100 persons per 6,400 acres is equal to 10 persons per square mile. Once set, the nomograph may be used to calculate densities and to reduce acres to square miles in one operation.

The next problem is the choice of a scale of densities; it is possible to obtain remarkably different impressions from the same statistics by changing the scale. This may be in arithmetical progression (0–100, 100–200, 200–300 per unit and so on), in geometrical progression (0–64, 64–128, 128–256 and so on), in terms of quartile groupings for a 4-category map, or at irregular intervals. The last is justifiable if it reveals significant features of distribution which would be lost by any regular scale. A dispersion diagram will help to show where significant groupings occur.[1]

The Drawing of Choropleth Maps. The boundaries of the particular administrative units which form the basis of the density values are first

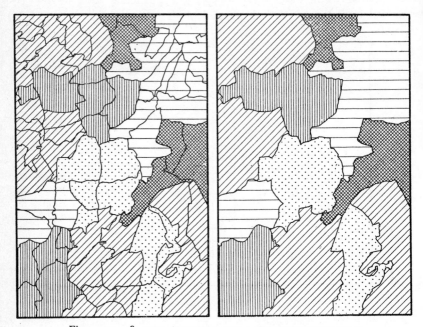

Figures 17, 18. THE SHADING OF A CHOROPLETH MAP

In Fig. 17, the boundaries of the administrative units upon which the values are based are shown in full, but in Fig. 18 only those boundaries are retained which delimit the areas of the same value.

drawn in pencil. A code number or letter (such as 1–5 or A–E in a five-density scale) is inserted in each administrative unit. When completed, it will be helpful to tint on the draft each category in crayon. Then the final drawing is made, outlining by a thin line each of the value-areas 1–5, and ignoring all other boundaries; as a rule, their

[1] A thorough examination of this problem was made by G. M. Schultz, 'An Experiment in Selecting Scale Values for Statistical Distribution Maps', *Surveying and Mapping*, vol. 21, pp. 224–30 (Washington, D.C., 1961). She produced a comparative series of choropleth maps with different scale values, with 3, 4, 5 and 6 categories, and examined the results in the light of the desired objective. She stresses the value of frequency distribution, and of quartile groupings for a 4-category map.

presence spoils the density impression of the shading (cf. Figs. 17, 18). Occasionally, however, it may be essential for the map to show specific values for each administrative unit, in which case the boundaries may be retained.

SHADING AND STIPPLES

Shading Ranges

Tints of shading must be carefully chosen to distinguish a specific distribution on a non-quantitative map, or to bring out and emphasize the isopleth intervals, or to distinguish the various density ranges on a choropleth map. It must be emphasized that it is much more logical to shade uniformly on a choropleth map an area which has an average density of, say, 100–200 units per average area, than to shade uniformly on an isopleth map the area between the 100- and 200-foot contours, or between the 50° F and 60° F isotherms. In the first case, it is the actual area that matters, but in the second case the shaded spaces may distract attention from the isopleths themselves, and produce an effect of stepped areal distribution which does not exist.

To achieve a graphic effect, the density of the shading should be intensified with the increases in distributional values. Yet one frequently sees on a map, for example, two different density values represented by horizontal and vertical lines, each the same thickness and distance apart. This gives to the eye an identical, instead of a contrasting, shading density. Figure 19 illustrates several alternative graded density ranges (examples A–F), of four, five and six depths, with and without black and white as culminating densities.[1]

It must be emphasized that it is not always possible to judge the success of the map from the appearance of the shading range in a few small panels. It is useful to shade one or two sample portions of the actual map, choosing areas both of uniform and of diverse distributions. If a series of maps is to be constructed, it might well be worth drawing one map as a trial, and if it is to be reduced either photograph it down or have a trial block made, before embarking on the series. The maps

---

[1] An extremely detailed review was made by G. F. Jenks and D. S. Knos, 'The Use of Shading Patterns in Graded Series', *Annals of the Association of American Geographers*, vol. 51, pp. 316–41 (Lawrence, Kansas, 1961). They discuss the types of shading, their reduction and reproduction limits, scales of the grey spectrum, texture types, the subjective impression (whether pleasing or otherwise), etc.

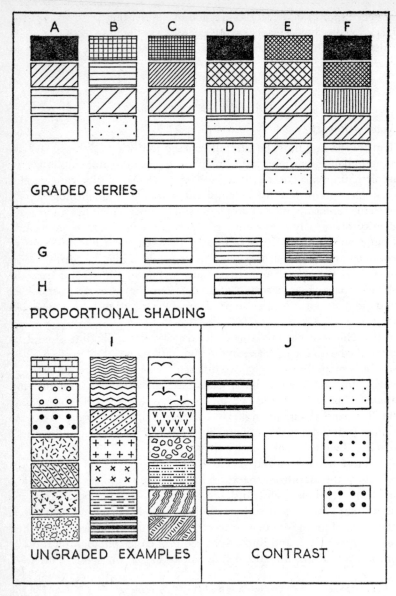

Figure 19. SHADING SYSTEMS

reproduced in this volume are intended to serve as a guide to effective shading systems (see, for examples, Figs. 116–19).

Care should be taken to provide sufficient shading contrast; a balance must be maintained between dark and light tints, and any suggestion of a monotonous grey uniformity over the whole map is to be avoided. Solid black, used with discretion, helps to clarify a map, but the chief difficulty is to avoid the obliteration of line-work, lettering and other detail. Panels may be cut in the black areas or heavier shadings in which essential lettering has to be placed, but if there is much detail, line-shading must be replaced by some faint dot-stipple. The lowest value is commonly represented by blank areas. This is not altogether satisfactory, for it does give an impression of no value at all instead of a low density, and moreover it is often essential to leave the sea un-shaded (occasionally the sea areas may be ruled horizontally with effect, as in Fig. 113).[1]

Two shading ranges on Fig. 18 have been devised to provide an arithmetically correct proportion between the shading and the relative value each represents. This relation can be used when the range of values is not too great. Thus on a four-grade range (say, 0–100, 100–200, 200–300 and over 300), horizontal lines may be drawn 10 mm, 5 mm, 2·5 mm and 1·25 mm apart (example G). Alternatively, the lines may be at a uniform distance apart, but each successive line is ruled twice as thickly as its predecessor (example H). However, as a general rule the compiler is aiming primarily at a striking effect, and this can be achieved without the finicky measuring necessitated by proportional shading. If the compiler ensures that the visual impression of each successive shading is approximately twice as dense as the preceding one, the result should be satisfactory. In addition, to ensure a clear visual impression between two successive shades, change the slope of the shading. Thus cross-hatching may be succeeded by closely ruled diagonal lines, and then in turn by even more widely spaced vertical lines. This will help enormously in emphasizing the pattern of the shaded areas. Care must be taken, when shading a map which is to be reduced, that lines are not ruled too closely, because blurring will take

[1] An analysis of ranges of tints according to scales of values is given, with examples, by R. L. Williams, 'Map Symbols: Equal-Appearing Intervals for Printed Screens', *Annals of the Association of American Geographers*, vol. 48, pp. 132–9 (Lawrence, Kansas, 1958). Some useful statistical summaries of practical tests in the efficacy of various ranges are given. A criticism of this article by A. H. Robinson appeared in *Annals*, vol. 49, pp. 457–60 (1959), and a reply by R. L. Williams in *Annals*, vol. 50, pp. 487–91 (1960).

place either due to halation from photographic reduction, or when the line-block is inked; cross-hatching is particularly susceptible. For a reduction of two and a half times, lines should never be ruled more closely than at millimetre intervals. Double cross-hatching, such as a diamond pattern superimposed over squares, should as a rule be avoided for line-blocks; in any case, such elaborate shading usually takes more time than the results justify.

It is a moot point whether pecks and dots should be used in the same density range as line-shading. The effect of patches of dots between areas of line-shading may not always be pleasing; on the other hand, used judiciously dots may take a useful place at the lower end of a range (Fig. 146). Sometimes one series of lines and another of dots may be used to differentiate between two distinct features; thus on a map representing distribution changes over a period of time, the scale of increases can be represented by lines of varying thickness, of decreases by dots of varying size, and of little or no change by white spaces (example J, and see also Figs. 116–19). A few ungraded shading systems are included on Fig. 19 for use in such non-quantitative distribution maps as land utilization and geological maps (example I).

Line-shading Technique

When the pencil draft is complete, showing the isolines or the limiting administrative boundaries, and the range of tints has been chosen, the final draft may be prepared in ink. First draw in the coast, frontier and boundaries of shading. Then, using a ruling-pen set to the exact line thickness, rule in each shade in turn. It is extremely difficult to obtain absolute uniformity; lines must be strictly parallel, and the ruling-pen should never be allowed to run dry. A single line, slightly fainter or heavier than its neighbours, shows up most glaringly, especially upon reduction. A piece of graph-paper, over which the tracing is pinned, will help to maintain parallelism if a large area has to be shaded, or a parallel rule can be used. Different straight-line shadings in adjoining areas should be 'staggered' and not ruled continuously across the boundaries, except where a gradual grading of related shadings is deliberately required. Pecked lines may be drawn direct, but it is often easier to rule a continuous line, and then either scratch out or paint out with process white the blanks between the pecks. Various automatic shading machines, such as the *Autoliner*, can be used to attain complete uniformity of spacing in line-shading.

Hand-stippling Technique

Uniformity is extremely difficult to attain when stippling is done by hand. Where even shading should appear, one often gets the appearance of closer 'drifts' and conversely of blank patches, or the eye may pick out whorls and other patterns. Here again, the placing of the tracing over graph-paper will help, for the dots can be inserted over the exact points when the graph-lines intersect. On the other hand, when skilfully done, a casual placing of the dots may give a more even impression than would a strict alignment of the dots, but it needs considerable practice to achieve this.

When a stippled map is to be reduced, care must be taken not to make the dots too small. Minute dots are particularly vulnerable to the acid-etching process of the block-maker, and often are found to have vanished when the map is printed. Again, if the dots are placed too closely, blurring is likely, particularly if the printing paper is of not too good a quality. Here again trial and error is essential to deter-mine working standards.

Mechanical Stipples

There are various methods by which shading and stippling may be applied mechanically to a map which is to be reproduced as a line-block. One method is to stick on to the manuscript map pieces cut to shape from a cellophane sheet, on which the various stipples are already printed. One variety, with the trade-name of *Contact* and made in the United States, consists of a wide range of screens, patterns and symbols, printed on *Dupont Mylor*. This can be easily cut out and laid on to the map, and will not dry out, shrink, tear or crack. Another is *Zip-a-Tone*,[1] for which a chart is available depicting 180 different shadings and stipples, and another gives 99 special map-symbols. In addition to black, *Zip-a-Tone* is available in 27 colours. The *Zip-a-tone* is cut with a special needle which a draughtsman can use as if it were a pen, holding it at an angle of 60 degrees. The reverse side of the screen is waxed, and adheres firmly and permanently to the drawing to which it is applied.

In making a map, *Zip-a-Tone* is placed over the specific areas, and the surface gently rubbed to make the screen adhere. Parts not

---

[1] The English agents for *Zip-a-Tone*, which is made by the American firm Para-Tone Inc., are Hunter-Penrose-Littlejohn Ltd., 7 Spa Road. Bermondsey, London, E.16.

wanted around the margins are cut away with the needle. Then the draughtsman goes over the *Zip-a-Tone* with a bone burnisher, applying sufficient pressure to ensure permanent adherence.

One of the most successful methods of mechanical stippling is the use of *Letratone*; this affords a range of 200 line and dot patterns on black and white adhesive film, which can easily be cut out and stuck down on the area to be shaded. Another range is *Formatt*, a type of acetate cut-out sheets, with 236 different shading designs.[1] The sheets are backed with a heat-resistant adhesive which holds at a touch when applied to the area to be covered with the chosen shading, but bonds tightly when burnished. Other makes of adhesive pre-printed materials are *Craf-Tone* and *Artype*.

For maps to be printed in journals and books, the system of mechanical stipples applied by the block-maker may be used. *Tint-books* showing a wide range of stipples (known to the trade as 'tints') are available, each of which has a standard number. The compiler of a map must complete all the line-work and lettering, and then add a fine line surrounding the area to be stippled. This area should be lightly shaded blue (which does not photograph) on one side or other of the tracing, unless the areas concerned are particularly complicated in shape, when a transparent overlay on which the areas are separately shaded may be drawn, carefully ensuring an exact register with the original. The reference number of the tint required must be added in the margin. If straight-line tints are desired, the angle at which the shading is to be laid must be indicated in the margin, either by ruling blue guide-lines or by specifying the exact angle to the vertical margin. Occasionally a graded series may be required, as on a relief map; this must be very carefully chosen from the available tints.[2]

Care must be taken with line-work and lettering when it is known that mechanical stipples are to be applied. In general, lines should be considerably thicker and lettering written more openly than usual. Many stipples simply obliterate underlying detail.

As a rule, this mechanical stippling should be used only when absolutely necessary. It is, of course, very expensive in proportion to the

[1] Produced by Graphic Products Corporation, 3810 Industrial Avenue, Rolling Meadows, Ill. 60008, U.S.A. A 16 pp. catalogue illustrates each of the available patterns.

[2] G. F. Jenks and D. S. Knos (op. cit., 1961) give on p. 332 a reference chart showing the percentage of area inked, from 0 (white) to 100 (black), through thirty-seven ranges of mechanical stipple, with accompanying densitometer rating, screen texture. pattern number and manufacturer.

cost of the block, and results are not always satisfactory. The most successful and commonly used mechanical stipple in the Lascelles series is No. 526, which provides an even grey stipple, through which lettering and line-work appear clearly; it uses some 10,000 dots per square inch without blurring, even on quite poor paper. This admittedly could never be attained by hand.

The stipples are applied by the block-maker, after the photographic reduction of the manuscript map, directly on to the block, but before the etching process. It is therefore a highly skilled and expensive operation. The tint specified is cut out to the exact shape and applied to the block, an acid-resisting ink is rolled through the tint-sheet on to the surface of the block, which thus applies the tint, and the block is then etched in the usual way.

COLOUR

The Use of Colour

In recent years there has been much research into the use of colours in the production of maps, particularly those depicting relief (see pp. 96 and 100).[1] There are what may be called the well-established conventional aspects of usage, the familiar associations; examples include blue tinting for the sea; stepped green, yellow, brown for relief; various land-use conventions such as brown for arable, green for forest; in maps showing temperatures the association of red with warmth and blue with cold; and the like. There is the association of magnitude with density of colour, for example, areas of heavy precipitation are often shown in dark blue, areas of light rainfall in pale blue, yellow or white. A. H. Robinson has discussed the three main colour dimensions. The first is *hue*, that is, the actual colour, the spectral colours of the rainbow and all the vast number of variants (as is shown, for example, in the *British Standards* colour range). The second is *value*, that is, the sensation of darkness (high value) and lightness (low value). And there is *chroma*, the actual reaction of the eye so that some colours are brilliant

[1] J. S. Keates, 'The Perception of Colour in Cartography', *Proceedings of the Cartographic Symposium*, pp. 19–28 (Edinburgh, 1962). In volume 7 of the *International Yearbook of Cartography* are published a series of papers on the subject of colour in cartography, originally delivered at the Third International Conference on Cartography at Amsterdam, 1967. These were: (i) A. H. Robinson, 'The Psychological Aspects of Colour in Cartography'; (ii) A. Heupel, 'Farbe und Topographische Karten'; (iii) C. Traversi, 'Emploi de la Couleur Verte dans les Cartes Topographiques'; (iv) R. Núñez de las Cuevas, 'Color in Topographical Maps'; (v) A. Makowski, 'Aesthetic and Utilitarian Aspects of Colour Cartography'; (vi) V. N. Philin, 'Colour Metrics in Cartography'.

and intense, others are dull, pastel, even 'washed out'; this, in fact, is
the actual amount of the hue in the colour. One difficulty is that not-
withstanding the generalizations made by psychologists and other
testers, there is an enormous amount of individual subjective choice,
whim, discrepancy, approval or the reverse, and in choosing colours
for maps the cartographer must seek to attain such practical aims as
clarity, legibility, contrast and balance, and hope that aesthetically
the result will be acceptable. In the preliminary compilation of an
atlas, for example, it is necessary not only to produce a whole series of
possible colour ranges but also to make several trial maps and choose
between the alternatives.

The printing and reproduction of colour-maps (such as those in
atlases), using half-tone tints, is outside the scope of this work, but
colour can be used in a limited way, on line-blocks, either by stipple or
by line.[1] If colour stipple alone is required, a single drawing is sufficient,
on which the limits of the coloured areas are indicated by black lines,
whether contours or any other isopleths, or the boundaries of administrative areas on a choropleth map. The areas to be coloured are marked
with blue pencil, as for stipples on an uncoloured map, and the various
tints are specified in the margin, with the name of the colour required
(e.g. 'Brown' for relief maps) placed beside them. The block-maker
produces two blocks, one carrying the line-work in black, the other the
stipple to be printed in colour; two printings are of course necessary.

If, however, it is desired to have in colour line-work, lettering, or,
in fact, any other detail than a mechanical stipple, there are two possible methods. In the first, two or more drawings are needed, one for
black and one for each colour; from one is made the black block, from
the others the particular colour blocks, and the separate drawings are
checked carefully for exact register. Alternatively, the map is drawn
in black and primary colours, and the several colours are separated
photographically by filter, thus enabling separate blocks to be made
with a high accuracy of register. The maps can of course be printed
in any colour.

These line-block colour-maps can appear in the text. This means

[1] For examples of the use of colour for textbook illustrations, see John Warkentin
(ed.), *Canada – Geographical Interpretation*, prepared under the auspices of the Canadian
Association of Geographers (Toronto, 1967). This has good examples of compound
graphs, columnar diagrams, colour wash and variegated symbols. See also Jan O. M.
Broek and J. W. Webb, *A Geography of Mankind* (New York, 1968), in particular, use of
colour for cartograms and cultural distributions, many of the maps being on a world
scale; and also population graphs and diagrams.

that the sheet (i.e. 16 pages of text) will be run through the printing-press once for each colour, so for economy as many as possible of the intended colour-maps should be placed on the same sheet. The method is obviously uneconomical for a single map. Alternatively, the colour map can be printed on a single sheet of glossy art paper and 'tipped in' (i.e. gummed in) by the binder. It must be emphasized that these colour maps are expensive luxuries, and should be used only when merited.

Colour, however, can be employed very effectively in the cartography notebook or in dissertations where only two or three copies are required and the maps can therefore be hand-tinted. The use of colour must not be overdone, however, as gaudiness can produce a most crude effect, and heavy colours will overshadow other parts of the map. A pastel shade, emphasizing but not obliterating the line-work, should be aimed at.

Colour-washes

The line-work must be finished in Indian ink, allowed to dry thoroughly, any pencil lines removed, and the surface of the paper dusted. Sufficient colour should be mixed for the whole map, and if possible a sample tint should be applied and allowed to dry for inspection. The chief difficulty is in laying a uniform wash over a large area; small patches of colour are comparatively easy to apply. The map should be put on a sloping board, and the wash laid on from the top downwards, brushed on in broad continuous free-flowing sweeps, using the largest possible brush. The wash should be applied as rapidly as possible, consistent with precision, so that the edge never dries out. When the lower margin of the map is reached, remove the excess colour by running a dry brush along it. One practical suggestion is to remember to fill in the key-panel while using each particular colour; it is often impossible to match shades afterwards.

Allow the colour-wash to dry thoroughly before applying an adjoining different colour or a further coat of the same colour if relative density layer-tinting is being attempted (see p. 96). One is frequently disappointed with layer-tinting; the result appears patchy and uneven, and touching-up usually makes things worse. Poor paper, uneven application, too heavy previous erasing which may have marred the surface of the paper, poor quality colours (especially greens), all these may account for the unsatisfactory results.

LETTERING

A considerable proportion of the information presented on a map is conveyed by the lettered names. Moreover, the standard of appearance of the finished map depends to a large extent on the quality of the lettering; all too often fine line-work is marred by unpleasing lettering, or essential detail is obscured. Lettering is a fine art, and to attain real proficiency demands a long and patient training, so obviously a student cannot spend much of his limited time with the object of becoming such an expert. His aim should be to produce simple, quickly written, easily legible, yet reasonably attractive lettering. Using various aids, most students can achieve an adequate standard in a relatively short time.

There are many works on lettering as a fine art.[1] A study of the methods by which such topographical maps as the various Ordnance Survey series[2] or atlas maps are lettered may be helpful, if only to indicate the lofty standards at which to aim. But the problems of lettering maps to be drawn as exercises or to illustrate dissertations or even for line-blocks, are of a different order. This section is intended to do no more than to summarize the simpler methods available, and to

[1] Some useful works on lettering are as follows:

J. G. Withycombe, 'Lettering on Maps', *Geographical Journal*, vol. 72 (London, 1929).

C. B. Fawcett, 'Formal Writing on Maps', *Geographical Journal*, vol. 95 (London, 1940).

F. Debenham, *Exercises in Cartography*, chap. 8 (London, 1937), which is written especially for the student draughtsman.

A very full treatise on lettering is G. Hewitt, *Lettering* (London, n.d.), written primarily for craftsmen and professional letterers. It is, however, copiously illustrated, and the student cartographer can derive much benefit from it.

P. Buhler, 'Schriftformen und Schrifterstellung', *International Yearbook of Cartography*, vol. 1, pp. 153–81 (London, 1954), deals with lettering styles, design, freehand lettering, the use of templates and lettering machines, etc., with many illustrations.

C. E. Riddiford, 'On the Lettering on Maps', *Professional Geographer*, vol. 4, no. 5, pp. 7–10 (New York, 1962).

A. H. Robinson and R. D. Sale, 'Cartographic Typography and Lettering', chap. 12, *Elements of Cartography*, pp. 273–91 (New York, 1969); A. Nesbitt, *The History and Technique of Lettering* (New York, 1957); and A. G. Hodgkiss, 'Lettering Maps for Book Illustration', *The Cartographer*, vol. 3, pp. 42–7 (Toronto, 1966); the last is an essentially practical survey, by a skilled practitioner, of the visual impact of lettering, its cost, and rapidity of execution of various methods.

[2] Current series of the O.S. maps are not lettered by hand, but names are set up in carefully designed type, and stuck on the face of the map in the correct position. Many other maps are lettered in this way. See J. S. Keates, 'The Use of Type in Cartography', *Surveying and Mapping*, vol. 18, pp. 75–6 (Washington, D.C., 1956).

give a few practical hints. Some of these principles, suitably developed, have been used to letter the maps which illustrate this book.

The Lettering-mask

Whatever method of lettering is employed, a number of general points must be borne in mind. When the line-work has been completed on the preliminary pencil-draft, prepare a lettering-mask to fit over the draft (mark the corner angles to give an exact register), which will indicate the positions of the various names. As far as possible, lettering should be placed horizontally for ease of reading; it ought to be possible to read comfortably all names on a map without turning it about. River names should be curved to follow approximately their courses; place each name on the north bank, unless the river is flowing north–south, when it should be placed along the west. Examine carefully the position of any larger names, such as those of countries; they should interfere as little as possible with other names or line-work, but should be spread out sufficiently to occupy most of the area to which they refer. Never allow names to interlace, nor space them so widely that the map appears to be dotted with odd unrelated letters. Make sure that it is absolutely clear to which feature or symbol a particular name refers. It may be necessary to alter the position of some names to avoid overcrowding; the location of a town-name, for example, can be transferred to the other side of its symbol, as long as it still clearly refers to it and to no other.[1] When the mask is fully lettered, examine critically the appearance of the lettering as a whole, and see if a harmonious and balanced composition has been attained relative to the line-work. If too many names are included, the clarity of the map will suffer.

Next analyse the names – countries, oceans, mountains, rivers and towns – and decide on the form for each group, differentiating these on the mask itself. Thus countries can be shown by roman capitals, towns by roman lower case, seas by italic capitals, and rivers and lakes by italic lower case. Italics lend themselves surprisingly well to the curving necessary along rivers, although stencil italics are difficult

[1] Attention has been paid to this important topic by E. Imhof, 'Die Anordnung der Namen in der Karte', *International Yearbook of Cartography*, vol. 11, pp. 93–129 (London, 1962). His detailed survey examines the optimum position on a map for a name, for legibility, clearness and certainty, the problem of curving a name, spacing a name out over an area, avoiding crossing and overlapping names, the problems of names of peaks and passses, etc. He has a number of striking comparisons of good and bad placing of names.

to place on curves. When a series of maps is to be lettered, these contrasting forms should of course be standardized throughout.

Finally, the completed mask can be checked at this early stage for actual spelling mistakes, and also to ensure consistency with the text and with the policy adopted where there are alternative place-name forms.

Alignment, Size and Spacing of Letters

The next step is to draw thin guide-lines on the actual map for the letters; these may be in pencil, and two pencils can be tied together at the correct spacing. More conveniently, a celluloid stencil may be cut, providing several parallel lines at different intervals, and other lines at right-angles and at various slopes. Yet another device is to place a piece of blue carbon-paper, face up, under the tracing, and draw the guide-lines with a pair of dividers; this will provide fine blue parallel lines which will not photograph if the map is to be made into a line-block. The beginner must draw three lines if he is using lower case letters, to delimit the base of the letters, the top of the lower case, and the top of the capitals. If stencils are being used, guide-lines are unnecessary except for the base-line of curved names. For horizontal names, the stencil is slid along a T-square or straight-edge in order to ensure perfect alignment. In addition to the horizontal guide-lines, it is very helpful to draw vertical or oblique guides in order to maintain a constant slope. With some experience, this will prove to be unnecessary, except at the beginning and end of a name in sloping letters.

The actual widths apart of the horizontal lines (determining, of course, the size of the lettering) is decided by a consideration of the number of names to be inserted, by the scale and extent of the map, i.e. the room available, and by the amount of reduction to be applied. The last is most important; frequently one can see a printed map, on the original of which the lettering had obviously been drawn much too small for the particular reduction, so that blurring and indistinctness resulted.[1] It must be appreciated that if the map is to be reduced, the letters must be proportionally wider and more open, the spaces between greater, and the fine lines less fine, than if no reduction is intended. Italics especially must be made much rounder, and acute angles widened.

[1] A. H. Robinson, 'The Size of Lettering for Maps and Charts', *Surveying and Mapping*, vol. 10, pp. 37-44, (Washington, D.C., 1950), deals with the questions of appearance, readability and reduction.

When guide-lines have been drawn in for every name, and any awkward or conflicting alignments adjusted, the names may be pen-cilled lightly in to obtain a correct spacing of the letters. This can only be done by eye; the space occupied by the different letters varies so much that mechanical spacing produces a displeasing effect.

Styles of Lettering

A style of lettering should be adopted which can be done naturally, and which is appropriate to the specific purpose. Find the particular nib which suits this style, and practise so that some degree of uniformity and speed is attained. Manuscript lettering may be divided into two groups. The first is lettering built up by multiple strokes of a broad stiff pen; the second is script lettering, in which each line consists merely of a single stroke of a steel or quill pen. The first method should be used only for titles and large names as it is a very slow process. A number of alternative styles is shown in Fig. 20.

In all cases, the form of the letters should be kept simple, neither distorted nor elaborated. Too severe and regular a style tends to make slight discrepancies stand out glaringly, but highly decorative styles, such as Old English (as conventionally used for antiquities) are hard to draw and are not very legible, nor do they reduce well. The extensive use of 'serifs' (the little ticks at the beginning and end of straight lines in letters) is generally to be avoided, as being time-wasting, and, moreover, if they are too exaggerated, they may mar the distinctive forms of the letters themselves. Used with discretion they may help to draw attention from irregularities in a long straight line of letters.

Light and heavy lines should be kept consistent in thickness by maintaining a constant angle at which the nib is held to the paper. A stiff broad steel nib will be found most convenient; a springy nib makes uniform line thickness very difficult to attain. Some final hints are to keep the nib clean by wiping it on a piece of linen at frequent intervals, and to rest the hand on a piece of paper to keep the map unsoiled. When finally inking-in the letters, insert them before the line-work, so that it will not be necessary to erase any detail to clear the letters.

Quill-lettering

A quill may be used very effectively instead of a steel nib for script lettering. The general effect is pleasing, while irregularities are not obvious, and with some practice a fair speed can be attained. It is possible for any student to produce quite passable results. It must be

MACEDONIA

Finland

ARABIA

Liverpool

BRADFORD

Warrington

MISSISSIPPI

River Amazon

PRESTON

Buntingford

PYRENEES

Brahmaputra

LLANGOLLEN

Finchingfield

Figure 20. LETTERING STYLES

The first four examples (capitals and lower case, in roman and italic styles) were built-up with a steel nib; the second four examples were written with a quill pen; the third group was drawn with the help of stencils, and then the ends of the letters were either trimmed with a razor blade or 'filled-in' to square them off. The last two examples were drawn with stencils, but using a finer nib than the makers specify.

admitted, however, that quill-lettering is most effective when not intended for reduction. Often a beautifully lettered map, so attractive in its original form, is most disappointing upon reduction, for angles tend to blur, and the delicate differences in stroke-width are lost.

Lettering-guides

The use of guides, such as stencils or templates, for lettering has much to commend it. It enables complete uniformity to be obtained, and where reduction is intended a little experience will readily show the exact size of lettering to be used for that particular reduction. Guides can be obtained in a range of sizes, for upper and lower case, roman and italic styles, and also for numerals. A pen of the hollow tube type should be used, so that by keeping the cylinder vertical the outlines of the letters can be followed exactly, and moreover the line thickness can be kept constant. Some guides specify the size of pen to be used; it has been found by experience that for clarity and better proportions, especially with smaller letters, a pen one size less than that specified should be used. The ruling of a base-line and correct spacing by eye are as essential as for freehand lettering, especially when the name is to be lettered on a curve. A number of types of pen and stencil are available, notably *Leroy, Wrico, Varigraph, Normograph* and *Speedball*.[1]

Stencil lettering tends to leave rounded, sometimes 'blobby', ends to the strokes. This can be removed by 'trimming' the letters with a keen blade after the ink has dried, so giving a clean-cut precise finish to the lettering. However, if vertical strokes are trimmed, they will be shortened, thus throwing those letters out of alignment with the rest; the corners must be filled-in with a fine nib to produce a straight-edge base and so keep the correct height of each letter. If working on a surface such as drawing-paper, all corners and ends must be squared by filling-in, as trimming with a blade will spoil the surface of the paper.

The mechanical perfection of the guided letters, together with the individual judgement and taste needed for effective alignment and spacing, combine to make trimmed guided lettering a very convenient, legible and effective mode of lettering for maps intended to be reproduced as line-blocks.

[1] R. F. George, *Speedball Textbook* (various editions), published by the Hunt Pen Co., Camden, N.J.

Dry-transfer Lettering

There is no doubt that lettering of maps has been revolutionized by the introduction of instant dry-transfer lettering, which uses self-adhesive characters printed on the underside of transparent acetate sheets. Such a method has been developed by the firm of *Letraset*,[1] which incidentally received the Queen's Award to Industry in 1966. The size of letters available ranges from 120-point (about 55 mm high) to 6-point (about 2 mm), and there are about eighty styles of type; some founts are familiar, such as 'Times Bold', 'Baskerville', 'Clarendon', 'Folio' and 'Old English', while others have been specially designed. In addition, Greek, Hebrew, Cyrillic and Arabic characters are available, as well as a range of numerals. Great care must be taken in arranging the position of the name and in spacing the lettering; ruled blue guide-lines are helpful. The protective blue backing-paper is removed from the sheet of letters and the first letter is positioned. The draughtsman then shades across the letter from top to bottom with a ball-point pen, using moderate pressure; when the sheet is slowly lifted, the letter will be left on the surface of the map, and the next one is selected. Finally, the blue protective backing-sheet is placed over the complete name and is rubbed hard to ensure maximum adhesion. This form of lettering, carefully executed, will give a really professional finish to maps and diagrams.

Mechanical Lettering

It is an increasingly common practice to use what is sometimes called 'stick-up lettering', whereby the names are set in printer's type,[2] and then affixed to the map. Some firms, given the desired list of names, can supply them printed on such media as matte or glossy acetate, or on such film as *Vinylite*, with a wax backing. They may be of the correct final size, or larger where the map is to be photographically reduced. The names are simply cut out and stuck down in the correct position

[1] Letraset Ltd., St. George's House, 195/203 Waterloo Road, London, S.E.1.

[2] A valuable summary, with illustrations of the different founts of type, their features and their value in cartography, is given by A. G. Hodgkiss, 'Typography for Carto-graphers', *Bulletin of the Society of University Cartographers*, vol. 3, no. 2, pp. 53–9 (Liverpool, 1969). See also *Type Face Terminology* (published by F. C. Avis, 26 Gordonbrook Road, London, S.E.4), which defines and illustrates over 200 terms relating to type design. A valuable discussion of various available printed type-faces is given by R. A. Gardiner, 'Typographic Requirements of Cartography', *Cartographic Journal*, vol. 1, pp. 42–4 (London, 1964). He discusses a standard work by Sir Cyril Burt, *A Psychological Study of Typography* (London, 1959).

on the map. One disadvantage of this method is its rather artificial appearance, as compared with lettering spaced by eye, and the names cannot readily be curved.

A more complex and sophisticated technique is known as *photo-typography*. The general principle is that the names are set up, letter by letter, using a manually operated keyboard; each letter is photographed in turn on a film, and then is automatically developed, fixed and dried. These names may be either positive (to be stuck directly on a map,

HEADLINER

YORKSHIRE

Urban Areas

H o l d e r n e s s

M i s s i s s i p p i

KINGSTON UPON HULL   River Humber

E n n e r d a l e   H u m b e r s i d e

R i v e r   T r e n t   **Kuala Lumpur**

B E V E R L E Y   *Geography*

Figure 21. AN EXAMPLE OF PHOTOTYPOGRAPHY
The names were set up by the *VariTyper Headliner 820 Photo Composing Machine.*

using an adhesive wax) or negative (to be stuck into a 'window' on a scribed negative sheet). While some devices space the letters automatically, other simpler types necessitate spacing by eye. A useful assistance in this connection is the *PPE Typographical Slide Rule*,[1] which shows the number of characters, of forty-four type faces and all point sizes, which can be contained within a given distance. One great advantage of phototypography is that the size of lettering can be readily varied by enlargement or reduction. Some recently developed machines can position the names in their correct locations on the actual

[1] Produced by Krisson Printing Co. Ltd., 184 Acton Lane, London, N.W.10

map, using a system of co-ordinates.[1] Mention may be made of two
very efficient lettering machines. The new model *KC5* of the *Photonymo-
graph* (developed from a machine which appeared in its earliest form
over thirty years ago) is made by Barr and Stroud, Ltd., and the
Monotype Lettering Machine is produced by the Monotype Corporation
of London. In the latter, the characters are selected manually by
dialling, after which the whole operation is automatically controlled;
it can print characters on either film or paper at a rate of up to sixty
per minute in type sizes from 5 to 90 point. Figure 21 shows various
kinds of printed letters, which have been set up by the *VariTyper Headliner 820 Photo Composing Machine*.

REDUCTION AND ENLARGEMENT

Reduction and enlargement of maps can be carried out most expedi-
tiously by photographic means (see p. 68). However, various graphical
and instrumental methods are available, and their use forms a valuable
cartographical exercise.

Graphical Methods

The Method of Squares. The original map is covered with a grid of
unit squares, either by ruling faint lines on the map itself, or by laying
over it a suitably ruled piece of tracing-paper, or a celluloid grid can be
used. The closer the grid, obviously the more accurate will be the
result; a 1-inch map could carry a $\frac{1}{4}$-inch grid. Rule another network
of squares, enlarged or reduced as desired, and copy the detail, square
by square, by eye on to the drawing-paper, noting particularly any
important intersections of detail with the grid lines.

It is, of course, the change of scale of the side of the square that produces the desired amount of enlargement or reduction. To enlarge three
times, for example, the side of the square on the drawing-paper will be
increased three times, that is, the area will be enlarged nine times.

Care must be taken over the size of the conventional symbols. As a
rule, unless the enlargement is very great, road widths and most
symbols such as churches should not be enlarged; on the One-inch
Ordnance Survey map most symbols are already exaggerated. Nor
should lettering be increased proportionally. The finished enlargement

[1] The student will find useful *An Experiment in Cartographic Typography*, produced by
the Oxford University Press and the Monotype Corporation Ltd., Monotype House,
43 Fetter Lane, London, E.C.4.

ORIGINAL

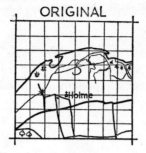

REDUCED TO ONE HALF

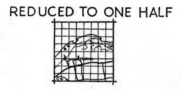

ENLARGED TWICE

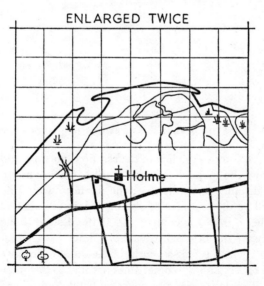

Figure 22. REDUCTION AND ENLARGEMENT BY SQUARES

should have as much balance as the original. Conversely, some detail on a reduction must be simplified, generalized or even omitted. Examples of enlargement and reduction by the method of squares is shown in Fig. 22.

The Method of Similar Triangles. This may be used for the reduction or enlargement of any narrow area, such as a length of road, railway or river, which would otherwise present considerable difficulties. Rule a

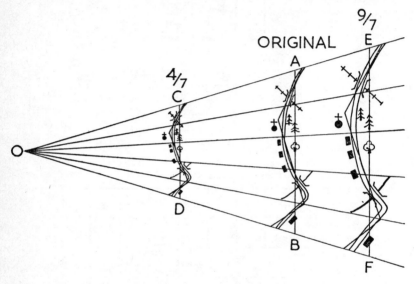

Figure 23. REDUCTION AND ENLARGEMENT BY SIMILAR TRIANGLES

straight line *AB*, across the line of the required portion, and rule guidelines to follow the major bends and curves. Choose any point *O* at a convenient distance from the line; the further away *O* lies, the more accurate will be the proportions obtained. Join each end of the section and any other significant points to *O*. If the desired reduction is, say, four-sevenths of the original, divide *AO* into seven equal parts, and find a point *C* which is four-sevenths of the distance from *O* to *A*. Draw *CD* parallel to *AB*, and also other lines parallel to the guide-lines along the section. The position of all important bends and other features will now be fixed along *CD*, and other detail can be drawn in by eye (Fig. 23). If enlargement is desired, produce lines *OA* and *OB* proportionally, and find the position of *EF*, then draw in detail similarly (Fig. 23).

Instrumental Methods

Proportional dividers can be used to simplify the copying of detail by the methods of squares or similar triangles. It is an instrument consisting of two bars, pointed at both ends, which act as dividers. These bars fit diagonally across each other, and are held together by a screw which slides in a groove down the centre of each bar. The screw is set to the proportion required, according to a scale-line on the top bar, and screwed down tight. If any distance is stepped off with the dividers at one end of the instrument, the dividers at the other end will give the same distance increased or decreased proportionally according to the scale-setting (Fig. 2).

Two instruments are in common use for redrawing maps on a different scale, the pantograph and the camera lucida. The *pantograph* (sometimes spelt pantagraph) consists in its simplest form of four metal arms of equal length, loosely jointed at all corners but one, which is fixed to a weighted stand. A pencil is inserted in the corner diagonally away from the fixed corner. A cross-bar, the position of which can be adjusted to determine the scale-factor, can be moved along parallel to two of the sides. This cross-bar carries a second pencil held in a marker, which, when the cross-bar has been moved to the correct scale-factor, lies on the diagonal from the fixed corner. Thus if the line-work to be copied is traced carefully round by the marker on the corner, the pencil on the centre of the diagonal will draw the same pattern on the reduced scale. Conversely, if the centre marker is used to trace the line-work, the pencil on the corner will reproduce the same pattern on an enlarged scale. This instrument is quite useful for reduction, but any enlargement tends to emphasize and exaggerate inaccuracies.

More elaborate models are available, with little wheels for smooth movement, or with refinements to guide the trace-point more smoothly, as in such patterns as the Coradi suspension model where the corners are held by taut wires from an upright arm to reduce friction. There is no point in describing the complications of these models; if the student wishes to use them, some practice following the maker's instructions will soon familiarize him with their operation. At one time, a pantograph was used very extensively by draughtsmen, but its use has declined since photographic methods have been developed.

The Camera Lucida depends on optical principles. The map to be copied is fixed in a vertical plane illuminated as brilliantly as possible, while the drawing-paper is laid in a horizontal plane. A prism, mounted on an arm, is placed so that when the eye is put to it an image of the map appears vertically below the prism over the drawing-paper. The disadvantage of this is that if the eye is moved relative to the prism, the image will move relative to the pencil, so that distortion is difficult to avoid. The method should only be used for the very approximate copying, enlarging or reducing of a simple map; a large wall-map, for example, can be copied conveniently. To enlarge or reduce the original, the position of map and drawing-paper relative to the prism must be adjusted; if these distances are equal, the image will be the same size as the original, while if the drawing-paper is nearer, the image will be reduced proportionally. Reduction is more satisfactory than enlargement, since the latter obviously increases errors.

Photographic Methods

A copying camera is a most essential part of a studio equipment. A half-plate camera, equipped with a ground-glass screen, and having at least a double extension and a 7–8-inch lens, can be arranged to move on runners about 12 feet long. By this means, maps up to about 4 feet square can be copied, and printed to any reduction. An enlarger capable of enlargements up to about 20 by 16 inches will enable negatives of photographed maps to be enlarged to a scale convenient for fine-drawing. Field-work plotted on a series of large-scale maps can be reduced photographically to transfer to the desired smaller-scale map. Moreover, if a map has to be made into a much smaller line-block, it is useful to have it photographed as an indication of the clarity of the lettering and shading.

MAP DUPLICATION AND REPRODUCTION

Printed maps are produced in large numbers by government agencies and publishing houses by methods described briefly below (p. 71). It should be noted that a strict copyright obtains in such printed maps (see p. 87). It is, however, often necessary to duplicate maps cheaply and in relatively small numbers for use in the classroom, cartography laboratory, planning office and architect's studio, or to provide base-maps for a series of thematic plotting exercises, for fieldwork, etc., and in any circumstance where the expense of a printed map would not be justified, e.g. for a local large-scale survey.

These low-cost, non-printing reproduction processes may be broadly divided into two groups: *stencil* and *photographic* methods.

Stencil Reproduction

The map is drawn as a positive on a special waxed stencil sheet, using a pointed stylo; names may be lettered by hand or typed in position, although the latter looks very artificial (except in legends). The stencil is then placed in a duplicating machine (such as *Gestetner* and *Roneo*), either hand or electrically operated, and as many copies as required are run off. Some types have a counting device which indicates the length of the run. It is, however, impossible to achieve a high standard of draughtsmanship when working directly on to a stencil with a stylo, though mistakes can be corrected by painting-out with an opaque material and redrawing. Recently photographic methods have become available which permit stencils to be produced automatically from a positive original drawing, with a very great resultant improvement in quality.

Photographic Methods

The *Dye-line* process is now widely used, especially by the drawing departments of local authorities and by commercial offices. A transparent sheet on which the map to be copied is drawn or printed is exposed over sensitized printing paper, and developed to give a print on the same scale as the original. Plans up to 1 m wide can normally be reproduced. The 'negatives' can be hand-drawn on tracing paper or transparent plastic film, or can be prepared photographically from opaque originals; of particular value are the transparencies prepared by the Ordnance Survey from their maps of all scales, though these are for official use only.

Two basic types of dye-line process are in use. Dry-developing machines have a high capacity and can produce not only prints but also new negatives. These are particularly useful as they can be amended by scratching out parts of the image on the underside of the film with a razor blade or lance eraser, then adding new information on the untouched top surface, so that a series of maps showing, for example, the successive development of a town can be produced with a minimum of labour. Prints from dry-developing processes do, however, have some disadvantages; they do not take colour-washes or waterproof ink at all well, and they fade with prolonged exposure to light. Prints made by semi-dry processes are rather easier to use, though they are not entirely stable. The difference between the two is that dry-developing involves the use of ammonia in the treatment of the image;

this uses, incidentally, a much more expensive machine. In the semi-dry process the print passes between rollers dampened with developer.

The main difficulty experienced in preparing art-work for dye-line printing is that the heat of the machine and the friction of the paper tend to produce heart-breaking damage. *Zip-a-Tone* stipples and *Letraset* lettering are particularly vulnerable, even if 'fixed' with protective sprays. Where these methods have been used it is wise to risk the original only once to prepare a working negative, and even then it is preferable for this to be done immediately after the machine has been switched on before it becomes too hot.

The *direct contact negative* (usually known as a *blueprint*) is made by exposing a translucent drawing over special iron-sensitized paper, which when developed and fixed reproduces the detail in white upon a blue background. This is more usually used for engineering drawings, however, than for cartographic work.

Direct photo-copying (such as *Photostat*) involves the exposure of the original drawing by way of a lens on to sensitized paper, which is wet-developed to produce a paper negative with white detail on a black background. The procedure is repeated with this negative to obtain black positive detail on a white background.

Direct photography, using a copying camera on to film, enables prints to be made, enlarged or reduced as desired, on to dimensionally stable draughting-film if further addition of material is required during map compilation. Most of the large photographic firms make suitable draughting-film.

One of the world's most successful duplication systems is *Xerography*, developed in the United Kingdom by *Rank-Xerox* in association with the parent *Xerox Corporation* of the United States. This is a dry process which does not involve chemical solutions; it can be carried out automatically, and is extremely rapid. Above all, it enables any type of paper to be used, in contrast to photo-copying which employs photographic paper; this is difficult to colour, has a pronounced texture, and tends to deteriorate on exposure to light. The greatest drawback to *Xerox* copying is its apparent inability to reproduce areas of solid black, which tend to 'ghost' in the centre. The principle used is that of the photoconductivity of the image on to a sheet of paper, which is then heat-fixed. Many institutions, libraries, etc., have *Xerox* apparatus; some even have automatic coin-operated machines for use by students. Care must be exercised in the copying of copyright material.

An alternative form of *Xerox* equipment is intended primarily for the production of plates for offset-litho printing (see p. 72), but is also particularly useful for the rapid and accurate enlargement or reduction of plans. The apparatus resembles a large camera, which is focussed to give the correct size of image on a sensitized plate, from which it is transferred to ordinary paper or transparent film. The biggest limitation to this process is that the size of the plan produced is restricted to approximately 250 by 500 mm.

Mention should be made of the range of *proofing-films*, which enable a cartographer to make contact film positives or negatives without requiring a dark-room, using various emulsions, developers and bleaching solutions. In some cases the sensitizer emulsion in various colours can be applied by hand to a plastic sheet as required. Proofing-film is especially valuable for checking the content of a map before it is sent to the printer, particularly where several colours are being used.

Finally, mention should be made of a number of highly sophisticated machines which have been produced, notably in the United States. An example is the *Electrofax Map Copier*, used by the U.S. Navy to make 5-colour charts by means of an electrostatic printing operation from 70-mm miniaturized film.[1]

The Production of Printed Maps

It is useful for the student to have an outline knowledge of the methods by which maps can be reproduced by printing, especially if he intends to seek employment in such work, or if he hopes to produce maps to illustrate books or periodicals.[2] After the invention of printing in Renaissance times, two main methods were used: the *woodcut* and the *copper engraving* (or *intaglio*). In the former, the outline was drawn on a block of hard wood, and the intervening areas were cut away so that the lines stood up above the surface to take the printer's ink. In the latter, the detail was cut into a sheet of copper with graving tools, so that when the plate was inked and then wiped clear, the ink remained only in the grooves, to be transferred to the paper pressed against it. One disadvantage of a copper-plate is its softness, so that only a relatively few copies can be taken from it before its fine lines begin to blur. (The British Ordnance Survey for its earlier editions overcame

---

[1] F. E. Shashoua, 'The Electrofax Map Copier', *Cartographic Journal*, vol. 1, no. 1, pp. 24–32 (London, 1964).
[2] Invaluable summaries of the whole field of map printing is given by W. G. Clare, 'Map Reproduction', *Cartographic Journal*, vol. 1, no. 2, pp. 42–8 (London, 1964); and by J. J. Ovington, 'An Outline of Map Reproduction', *Cartography*, vol. 4, pp. 150–5 (Melbourne, 1962).

this difficulty by making electrotypes from the copper plates.) The advantage is that copper-plates can easily be revised and corrected, by beating out the surface and re-engraving, and it is still used by the British Admiralty Hydrographic Department; in this case, however, the engraved plate is used to make only one print, which is then reproduced lithographically (see below).

Towards the end of the eighteenth century came the invention of *lithography*, whereby the map was drawn in greasy ink mirror-wise on a smooth slab of stone, against which the paper was pressed. From these elementary methods have evolved the three main types of map reproduction: photolithography, photoengraving and photogravure.

In *photolithography* the map, drawn on paper, card or plastic sheet, is photographed down to the desired size, and the negative is exposed over a sensitized zinc or aluminium plate. After washing, the image remains on the plate (chemically strengthened in various ways), and is receptive to greasy ink. Thus the image and the non-image areas are on the same plane (i.e. the surface of the plate), but the former is ink-receptive while the latter, kept moistened with water, is ink-repellent. The plate is then curved around the cylinder of a rotary press, the inked image is transferred to a rubber roller, and from there to the printing paper. This is known as *offset-litho* (or rotary offset printing), and maps can be printed very rapidly (up to 10,000 per hour), using separate images, plates and rollers for each individual superimposed colour. 'Colour separation' (i.e. the production of the individual plates for each colour from the original) can be carried out in several ways, normally using a separate drawing for each colour, with 'register marks' (crosses) printed on each, so that they coincide when overlain.[1] Offset-litho is mainly used for sheet-maps by government agencies, where very long runs are involved, but it is also increasingly used in small printing works, such as those of local authorities, where plans and text can be reproduced in a single operation, without the time-consuming need to use separate founts of type. If only short runs (up to about 1,000 copies) are required, and where only black and white reproduction is required, a *Xerox* positive on a special waxy paper may be used as the printing plate.

[1] An interesting discussion of these problems of colour map reproduction is given by D. W. Gale, 'Register Control in Map Reproduction', *Cartographical Journal*, vol. 2, no. 2, pp. 68–74 (London, 1965). Miss A. F. Coleman discusses problems of colour-printing in 'Some Technical and Economic Limitations of Cartographic Colour Representation on Land Use Maps', *Cartographic Journal*, vol. 2, no. 2, pp. 90–4 (London, 1965).

Maps and diagrams in books and periodicals are commonly produced by *photoengraving*. The map is transferred photographically to a
metal plate, and the image has been emphasized and strengthened by
an application of an acid-resistant resin, the interline areas are etched-
out with acid. The plate is fastened to a block of wood, so that the
surface of the upstanding image is at exactly the same height (0·918 in.)
as the printer's type, which is convenient when a book or periodical
consisting of text with illustrations has to be printed by *letterpress*.
Halftones (plates) are produced by using a 'screen' of dots of varying
coarseness to depict the detail. In the *photogravure* process the material
is photographed using a screen to convert all the detail (line-work,
lettering, stipple) into a series of minute square 'cells'. These are transferred to the printing-plate on which the cells are etched to varying
depths, so holding when inked a varying amount of ink; shallow cells
produce light tones, deeper cells darker ones. The result is coarser and
less regular than that produced by other reproduction methods; fine
lines and shading must be avoided in the drawing. The process is
mainly used for 'long-run' magazine printing.

MEASUREMENT OF AREA AND DISTANCE

It is occasionally useful during a course of cartographical exercises to be
able to measure the area of any unit on the map with a fair degree of
accuracy. Of course, when using administrative divisions as a basis for
computations, the exact areas can be read off from the cadastral
survey records, from large-scale maps on which areas are printed,
or from census volumes. But when the student is dealing with a non-
administrative unit, particularly in connection with land-forms, it
may be necessary to compute the area; for example, the area within
specific contours must be known in order to draw a hypsometric curve
(see p. 151). There are several graphical methods, of a greater or less
degree of accuracy, and various instruments which may be used.[1]

The Method of Squares

A somewhat tedious procedure is to cover the area with unit squares,
either by tracing the outline on to graph-paper, or by superimposing

[1] A definitive survey is given by M. Proudfoot, *The Measurement of Geographic Area*
(U.S. Bureau of the Census, Washington, D.C., 1946). See also J. W. Gierhart,
'Evolution of the Methods of Area Measurement', *Surveying and Mapping*, vol. 14, pp.
460–5 (Washington, D.C., 1954).

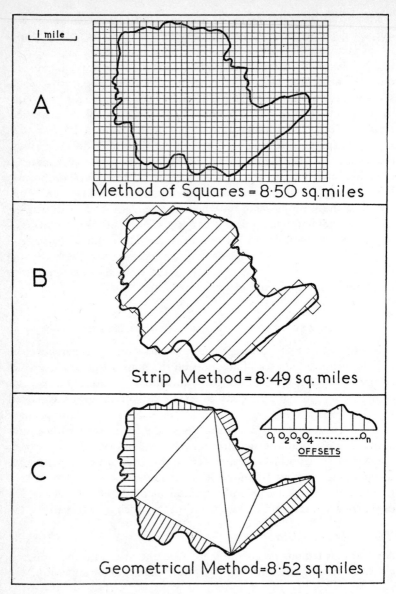

1 mile

A

Method of Squares = 8·50 sq. miles

B

Strip Method = 8·49 sq. miles

C

O_1 O_2 O_3 O_4 - - - - - - - - - - O_n
OFFSETS

Geometrical Method = 8·52 sq. miles

Figure 24. MEASUREMENT OF AREA

the graph-paper on the map over a brilliantly lighted tracing-table. Count the large squares, then the small; where the outline crosses a small square, include it if more than half its area lies within the outline. Apply the scale-factor to convert the area on the map into the area on the ground. Areas on a 1-inch map can be readily measured on inch (and tenths) graph-paper (Fig. 24, example A).

The Strip Method

This is a more speedy method, but is not as accurate. Rule a series of parallel lines a unit distance apart, either upon the face of the map or on tracing-paper. The smaller the unit, the more precise will be the measurement, but it should be some convenient unit for the scale of the map; on a large area of a 1-inch map, place the lines an inch apart. Rule vertical lines at each end of every strip to convert them into rectangles; the vertical lines should be placed as 'give and take lines' across each portion of the boundary so as to exclude as much area as they include (Fig. 24, example B). Add the lengths of all the strips, which will give the total area in square units, and apply the scale-factor.

The 'Dot Planimeter'

A method of calculating area with speed and accuracy was devised by W. F. Wood,[1] based on the principle of an areal grid, but using dots instead of squares. A master-planimeter is made by covering a sheet of tough tracing-paper with evenly spaced dots; the accuracy of the calculation will depend on the degree of fineness or coarseness of the dot-grid. The dots should be alternately open or solid, or red and black, both in the vertical and horizontal dimensions. The planimeter is laid over the area to be measured and the dots falling within its boundary are counted; if the boundary touches one type of dot it is included, if the other it is omitted. The number of dots is then multiplied by the area-factor each represents. The process can be speeded by adding a line-grid enclosing blocks of ten and a hundred. This process is certainly easy, quick and quite accurate (Fig. 26).

The Blakerage Grid

This type of dot planimeter, developed by R. N. E. Blake, consists of a grid with 100 dots evenly spaced within each 40 mm. square. On a scale of 1:2,500, 1 dot will equal 0·01 hectares (0·025 acres); on a scale

[1] W. F. Wood, 'The Dot Planimeter: A New Way to Measure Map Area', *Professional Geographer*, vol. 6, no. 1, pp. 12–14 (New York, 1954).

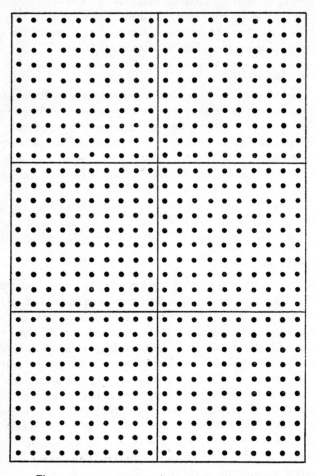

Figure 25. THE BLAKERAGE GRID DIAGRAM

of 1:25,000, 1 dot will equal 1·0 hectares (2·5 acres); and on scale 1 : 250,000, 1 dot will equal 100 hectares (250 acres). Thus by laying the grid over a map of the particular scale, the dots falling within a particular unit can be readily counted and thus its area determined (Fig. 25).

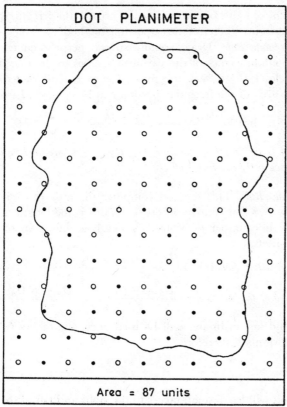

DOT PLANIMETER

Area = 87 units

Figure 26. DOT PLANIMETER

Based on W. F. Wood, op. cit. (1954).
   Where the boundary touches the open dot, it is included; where it touches the black dot, it is omitted from the count.

Geometrical Methods

When an area has a relatively simple outline, divide it into triangles occupying as much of the figure as possible (Fig. 24, example C). Their areas can be computed by either of the formulae:

$$Area = \frac{Base \times perpendicular\ height}{2} \tag{1}$$

$$or \quad Area = \sqrt{s(s - a)\ (s - b)\ (s - c)} \tag{2}$$

where a, b, c are the three sides, and $s = \dfrac{a + b + c}{2}$

The problem remains of computing the area of the irregular portions along the margins; there are three possible methods.

Mean Ordinate Rule. Draw equally spaced perpendiculars (offsets) from the bounding lines of the various triangles to the margin of the area; the closer the offsets, the more accurate, if more tedious, the result. The area of the irregular portion can be computed from:

$$Area = \frac{l(O_1 + O_2 + \ldots O_n)}{n}$$

where l is the length of the line, O_1, $O_2 \ldots O_n$ *are the lengths of each offset, and n the number of offsets.*

Trapezoidal Rule. This method computes the area of each division bounded by two consecutive offsets. Its width is the length of the line divided by the number of offsets, its length is the mean of the two bounding offsets.

$$Area\ of\ the\ first\ two\ divisions = \frac{(O_1 + O_2)}{2} + \frac{(O_2 + O_3)}{2} \times \frac{l}{n + 1}$$

$$Area\ of\ the\ last\ two = \frac{(O_{n-2} + O_{n-1})}{2} + \frac{(O_{n-1} + O_n)}{2} \times \frac{l}{n + 1}$$

The first and last ordinates will be used once, the others twice. The complete formula is therefore:

$$Area = \frac{O_1 + 2(O_2 + O_3 + \ldots O_{n-1}) + O_n}{2} \times \frac{l}{n + 1}$$

This method will leave small pieces to the right and left of the last offset; the area of these can be estimated, or weighed against similar pieces elsewhere.

Simpson's Rule. Without entering into the geometrical complexities of the principles involved, the application of this formula will give a more accurate result than will the preceding methods. Divide up the bounding line into an odd number of offsets, so that there will be an even number of unit areas. The formula is:

$$Area = \frac{l[O_1 + O_n + 2\ (sum\ of\ odd\ ordinates) + 4\ (sum\ of\ even\ ordinates)]}{3\ (n + 1)}$$

It is to be doubted whether the mathematical tedium of this method is worth the slightly more accurate result.

Instrumental Methods

The most convenient instruments which can be used for the measurement of area include the computing-scale and several planimeters of varying degrees of refinement.

Computing-scale. The computing-scale is a simple instrument which in effect applies cumulatively the strip method without the necessity of ruling vertical bounding lines or of measuring separately the length of each strip. It consists of a hard-wood rule, calibrated on its four edges (obverse and reverse sides) for the 1/2,500, 6-, 1- and ¼-inch scales, with a cursor which can slide from one end of the scale to the other. Rule a series of equidistant parallel lines over the area, and place the scale along the first strip, with the index wire over the position where the first vertical bounding line would fall, touching the zero of the scale. Slide the cursor along to the right-hand edge of the first strip, then move the rule to the second strip, and repeat this operation. When the cursor arrives at or near the end of the scale, note down the reading and start at zero once more. When the cursor arrives at the right-hand edge of the last strip, note the final reading, and add to it all the previous ones. Some computing-scales are made to read off the area in acres for a particular scale, while others give an answer in square inches which must be converted into actual area measurements by applying a scale-factor.

Planimeters. Several makes of planimeter are available[1] varying from a simple form of tracer-bar, known as a *hatchet-planimeter*,[2] to delicate instruments fitted with recording dials, known as *wheel-planimeters*. It is not easy to describe the theory of these instruments, but if the maker's instructions are carefully followed, a short period of practice will soon familiarize the student with their operation. The principle of each model is that a point is carefully traced round the perimeter of the area to be measured. In the case of a wheel-planimeter, the dial records

[1] J. W. Gierhart, 'Evaluation of Methods of Area Measurement', *Surveying and Mapping*, vol. 14, pp. 460–9 (Washington, D.C., 1954), surveys the use and value of various types of planimeter. A clear exposition of the theory of these instruments is given by F. Debenham, *Exercises in Cartography* (London, 1937).
[2] R. L. Williams, 'The Hatchet Planimeter', *Professional Geographer*, vol. 6, no. 2, pp. 14–16 (New York, 1954).

the distance travelled while tracing the perimeter, and this figure, multiplied by a known constant for the instrument, gives the area. Some models have a variable tracer-arm, which will allow direct measurement in any unit, others have a fixed tracer-arm which will give the area on the paper in square inches, and therefore a scale-factor must be applied.

In recent years a number of increasingly complex instruments have been evolved for area measurement.[1] Until recently the United Kingdom Ordnance Survey measured the areas of 'parcels' on their 1 : 2,500 plans by means of grid-squares and a computing scale. In 1959 an automatic reading planimeter was introduced which measures areas electronically, the results of which are transferred to punched cards. This has recently been replaced by a larger tape-operated installation. The areas are printed out by Monophoto Filmsetters to provide positive figures of acreages, which are mounted within each 'parcel' on the drawn plans by wax adhesives.

Measurement of Distance

It is frequently necessary to measure the length of some irregular line on a map, such as a road, railway or river. If the line is not too irregular, a number of short straight portions can be stepped off successively with dividers, and summated. Alternatively, the end of a piece of fine thread is placed at the starting-point and then laid along the line, carefully following each curve. Again, a small toothed wheel fitted with a recording dial, known as an *opisometer*, can be run carefully along the line, the total length given on a dial is read off in inches or centimetres, and this is converted into actual lengths by applying a scale-factor. In each of these cases, it is well to measure the line twice, once from each end, and calculate the mean of the two results; this will balance 'inside' and 'outside' curves.

Each of these methods measures the length of the line as its projection on to a plane surface, which in hilly country will be considerably shorter than its actual length. If for some special reason a very accurate measurement is desired, it may be necessary to construct a profile of the road with no vertical exaggeration, and then to measure the actual profile.

[1] C. J. McKay, 'Automation Applied to Area Measurement', *Cartographic Journal*, vol. 3, no. 1, pp. 22–5 (London, 1965).

AUTOMATED CARTOGRAPHY (Figs. 185, 221)

In recent years enormous developments have taken place in the production of maps and graphs ('pictorial computer output') from statistical information handled by a computer and presented in terms of point, line and area. The student is referred to the copious bibliography[1] should he wish to pursue the fascinating applications, and he should certainly be aware of the exciting implications, which may well concern him in post-graduate work or should he seek to make a career in a map-producing agency or house (as, for example, at the Clarendon Press, Oxford), or in planning.

In general, a storage system ('data bank' or 'library')[2] enables all spatially mappable data to be stored either as strings of line information in the form of co-ordinates on paper or on magnetic tape (for producing continuous lines), or as points (pairs of co-ordinates either on punched cards or tape). This information, known as the 'input', is manipulated according to the particular programme written for the computer and presented in terms of 'output'. This 'output' is linked to some type of printing device, a *computer-output printer*; various types include the co-ordinate-plotter (which prints any kind of symbol at the exact position specified by the pair of co-ordinates), the line-plotter, the continuous curve-plotter and the graph-plotter. The lines, in fact, are drawn by plotting a large number of points so closely spaced that they are visually contiguous (at least 100 pairs of co-ordinates per

[1] A few of the more accessible sources are: (i) W. C. Cude, 'Automation in Mapping', *Surveying and Mapping*, vol. 12, pp. 413–36 (Washington, D.C., 1962); (ii) J. S. Keates, 'Digital Drafting: Some Further Developments', *Cartographic Journal*, vol. 2, no. 2, pp. 60–1 (London, 1965); (iii) M. S. Monmonier, 'The Production of Shaded Maps on the Digital Computer', *Professional Geographer*, vol. 17, pp. 13–14 (Lawrence, Kansas, 1965); (iv) G. Petrie, 'Numerically Controlled Methods of Automatic Plotting and Draughting', *Cartographic Journal*, vol. 2, no. 3, pp. 60–73 (London, 1966) (this describes some of the machines in use, with photographs); (v) J. D. Porteous, 'Computer Graphics: a Further Note', *Area*, no. 3, pp. 37–9 (London, 1969); (vi) K. E. Rosing, 'Computer Graphics', *Area*, no. 1, pp. 2–7 (London, 1969); (vii) R. A. Siders, *et al.*, *Computer Graphics* (London, 1966), a comprehensive textbook; (viii) W. R. Tobler, 'Automation and Cartography', *Geographical Review*, vol. 49, pp. 526–34 (New York, 1959); (ix) W. R. Tobler, 'Automation in the Preparation of Thematic Maps', *Cartographic Journal*, vol. 2, no. 1, pp. 32–8 (London, 1964) a most valuable summary, with a bibliography of fifty-one items; (x) N. L. G. Williams, 'The Oxford System of Automatic Cartography', *Cartography*, vol. 16, pp. 17–20 (Ottawa, 1966); and (xi) P. C. F. Wolfendale, 'Machine Accuracies in Automatic Cartography', *Cartographic Journal*, vol. 4, no. 1, pp. 24–8 (London, 1967).

[2] S. M. Howard, 'A Cartographic Data Bank for Ordnance Survey Maps', *Cartographic Journal*, vol. 5, no. 1, pp. 48–53 (London, 1968).

centimetre); coastlines and contours are thus drawn. So far the most successful results have been obtained in the production of point-symbol, isopleth and choropleth maps, and various systems have been developed. Mention may be made of *SYMAP*,[1] an abbreviation for 'Synagraphic Mapping System', developed at Harvard; this uses a standard line printer operated by the Fortran IV machine language (see p. 466). Another is *LINMAP*, developed by the Urban Planning Directorate of the United Kingdom Ministry of Housing and Local Government, a representation of a value in an area by means of a single symbol located in that area, so as to produce rapidly maps involving vast quantities of census data (see p. 465). Broadly speaking, a grid with a mesh fine enough to recognize critical data is superimposed over the area to be mapped, and either a dot or a symbol is placed according to instructions at the grid intersections, or average values are located within each grid-cell so as to produce patterns of polygons. An invaluable development is the *proximal map*, where 'value-areas' are assigned according to their proximity to each datum-point, what has been termed 'nearest neighbour technique'. Computer programmes can also be written for map-lettering (see p. 64); one machine, developed in America, not only produces names with their co-ordinates on tape, but locates and prints them on the map, accurate to within $\pm$ 0·0025 inches, at a rate of about 1,000 per hour.

Conversely, it is possible to convert data presented in the form of graphs, charts, drawings, photographs, etc., into digital form for subsequent data-processing by a computer. This is possible by using such complex equipment as the 'd-Mac Pencil Follower', which consists of a reading table on which is placed the pictorial information to be analysed. By following the trace with the 'Reading Pencil' connected to an automatic mechanism beneath the table surface, each position of the Pencil is digitized.

While the earliest automated printing was admittedly somewhat crude and rough, maps of increasingly attractive appearance are being produced. However, the machine cannot as yet compete with the human cartographer in aesthetic terms; its value, especially for research purposes, is the rapid production of a map from 'massive packets of data', or in the words of J. D. Porter, '. . . the speedy output of basic material for geographical interpretation' and for analytical research.

[1] J. C. Robertson, 'The Symap Programme for Computer Mapping', *Cartographic Journal*, vol. 4, no. 2, pp. 108–13 (London, 1967).

MODELS

Reference may be made briefly to an aspect of geographical research which has developed strikingly, the use of models[1] and analogies. These techniques are by no means new, for in fact maps are themselves models, known in the jargon as *'iconic'* or *'representational'* models. Moreover, as C. Board[1] points out, since they contain 'the essence of some generalization about reality', they are also *'conceptual models'*. However, in recent years the use of models has been elaborated and systematized, notably through the work of R. J. Chorley and P. Haggett.[2]

In its simplest form, a *relief* or *terrain model* affords a three-dimensional reproduction of the landscape, with length and breadth to scale, though altitude is of necessity exaggerated (see p. 120). The material may be of plaster, potter's clay, superimposed layers of paper, pulp, card, hardboard or plywood, or various plastics;[3] the last may be moulded, or produced by a vacuum-forming process. One example of these models was by W. V. Lewis and M. M. Miller, who used a kaolin mixture to simulate the formation of crevasses in a glacier.[4] These are sometimes described as *hardware models*;[5] the kaolin glacier is in fact an *analogue hardware model*. These models may be prepared from topographical maps and/or air photographs.[6] Apart from illustrating landforms, they are widely used in civil engineering projects, planning,

[1] C. Board, 'Maps as Models', pp. 672–725, in R. J. Chorley and P. Haggett, *Models in Geography* (London, 1967); this contribution has a bibliograply of 124 items.
[2] R. J. Chorley and P. Haggett, *Models in Geography* (London, 1967, also available as three paperbacks), a massive work of 800 pp., with numerous contributors and a vast bibliography. A useful summary is by J. P. Cole and C. A. M. King, 'Models and Analogies', *Quantitative Geography*, chap. 11, pp. 463–520 (London and New York, 1968), with numerous examples and a select bibliography.
[3] See, for example, the use of polystyrene by W. L. Mowbray and M. Galley, 'Relief Models Using New Materials', *Geography*, vol. 53, pp. 308–9 (Sheffield, 1968).
[4] W. V. Lewis and M. M. Miller, 'Kaolin Model Glaciers', *Journal of Glaciology*, vol. 2, pp. 535–8 (Cambridge, 1955).
[5] R. J. Chorley, 'Models in Geomorphology', in R. J. C. and P. Haggett, *Models in Geography*, pp. 63–8 (London, 1967); and M. A. Morgan, 'Hardware Models in Geography', being pp. 727–74 of the same work, with a bibliography of seventy-five references.
[6] P. G. Mott, 'Topographical Model Making Using Air Photographs', *Cartographic Journal*, vol. 1, no. 2, pp. 29–32 (London, 1964), discusses a method of making a relief model from layers of board, the steps between the layers filled with a plaster-pulp mixture, and air photographs are actually mounted over the relief surface by 'skinning'.

architecture, for military purposes ('terrain intelligence') and geological studies (the rock-strata may be depicted in section on the sides
of the model. Such models are fundamentally static, but *working* or
dynamic scale-models may be of great value, for they enable a process to be
simulated and its results studied. For example, R. A. Bagnold used a
wind-tunnel in his investigation of sand movement and dune formation,[1] while others employ tanks for the study of waves, river erosion,
shoreline processes, etc. A working model of the tides in Southampton
Water and the Solent is now kept at the University of Southampton; a
tidal model of the Mersey is at the Hydraulics Experiment Station at
Wallingford, Berkshire; and numerous river models (including one of
Niagara Falls and another of the Arkansas River) are at the U.S.
Corps of Engineers Research Establishment at Vicksburg, Tennessee.
Of course, the problem of scale, dimensional similarity and the speeding up of the time element presents difficulties, but apart from their
visual impact these models can produce valuable quantitative experimental data.

The construction of models in a wider sense can be used to create a
bridge between observation and theory, to bring together factual information so as to form a working hypothesis which can be tested
against reality. Models can be created in any aspect of Geography, in
geomorphology (e.g. a model of the development of a complex drainage
system), climatology (e.g. a model of a frontal zone), agricultural and
industrial location (such as the famous pattern of Von Thünen's concentric rings of land-use, a classic and possibly the earliest model of
agricultural location), and in the form of regional models. They may be
presented in the form of graphs, profiles, networks, flow-diagrams and
circuits, and mathematical equations. Elaborate classifications of
models have been presented: they may be *mathematical*, where mathematical symbols are used to represent constants and variables, so to
produce model predictions which can be compared with real physical
situations; they may be *experimental* or *natural*; they may be *iconic* (i.e.
the static and working models described above), or *analogue*, *symbolic* or
simulation models, in which the actual properties are represented by
other different, though analogous, properties. Again, a *stochastic*
model is a locational model involving the laws of probability, though
taking into account chance process. Many of the less complex types of
model are in fact exemplified in this book.

[1] R. A. Bagnold, *The Physics of Blown Sand and Desert Dunes* (London, 1941).

TOPOLOGICAL MAPS

Topology is a branch of geometrical mathematics which is concerned with order, contiguity and relative position, rather than with actual linear dimensions. It is sometimes referred to as 'the rubber sheet geometry', since a pattern on such a sheet can be deformed, yet points on it remain in the same order or relationship; in point of fact, the scope of topology is really much wider than this. In contrast to a topographical map, which retains the familiar scale and orientation, a topological map, while retaining contiguity of relationships (such as boundaries, relative positions of towns, etc.), uses other criteria (area, annual precipitation, density of population, *per capita* income, communications systems, etc.) to determine the scale, that is, the information is subjected to a topological transformation. Such a diagrammatic map may clearly bring out novel relationships and patterns. For example, a topological map is seen in every London Underground train compartment, showing the correct sequence of stations, though not to scale and only diagrammatically orientated. Figure 190 in this book is an example of a topological map in which countries are depicted in proportion to the size of their populations. In more complex topological maps, systems of nodes (i.e. 'zero dimensional' points and dots), arcs ('single dimensional' lines) and regions (two-dimensional surfaces or spaces) are used to construct networks and inter-relationships of immense significance.[1] The linking of arcs will produce *networks* of various kinds, and the construction and analysis of *network models* (in terms of what is known as *graph theory*) is making an important contribution to locational theory.[2]

[1] J. P. Cole and C. A. M. King, *Quantitative Geography*, pp. 85–91, with many examples and a full bibliography (London, 1968).
[2] P. Haggett and R. J. Chorley, *Network Analysis in Geography* (London, 1969).

2

Relief Maps and Diagrams

An accurate topographical map can be used as the basis of much geographical work. The reading of a contour-map is not easy and needs considerable practice to enable the landscape to be visualized. In fact, the representation, recognition and description of relief features from their contour patterns, ranging from simple examples such as concave slopes, spurs and cols to complex land-forms, provides much of the content of map-work as generally understood. More advanced interpretation of topographical maps will help the student to examine and explain various geomorphological concepts. It is not going too far to say that the large-scale topographical map is secondary only to the ground itself in such work. As Professor A. A. Miller has said, 'I referred to the map as a tool; in reality it is a whole bag of tools containing more ingenious devices than a boy scout's knife, and if properly used it will open almost any geographical problem. . . .'[1] But it must also be emphasized that the map must never be divorced from the ground. As W. M. Davis once somewhat ironically wrote, the study of maps 'seems to lead different investigators to different results'.[2] It is all too easy to read too much into, and to deduce unjustifiably too much from the map, without careful ground corroboration.

A geographer should also be able to add to a published map any further material he requires;[3] he 'must supplement the map informa-

[1] A. A. Miller, 'The Dissection and Analysis of Maps', Presidential Address, 1948, to the Institute of British Geographers, published in *Transactions of the Institute*, no. 14, p. 2 (London, 1949).

[2] W. M. Davis, 'The Peneplain', *Geographical Essays*, p. 353 (Boston, 1910).

[3] R. F. Peel, 'Geomorphological Fieldwork with the Aid of Ordnance Survey Maps', *Geographical Journal*, vol. 114, pp. 71–5 (London, 1949). As he says (p. 17) '. . . even the best [map] cannot provide more than a partial definition of the ground, or one more accurate than its scale will permit'. The same writer strikingly illustrates his own precept in 'A Study of Two Northumbrian Spillways', *Transactions and Papers, 1949*, no. 15, pp. 73–89, the Institute of British Geographers (London, 1951). He published a series of maps and profiles, based on large-scale O.S. maps, with additional details surveyed in the field; they include two contour-maps, with an interval of 10 feet.

tion with his own measurement of valley profiles, hill-slopes, etc., using
levels, clinometers, aneroids, or field-sketching, according to the individual problem and the degree of accuracy required'.[1]

A detailed topographical map provides, then, much definite and
exact information which can be used as a basis for various purposes,
'a starting-point for further analysis'.[2] Significant contours can be
extracted; outlines can be traced as a basis for plotting field information; gradients, slopes and relative relief can be calculated; and profiles
can be drawn.

One word of caution should be noted. A strict copyright[3] exists in
all official topographical and geological maps published by government
agencies, whether British or foreign. Permission is required for reproduction for publication of a portion of such a map; this as a rule is
readily forthcoming, with the proviso, in the case of the British Ordnance Survey, that 'Crown Copyright Reserved' must be printed below,
and a fee is required. More doubtful is the common case when certain
detail is extracted to be used as a base for a newly compiled map. If
it bears any resemblance to the original source, it is safer to consult
the Ordnance Survey or other agency responsible.

THE DEPICTION OF RELIEF

Since the earliest days of map-making, the depiction of relief has been
one of the major problems of cartographers, for it involves the representation of three dimensions upon a plane surface. From the primitive
efforts, using crude pictorial symbols in profile or the so-called 'hairy
caterpillars', to modern colour-printing, which employs several
methods in careful conjunction, is a long story of trial and experiment,
and of increasing technical efficiency.[4]

---

[1] A. A. Miller, op. cit., p. 2 (1949).   [2] A. A. Miller, op. cit., p. 4 (1949).

[3] The question of copyright is discussed by C. B. Hagen, 'Maps, Copyright and Fair
Use', *Special Libraries Association*, Geography and Map Division, *Bulletin* no. 66, pp.
4–11 (Ottawa, 1966).

[4] A very useful survey of modern aspects of the problem is given in 'Questions
générales concernant la représentation du relief au point de vue topographique et
morphologique', *Comptes Rendus du Congrès International de Géographie, Lisbonne, 1949.
Tome I. Actes du Congrès. Travaux de la Section I* (Lisbon, 1950). A brief but comprehensive survey is J. S. Keates, 'Techniques of Relief Representation', *Surveying and Mapping*,
vol. 21, pp. 459–63 (Washington, D.C., 1961). One of the world's leading exponents of
terrain representation in cartography is Professor Eduard Imhof, of Zürich, who until
1966 edited *The International Yearbook of Cartography*. He presented a remarkably concise
summary of methods of terrain representation (maps, panoramas, block-diagrams,

It is convenient to summarize in turn the main methods of relief depiction, indicating at the same time their application to the work of the geographer. It will be appreciated that the various methods, each with some advantage and usually with some limitations as well, may be combined with profit.

(a) *Spot-heights*

At various points on the map, heights are marked which have been carefully computed relative to a chosen datum.[1] This datum for heights above the sea is determined from a series of tidal observations, providing a 'mean sea level'. The datum for depths below the surface of the sea is usually taken on British charts to be the lowest low water springs, i.e. the worst water conditions for navigation. The chief merit of spot-heights is that they provide definite and precise information, their chief defect is that distributed over the map they give little or no visual impression of the general pattern of the relief.[2] Used in conjunction with other methods, however, they provide that exactness which is otherwise often lacking. Prominent summits should have their heights

---

stereoscopic images, anaglyphs, relief models, etc.) in 'Kartenverwandte Darstellungen der Erdoberfläche', *International Yearbook of Cartography*, vol. 3, pp. 54–99 (London, 1963), with many examples and illustrations. His *Kartographische Gelandedarstellung* (Berlin, 1965), is an elegantly produced volume of 425 pp., with 222 illustrations and 14 colour plates. This (in the words of a reviewer) is '. . . a culmination of a lifetime's experience and analysis of a subject to which the author is passionately devoted'. See also L. D. Carmichael, 'Experiments in Relief Portrayal', *Cartographic Journal*, vol. 1, no. 1, pp. 11–17 (London, 1964); and J. P. Curran, 'Cartographic Relief Portrayal', *The Cartographer*, vol. 4, pp. 28–37 (Toronto, 1967), which forms a useful summary with illustrations.

[1] The Old British Datum was based on a series of short-term tidal observations carried out between 7 and 16 March 1844 at the Victoria Docks, Liverpool. Tidal observations for the ten days were taken at five-minute intervals for an hour about high and low water. Thus the Ordnance Survey obtained a datum, a Mean Sea Level, which held good until 1921, and indeed heights on some maps are still given in terms of the Old Datum, as the transfer is still in progress. When the O.S. decided in 1911 to re-execute the primary level network for Great Britain, it was also decided to obtain a new datum. Newlyn Tidal Observatory, on a pier projecting into the sea, had virtually an open-ocean site. From 1 May 1915 to 30 April 1921 the mean of hourly records was computed, and, after various corrections had been applied, a New Datum was determined as the basis for all heights in Great Britain.

See also S. D. Hicks, 'Sea Level: a Changing Reference in Surveying and Mapping', *Surveying and Mapping*, vol. 28, 285–90 (Washington, D.C., 1968).

[2] K. H. Huggins, 'The Scottish Highlands: A Regional Study', *Scottish Geographical Magazine*, vol. 51, pp. 296–306 (Edinburgh, 1935), uses a form of spot-height very effectively in an effort to demarcate the Highlands. On a map bearing only the 800-foot contour, he indicated with a dot each summit over 1,500 feet, to give a 'crude indication of the degree of relief' (p. 299).

marked, even on a small-scale map, and there should be a few heights in lowland areas and valley bottoms, so often ignored. Figures in areas of heavy shading should be inserted in white panels.

(b) Contour-lines

Contour-lines or contours (sometimes known as *isohypses*) are drawn on a map through all points which are at the same height above, or depth below, a chosen datum. Some contours are surveyed in on the ground, others are interpolated, partly from accurately determined spot-heights, partly by the eye of a skilled surveyor in the field.[1] These interpolated contour-lines are commonly known as *form-lines*, while the surveyed contours are thickened or emphasized in some way. Modern photogrammetric methods enable extremely accurate and rapid contouring to be carried out from air photographs, using complicated stereoplotting machines.

The Contour-interval. Occasionally a variable contour-interval may be used; it may be increased in mountainous areas above a certain height, or extra contours may be inserted, thus decreasing the interval, in lowland areas. Some authorities claim that this change of interval is permissible only on small-scale maps[2] (such as the International 1: Million series), or on atlas maps. On large-scale topographical maps the interval should if possible be maintained, even if the contours become crowded in mountain regions; this crowding, in fact, precisely indicates the steepness of the relief with some visual effect akin to that of hachuring. But such close contouring is seldom possible, or desirable, on manuscript maps. This problem of contour-interval[3] is akin to that of all isopleth or choropleth intervals (discussed on p. 41).

(discussed on p. 41)

---

[1] K. M. Clayton, 'A Note on the Twenty-five Foot "Contours" shown on the Ordnance Survey 1: 25,000 Map', *Geography*, vol. 38, pp. 77–83 (London, 1953). This contains much information about contours on British official maps, with a diagram (p. 80) of heights at which contours were drawn or interpolated.

[2] One of the most attractive and informative relief maps produced is that of Belgium, the *Carte Oro-hydrographique*, compiled by A. de Ghellinck, M. A. Lefèvre and P. L. Michotte, printed by the *Institut Cartographique Militaire*. It is tinted in eleven shades of green, yellow and brown, at 5, 20, 50, 100 and every 100 metres up to 700 metres.

[3] See (i) 'Selection of Contour Intervals' (A Panel Discussion), *Surveying and Mapping*, vol. 12, pp. 344–58 (Washington, D.C., 1952); (ii) G. D. Whitmore, 'Contour Interval Problems', *Surveying and Mapping*, vol. 12, pp. 174–7 (Washington, D.C., 1953), a summary restatement of the above discussion; and (iii) R. Finsterwalder, 'Zu den Schichtlinien der deutschen Karte 1: 25,000', *Die Erde*, vol. 3, pp. 36–43 (Berlin, 1951–2).

Significant Contours. The topographical map provides detailed information about the relief, for every contour, corresponding to the particular contour interval, is included. In drawing a relief map for some specific purpose, it is usually necessary to select certain significant contours, partly for clarity and emphasis, partly for ease of drawing and reproduction. The contours must be carefully chosen; quite a different impression can be given by using an alternative series.

Sometimes a single contour is in itself highly significant. The 200-foot contour in the London Basin, the 70-metre contour in north-eastern Belgium, the 800-foot contour in Scotland,[1] the 1,300-foot contour in the Ingleborough district,[2] all these illustrate and emphasize some interesting feature.

Generalized Contours. As a rule, contours should be very accurately traced. Occasionally, however, a clearer picture can be obtained when minor detail is ignored. S. W. Wooldridge, for example, produced a map of the Chiltern dip-slope in which he carried the contours, at 50-foot intervals, across the minor valleys which dissect the dip-slope, so as to link up the interfluves.[3] Needness to say, this practice should be used with great discretion.[4]

Extrapolation of Contours. Under the guidance of an experienced geomorphologist, it is sometimes possible to reconstruct cartographically erosion surfaces or platforms which have been largely destroyed by subsequent dissection, and to insert the 'restored contours' or *eohypses*. A. A. Miller[5] identified in the field the surviving portions of the 600-

[1] K. H. Huggins, op. cit. (1935).

[2] M. M. Sweeting, 'Erosion Cycles and Limestone Caverns in the Ingleborough District', *Geographical Journal*, vol. 115, p. 78 (London, 1950). Note the map of geomorphological features, outlining a striking erosion surface by means of the 1,300-foot contour.

[3] This map helped to bring out a broad bench, or flattening, between the 500- and 650-foot contours, with a steeper slope behind. Professor Wooldridge related this to the marine abrasion of Pliocene times. See S. W. Wooldridge and R. S. Morgan, *The Physical Basis of Geography*, p. 260 (London, new impression, 1946). *Note.* This (and the useful chapter containing it) was omitted from the 1959 and subsequent editions, entitled *An Outline of Geomorphology*. See also S. W. Wooldridge and D. L. Linton, *Structure, Surface and Drainage in South-East England* (London, 1955).

[4] A. J. Pannekoek, 'Generalization of Coastlines and Contours', *International Yearbook of Cartography*, vol. 2, pp. 55–75 (London, 1962). This has a number of examples showing the drawing of similar contours on different scales, stressing the degree of simplification required.

[5] A. A. Miller, 'The 600-Foot Plateau in Pembrokeshire and Carmarthenshire', *Geographical Journal*, vol. 110, pp. 148–59 (London, 1937).

foot plateau in south-western Wales and plotted their margins at the
point of the break of slope. These surviving portions of the plateau were
stippled. He then extended the system of contours by extrapolation to
show the probable original extent of the plateau.[1]

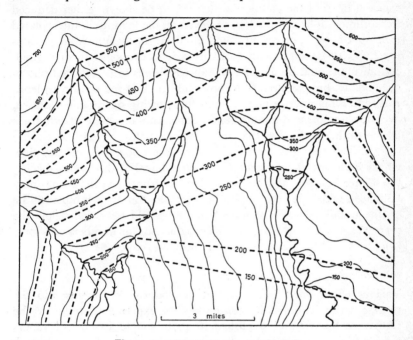

Figure 27. STREAM-LINE SURFACES

Based on G. H. Dury, 'Methods of Cartographical Analysis in Geomorphological
Research', *Indian Geographical Society: Silver Jubilee Souvenir*, pp. 136–40 (Madras,
1952).

Stream-line Surfaces (Fig. 27). In an upland area which has been deeply
dissected by rivers, extensive interfluves survive between the valleys.
An indication of the amount of denudation still to be achieved can be
obtained by interpolating the contours of the 'stream-line surface'. This
is done by plotting intersections of actual contours with the main
streams; thus the 600-foot stream-line contour can be inserted by

[1] Striking examples of the use of contour extrapolation in the elucidation of geomorphological problems are given by H. Annaheim, 'Studien zur Geomorphogenese
der Südalpen und Alpenrand', *Geographica Helvetica*, vol. 2, p. 90 (Bern, 1946); and
by D. L. Linton, 'Some Aspects of the Evolution of the Rivers Earn and Tay', *Scottish
Geographical Magazine*, vol. 56, pp. 1–11 (Edinburgh, 1940).

joining points where the main streams cross each actual 600-foot contour-line.[1]

The Technique of Drawing Contour-maps. The usual practice is to make a pencil draft of the contours which are to be shown, either deduced from photographs and models, surveyed on the land itself, interpolated from spot-heights, or traced from an official topographical map. Contours may be traced directly in ink from a clear topographical original. Indian ink should be used for a map which has to be reduced photographically, but coloured waterproof ink (such as red or brown) lends clarity to a manuscript map. A smoothly flowing steady line is drawn, avoiding tremors and minute bends. Sometimes it is preferable to trace in the drainage first, so that the contour re-entrants can be exactly placed. The contours are numbered on the upper side of each line, which indicates at first sight uphill and downhill directions, placing the figures in a row above one another. Alternatively, the figures may be placed within breaks in each line, but the top of each must still be on the uphill side. The contour-interval should be stated below the map. When the map is to be reduced, the contour numbers and spot-heights should be drawn sufficiently large to reduce clearly.

(c) *Elaborations of the Contour Method*

Various efforts have been made to introduce refinements into the representation of relief by contours. One experiment, tried by the Ordnance Survey and published in 1866, sought to make contours more striking by using white lines on north-western slopes and black ones on south-eastern slopes, superimposed over layer-tinting in grey. These contours were indeed so striking that they looked like a series of terraces.

The student can try the effect of 'illuminating' contour-lines with the aid of a tracing; go over the contours on the tracing, resting it on drawing-paper on a sheet of glass, with a hard pencil, which will make a slight indentation in the paper. Apply hill-shading (see p. 97) with a soft carbon pencil, which will make the contour-lines stand out in white. Finally, go over the contours on south-eastern slopes in Indian ink.

It is a criticism of contours that they give a somewhat smooth, rounded and rolling effect to the relief, while sudden changes of slope, sharp breaks or edges, and any interruptions may be obscured, unless

[1] G. H. Dury, 'Methods of Cartographical Analysis in Geomorphological Research', *Indian Geographical Society, Silver Jubilee Souvenir Volume*, pp. 136–40 (Madras, 1952).

it is craggy enough to be shown by rock-drawing (see p. 101). R.
Lucerna[1] superimposed heavy lines, known as 'edge-lines' or 'break-
lines', to indicate these *Kanten* or edges, hence the German name of
Kantographie and the American form Kantography. Some of these lines,
indicating deeply cut lateral valleys, hanging-valleys or rock-steps, ran
across the contours, others indicating rock-crests or river terraces were
drawn parallel to the contours. The contour-lines were usually interrupted or broken at these edge-lines.

A number of experiments have recently been made in Germany on
the effective presentation of relief by means of contours with additional
information.[2] These involve the careful selection of the contour interval,
which varies from map to map according to the nature and type of the
relief, so that the pattern of the contours may afford some degree of
pictorial representation. Kantography is used to indicate breaks of
slope by means of strips of superimposed shading, and various symbols
are lightly applied to show rock outcrops, scree and other surface
characteristics.

It is a useful field-exercise for students to take an outline contour-
map into some area of prominent relief, such as the Lake District.
Work systematically over the ground, fixing break-lines and plotting
them on the map; a major break can be shown by a heavier line than
that used for a minor one.

Tanaka Kitirô's Methods. Two original methods of relief depiction,
using contours as basic material, have been developed by Tanaka
Kitirô. These methods and the mathematics behind them are somewhat involved; they are described and very strikingly illustrated in
several detailed articles.[3]

[1] R. Lucerna, 'Neue Methode der Kartendarstellung', in *Petermanns Mitteilungen*,
vol. 74, pp. 13–18 (Gotha, 1928); and 'Kantographie', *Comptes Rendus du Congrès
International de Géographie*, vol. 2, sect. 1, pp. 23–4 (Leiden, 1938).

[2] L. Brandstätter, 'Schichtlinien und Kanten-Zeichnung: Neue Methode der
Geländedarstellung auf der Topographisch-morphologischen Kartenprobe 1 : 25,000',
Erdkunde, vol. 14, pp. 171–81 (Bonn, 1960). Two map excerpts are included.

[3] (a) Tanaka Kitirô, 'The Orthographical Relief Method of Representing Hill
Features on a Topographical Map', *Geographical Journal*, vol. 79, pp. 213–19 (London,
1932).

(b) Tanaka Kitirô, 'The Relief Contour Method of Representing Topography on
Maps', *Geographical Review*, vol. 40, pp. 444–56 (New York, 1950). Professor Tanaka
Kitirô is a member of the Faculty of Engineering of Kyushu Imperial University,
Japan. Note should be taken, however, of T. M. Oberlander, 'A Critical Appraisal of
the Inclined Contour Technique of Surface Representation', *Annals of the Association of
American Geographers*, vol. 58, no. 4, pp. 802–13 (Lawrence, Kansas, 1968), who stresses

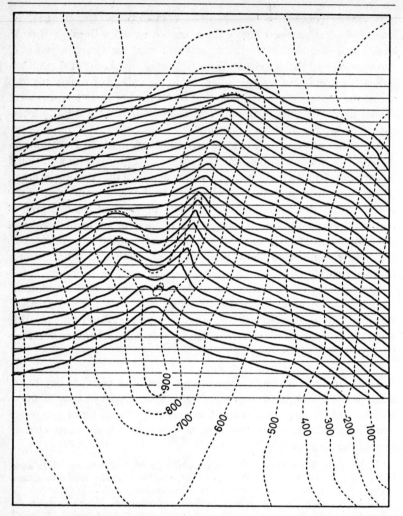

Figure 28. THE INCLINED CONTOUR METHOD OF RELIEF DEPICTION
Based on the Orthographical Relief Method, devised by Tanaka Kitirô (op. cit., 1932).

In the *Orthographical Relief* method, devised in 1931, he used 'inclined contours' to delineate the exact intensity of light and shade on the

the shortcomings and pit-falls of the method. One of his major criticisms is that the angle and interval of the inclined plane is rarely given with the map, which is fundamental, since the representation can vary drastically when the angle of the construction planes is changed.

ground surface (Fig. 28). These inclined contours, although mathematically determined, can be drawn mechanically and with precision
by a draughtsman, and no judgement, estimation or artistic skill is
necessary. It emphasizes steep slopes and striking relief extremely well,
but the multiplicity of lines gives a heavy, even an obscuring effect, to
the map. This is shown very obviously in the example specially drawn
by Ordnance Survey draughtsmen of the country round Perth.[1]

In 1950 Tanaka published in America his *Relief Contour* method,
which he devised in order to give the map an appearance of detailed
relief as produced by oblique lighting, while preserving the outlines
of ordinary contours. The exact thickness of each contour-line was
calculated, varying according to the degree of slope. Contours on slopes
away from the source of illumination were drawn in black, those facing
the source of illumination in white; these he called 'relief contours'. All
other detail – rivers, roads, cities, spot-heights and names – were drawn
in black over the relief contours. The map can be reproduced either
by a single half-tone block, or by two printing plates, one carrying the
black contours and all other detail, the second with a uniform neutral
background on which the white contours alone appear. The second
process gives sharper line-detail than does the half-tone.[2]

(d) *Layer-shading and -tinting*

It is possible to shade or colour each part of a map lying between two
particular contours in order that the distribution of high and low land
can be seen at a glance. This is known as layer-tinting, or as hypso-
metric-shading. It is obviously more useful for landscape of varied relief,
for there is little point in tinting with one colour the whole of a map
showing uniformly level country, whether on a plain or a plateau.
The principle of layer-tinting is in a sense misleading, since a single
shade between the 100- and 200-foot contours indicates a uniform
level, instead of a progressive change in height. Nevertheless as in the
case of all maps drawn on a similar principle (see pp. 47–55), the process is helpful if used with discretion.

[1] H. St J. L. Winterbotham, 'Note on Professor Kitirô's Method of Orthographical
Relief', *Geographical Journal*, vol. 80, pp. 519–20 (London, 1932).

[2] The student should examine the relief-contour map of the Kirishima volcanic
group near Kagoshima in the island of Kyushu, published in Tanaka's article, and
compare it with the reproduction of the official Japanese topographical map, on
which the relief-contour map was based. The latter gives a curious appearance, as
if it were built up of cardboard layers, but the relief effect, as applied to an admittedly
striking area of volcanic topography, is extraordinarily vivid.

Line-shading. As a rule, line-shading and hand-stippling should be avoided for manuscript relief maps. It is slow to execute, a 'stepped' effect is unavoidable, and intermediate contours, lettering and other detail are often obliterated. Mechanical stipples, printed in black, may occasionally be used (see p. 51–3), but a carefully graded series must be chosen.

Colour. Modern printing processes have enabled layer-colouring to be employed very successfully on atlas and topographical maps, using half-tone blocks; these are obviously outside the scope of this book. Hand-tinting may, however, be used on manuscript maps, or by means of mechanical stipples on line-blocks (see p. 54); if the cost is merited, a brown stipple can be extremely effective.

The range of colours for hand-tinting of relief maps must be carefully chosen. One alternative is to use a sequence of greens, yellows and browns in ascending altitude, possibly culminating in red, purple and even white in high country. The contour interval at which the colour-change is made must be carefully chosen. As many depths as possible of each colour should be used to give a gradual sequence and to avoid the stepped effect of sudden contrasts. A second method is to use one colour only, ranging from the faintest to the darkest possible density.[1] A third method is to merge or grade the successive tints, so that a stepped appearance is avoided; it lacks the absolute quantitative nature of colour used as contour-filling (although it can be super-imposed over contours), yet it is very effective, especially if combined with light hill-shading. The colours should be subdued in appearance, almost as if a pale grey wash were superimposed, so that a series might range from greenish-grey to greyish-brown. The key-panel consists of a single column, with merging colours, and with significant heights ticked off alongside.

A most revealing exercise is to tint two copies of an outline contour-map, using two different colour systems; sometimes quite striking contrasts in the general appearance of the map will result.

(e) *Hachures*

Hachures are lines drawn down the slope in the direction of the steepest gradient; conventionally, they are drawn more closely together where

---

[1] Examples include the purple layer-tinting on R.A.F. maps, the attractive greys of Bartholomew's *Road Atlas of Great Britain*, and the greys of the British Council's *Map of the Middle East.*

the slope is steeper. Another method employs the same number of lines per inch, but each one is proportionally thicker, as in the Lehmann system, where the exact thickness of the individual hachure is determined according to the angle of slope. Another adaptation is where the hachuring is assumed to be obliquely lighted, usually from the north-west, but this is only effective in regions of strong relief, with sharply defined ridges, as in Switzerland, where the method was developed on the Dufour map.

The chief disadvantages of hachuring are the lack of absolute information (to meet which numerous spot-heights have to be inserted), the difficulty of drawing hachures in the field unless one has a very good eye for country, and the problem of distinguishing directions of slope. Most hachuring on modern printed maps is in colour, usually brown, purple or grey. This removes the grave disadvantage of the obliterating effect of black hachures in hilly areas, shown, for example, on the sheets of the first and second editions of the One-inch series of the British Ordnance Survey. On the other hand, its chief advantage is that it enables minor but important details, lost on a contour-map within the contour interval, to be brought out, and sometimes it can show country of striking relief in a very dramatic manner (Fig. 29).[1]

(f) *Hill-shading*

Hill-shading, known in the United States as 'plastic-shading', aims at producing something of the effect of a relief model. It is imagined that such a relief model is brightly illuminated, either by a vertical source of light, or obliquely, usually from the north-western corner.[2] P. Richarme,[3] in fact, makes a relief model in plaster, illuminates it (considering the merits of both zenithal and oblique lighting, and a

[1] Compare the hachuring on (*a*) the *Topographischer Atlas der Schweiz (Dufour)*, 1 : 50,000; (*b*) the *Topographischer Atlas der Schweiz (Siegfried)*, 1 : 50,000; (*c*) the new *Landeskarte der Schweiz*, 1 : 50,000; (*d*) the *Carte de France au* 80,000; (*e*) the 1st, 2nd and 3rd Editions of the One-inch series, Ordnance Survey; and (*f*) the 5th (Relief) Edition of the One-inch series, Ordnance Survey. The last of these, with contours in brown, took the hachures from the copperplates of the 3rd Edition, and printed them in orange on north-western slopes and in grey on south-eastern slopes.

[2] P. Yoëli, 'Relief Shading', *Surveying and Mapping*, vol. 19, pp. 229–32 (Washington, D.C., 1959), pays special attention to vertical, oblique and combined lighting, analysing the effective results. See also R. Mean, 'Shaded Relief', *U.S. Aeronautical Chart and Information Service*, Technical Manual RM–895 (Washington, D.C., 1958); and L. J. Harris, 'Hill Shading for Relief Depiction in Topographical Maps', *Chartered Surveyor*, vol. 91 (9), pp. 515–20 (London, 1959).

[3] P. Richarme, 'The Photographic Hill Shading of Maps', *Surveying and Mapping*, vol. 23, pp. 47–59 (Washington, D.C., 1963).

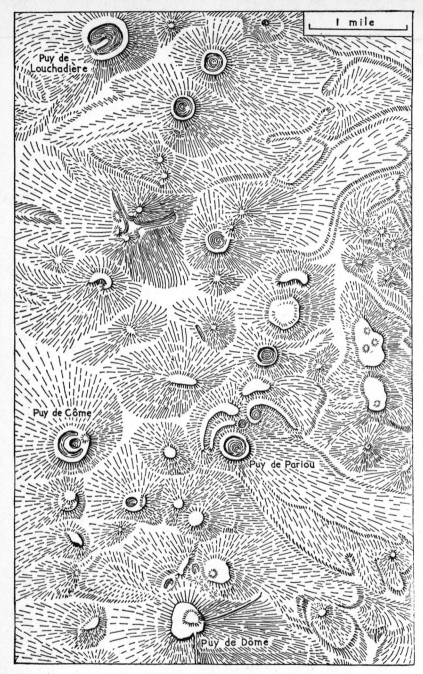

1 mile

Puy de Louchadière

Puy de Côme

Puy de Pariou

Puy de Dôme

Figure 29. HACHURING

Based on the 1 : 80,000 series, *type 1889, Service Géographique de l'Armée* (Paris).

mixture of both), and photographs it, using a special large camera. The major problem is the transforming of the perspective projection provided by photography into the required orthogonal projection. With a vertical light, the steeper the slope the darker the shadow, while ridge crests, plateaux, valley bottoms and plains remain in the light. With an oblique lighting, north-western and western slopes are unshaded, while the eastern and south-eastern slopes are in shadow. Lighting from the south gives a curiously 'photograph-negative' appearance, and is never used. Grades of grey are used for shading, but occasionally in regions of striking relief solid black shading can be used for the areas in shadow, producing a very effective, if somewhat stark, effect.[1] It is impossible, of course, to add any more detail or lettering over such shading. Some of the disadvantages inherent in this method are common to those of hachuring: lack of specific information (unless there are many spot-heights); doubt about which is uphill and downhill, spur or valley, plateau or plain; and confusion about the relative steepness of slopes, particularly with oblique lighting.

The printed topographical map can, however, make quite effective use of hill-shading, when it is applied in a subdued stipple in some neutral tone, and used in conjunction with contours, hachures or both.[2]

A series of interesting efforts to solve some of the problems of hill-shading has been made by P. Yoëli.[3] In effect he sought to produce an

[1] E. Eiselen, 'The Central Valley (of California) Project', *Economic Geography*, vol. 23, pp. 22–31 (Worcester, Mass., 1947), has four such maps, which show the relief of the state of California with remarkable clarity, emphasizing, of course, the longitudinal valley and the ranges on either side. Information concerning canals, reservoirs and irrigation projects is lettered in panels outside the shaded area, and located by means of arrows.

[2] The French *Service Géographique de l'Armée* employed on the 1 : 50,000 series, published in the decade before 1914, brown contours at 10-metre intervals, with a double system of hill-shading. The shadow of a vertical lighting was printed in brown, that of an oblique lighting from the north-west in purplish-grey. The *Nouvelle Carte de France*, on the same scale, published after 1922, used for economy only oblique hill-shading from the north-west in brown.

The Ordnance Survey has used the technique of hill-shading on four of the One-inch Tourist Series. On the Peak District sheet, a combination of layer-colouring and hill-shading above 900 feet O.D. is used. On the North York Moors sheet, layer-colouring is applied in tones of apricot and hill-shading in violet on slopes to the south and east above 400 feet. On the revised Lake District (1958) and new Lorne and Lochaber sheets (1959), pale green tinting was used for all land below 500 and 100 feet respectively, light brown tinting above, and contours at 50-foot interval. A most striking three-dimensional effect was produced by the superimposition of hill-shading in brown.

[3] P. Yoëli, 'Analytical Hill Shading', *Surveying and Mapping*, vol. 25, pp. 573–9 (Washington, D.C., 1965); and 'Analytical Hill Shading and Density', *Surveying and Mapping*, vol. 26, pp. 253–9 (Washington, D.C., 1966).

analytically determined shadow picture of an illuminated three-dimensional body by resolving the surface of the earth into small plane unit-areas or facets, to be prepared from stereoscopic air photographs. He then applied a computed density of tones to the hill-shading. Later he discussed the mathematical theory and the practical application of this method, with examples, using an electronic computer to calculate the value of each minute facet and an electronically directed light-ray on an emulsion-coated surface to give the particular shade of grey.[1]

In order to appreciate the problem of hill-shading, choose some clear-cut relief feature, such as a volcanic cone or a fretted ridge, draw a contour-map of it, and then try to produce a shaded map. The shading may best be done by using a very soft graphite or carbon pencil; apply light strokes on the slopes in shadow. The individual strokes may be obliterated into a uniform tone by rubbing with a stump of rolled paper, or even carefully with the finger-tip. The student may try the effect of a graduated colour-wash in brown or grey, but great care is needed to avoid patchiness. Damp the paper and before it dries apply all the colour, leaving no sharp edges.

Perspective Terrain Maps. This term has been given by F. Hölzel to a type of shaded relief map in which some 'bird's-eye impression' of the third dimension is achieved by perspective design.[2] The reader is referred to the examples shown in this article, notably to an impression of the country around Lake Como, drawn in both the commonly used 'central perspective' and his so-called 'progressive perspective'.

(g) Landscape Colour Maps

Some general considerations concerning the use of colour in cartography have been outlined on pp. 53-5. It has been employed on maps since early times, although until the end of the last century it was of course applied by hand. Since then, developments in colour-printing have proceeded apace, sponsored both by government topographical agencies and by the great atlas-houses. As J. S. Keates,[3] who has

[1] P. Yoëli, 'The Mechanisation of Hill Shading', *Cartographic Journal*, vol. 4, no. 2, pp. 82-8 (London, 1967).

[2] F. Hölzel, 'Perspektivische Karten', *International Yearbook of Cartography*, vol. 3, pp. 100-18 (London, 1963).

[3] J. S. Keates, 'The Small-scale Representation of the Landscape in Colour', *International Yearbook of Cartography*, vol. 2, pp. 76-83 (London, 1962). This includes a particularly effective colour-map of part of West Africa, published in an atlas by Thomas Nelson & Sons, Ltd., Edinburgh. See also P. Yoëli, 'Relief and Colour',

devoted a good deal of research to this topic, states, the combination of layer-tinting and hill-shading has reached its zenith. He goes on to discuss the principles of landscape colour maps, as follows: 'Landscape colour maps are not to be confused with colour air photographs, and do not attempt to reproduce the colours seen in nature. The colour map must select, identify and generalize in the same way as any other map of equivalent scale. Its purpose, like that of a geographical description, is to illuminate, to stimulate and to enable comparisons to be made. In this way, colour is used in an impressionistic, not a "realistic" manner.' After an interesting discussion on the psychology of colour association, Keates goes on to describe the problems of the compilation in manuscript of available information in the form of guide base-maps, the choice of media (coated plate or laminated board) and pigment (he prefers water-colours), the background delineation of guide-lines in pencil, the actual application of the colour (its intensity, brightness, shadows, emphasis), problems of retouching, and the combination of line-drawings depicting drainage, names, etc., with the half-tone colour-map. The net result is to emphasize, in a three-dimensional effect, the significant features of the landscape.

The student is referred to Keates' original paper, which includes an effective map-sample. It is clear that this method can be highly successful, but obviously calls for a considerable degree both of artistic sensitivity and skill, and of cartographic appreciation.

(h) Cliff- and Rock-drawing

It is very useful to be able to indicate on a map the occurrence of steep cliffs and rock-faces. This is done by wedge-shaped black lines, with the thin ends pointing down the slope, but it is very difficult to do on a manuscript map without giving a 'fringed' or 'tasselled' effect.[1]

Cartographic Journal, vol. 1, no. 2, pp. 37–8 (London, 1964); he discusses the possibility of combining 'natural colours' for vegetation (using both solid colour and symbols) with hill-shading in black and white.

[1] The Lake District and Snowdonia sheets of the One-inch (Tourist Edition) of the Ordnance Survey, some of the sheets of the Fifth (Relief) Edition and the more detailed 1 : 25,000 series, show crags very effectively. The Swiss maps use beautifully drawn symbols, particularly the 1 : 50,000 *Siegfried* maps and the new *Landeskarte der Schweiz* on the same scale. See also the superb map of Mt McKinley, Alaska, on a scale of 1 : 50,000, surveyed and edited by Bradford Washburn, printed in Bern (1960) by the Swiss Federal Institute of Topography.

(i) *Physiographic (Pictorial Relief) Maps*

A. H. Robinson[1] once wrote, 'The variously titled landform map or physiographic diagram is possibly the only type of map that can be claimed as a wholly original contribution of American geographic cartography.' Several American geographers, particularly E. Raisz,[2] have devised methods of showing physiographic features on small-scale maps by the systematic application of a standardized set of conventional pictorial symbols, based on the simplified appearance of the physical features they represent as viewed obliquely from the air at an angle of about 45 degrees. Some American geographers call this a 'morphographic' or 'morphologic' method. In principle, the method goes back to the primitive concepts of early maps, whereby relief features were shown obliquely and in some degree of perspective, instead of by vertical conventions. Many of these physiographic symbols are derived from block-diagrams used by such pioneers as W. M. Davis. In the article quoted, Raisz standardized the symbols to be used into a set of forty 'morphologic types', and added a further ten, based mainly on natural vegetation, to diversify the category of plains. Raisz has used the name *trachographic map* for one which shows specific ruggedness of land by means of symbols.[3]

The advantages of this method are most readily appreciated when it is used for semi-diagrammatic small-scale maps, such as military maps to show a campaign, or for teaching purposes, for they do give a good broad impression of the country.[4] Some American regional

[1] A. H. Robinson *et. al.*, 'Geographic Cartography', *American Geography: Inventory and Prospect*, p. 557 (Syracuse, N.Y., 1954).

[2] E. Raisz, 'The Physiographic Method of Representing Scenery on Maps', *Geographical Review*, vol. 21 (New York, 1931); and E. Raisz, 'Developments in the Physiographic Method of Representing the Landscape on Maps', *Comptes Rendus du Congrès International de Géographie. Amsterdam, 1938.* Tome 2. Travaux de la Section I. Cartographie, p. 33 (Leiden, 1938). See also E. Raisz, *Landform Maps: A Method of Presentation:* Part I of a Final Report of Contract Nonr (2333(00)) with the Geography Branch, Office of Naval Research (Washington, D.C., 1958). This paper describes in detail the method of drawing these maps with physiographic symbols, especially making use of aerial photographs. The paper is accompanied by a physiographic map of Mexico, on a scale of approximately 1 : 3 million, by an outline map of physiographic provinces, and a descriptive summary of the latter. A further account by E. Raisz, 'A new Landform Map of Mexico', is given in *International Yearbook of Cartography*, vol. 1, pp. 121–8 (London, 1961). Raisz's well-known sheet-map of 'Landforms of North America' on various scales has run to a number of editions.

[3] E. Raisz, *Principles of Cartography*, pp. 88–9 (New York, 1962).

[4] A valuable summary of landscape maps, using the term in its widest sense, is provided by A. G. Isachenko, 'Landscape Mapping (its Significance, its Present State

geographers, such as Preston E. James,[1] have made extensive use of the method. The chief difficulty is to lay these symbols, which in appearance are like block-diagrams and therefore are intended to be viewed obliquely, on to a map which has to be viewed vertically. The symbols can of course be superimposed on faint layer-tinting, or a number of spot-heights can be added.

Raisz suggests the name of '*land-type*' maps, where the black symbols representing land-forms are superimposed upon tints of brown, green, yellow, etc., for cropland, forest, grassland, etc. As he says in the paper cited below (1953, p. 501): 'It is high time to break with the conventional layer tinting for medium- and small-scale maps, with deceptive greens on the driest of deserts. Vegetation and cultivation are far more important than elevation above sea level, and if we follow in our symbolism the natural colouring we arrive at the land-type map, the map of the Air Age.' Information to be included in these maps is obtained from contour-maps of the largest possible scale, field-work and air-photographs.[2]

An interesting application of the principle consists of making a large-scale physiographic map of some striking piece of country, using various symbols. Trace a contoured base-map from a large-scale topographical map. Then, with the help of geological maps, oblique air photographs if available, and field observation, draw in pictorial

---

and its Tasks)', *Soviet Geography: Review and Translation*, vol. 2, no. 2, pp. 34–47 (New York, 1961).

[1] P. E. James, *Latin America* (New York, 1942); and *A Geography of Man* (Boston 1949). Note particularly in the latter the striking maps of the western Sahara (p. 41), the Tarim Basin (p. 52), South Africa (p. 56), India (p. 102), California (p. 124), Greece (p. 133), Italy (p. 148) and eastern North America (p. 235). Other very attractive examples are shown in articles by E. O. Teale and E. Harvey, 'A Physiographical Map of Tanganyika Territory', *Geographical Review*, vol. 31, p. 655 (New York, 1941); and by H. de Terra, 'Component Geographic Factors of the Natural Regions of Burma', *Annals of the Association of American Geographers*, vol. 34, p. 71 (Lancaster, Pa., 1944).

[2] E. Raisz, 'The Use of Air Photos for Landform Maps', *Annals of the Association of American Geographers*, vol. 41, pp. 324–30 (Lancaster, Pa., 1951); and 'Direct Use of Oblique Air Photographs for Small-scale Maps', *Surveying and Mapping*, vol. 13, pp. 496–501 (Washington, D.C., 1953); and 'Landform, Landscape, Land-use and Land-type Maps', *Surveying and Mapping*, vol. 6, pp. 220–3 (Washington, D.C., 1946). He defines these four categories of maps as follows: (i) *landform* – a map showing relief by means of physiographic symbols; (ii) *landscape* – an artist's representation in true colours, though conventionalized where necessary, such as a coloured air photograph with added symbols and lettering; (iii) *land-use* – a map showing categories of arable, pasture, woodland, built-up and industrial areas, etc.; (iv) *land-type* – a combination of these three.

symbols, using the contours as a location guide. The symbols need not be restricted to Raisz's chosen forty; special symbols can be devised for the particular land-forms. For example, a physiographic map of Craven in the West Riding of Yorkshire would show scars, gorges, dry-valleys, pot-holes and water-sinks, monadnocks, areas of clint pavements, areas of peat bog, millstone grit 'edges' and so on. A physiographic map of Snowdonia would contain an immense variety of striking phenomena – glaciated valleys, hanging valleys, cwms, llyns, rock-steps, upland moors and arêtes. The area chosen need not be viewed so that the top of the map is in the north; for Craven this might indeed be the most revealing aspect, but for Snowdonia a view south-westward across the Glyders towards Snowdon itself would be preferable. This work has much of the nature and quality of landscape-drawing and should preferably be carried out in the field.

The Method of A. H. Robinson and N. J. W. Thrower. A physiographic method of depicting the terrain was devised by the co-authors[1] '. . . to the end that more of those who have the requisite scientific training but not the artistic may be able to produce acceptable landform drawings' (op. cit., p. 509). The difficulty is that if a land-form is drawn in perspective, the vertical dimension must occupy horizontal space, and therefore much of the feature must be drawn in the wrong place on the map. On a small-scale map the deviation from the correct position of any feature is not serious; the realistic and artistic appearance of the land-form depiction is the first desideratum. The basis of the method is Tanaka Kitirô's concept of the 'inclined contour' (see p. 94). The authors covered a contour map of the desired area with equally spaced horizontal lines, then drew in the inclined contours by joining the intersection of the lowest contour and the lowest horizontal line with the intersection of the next contour and the next horizontal line. This in effect is projecting the contours orthogonally with a high upward oblique angle of view, so that '. . . a perspective-like profile can be placed on a map without violating the planimetry of the map' (op. cit., p. 512). The trace of the inclined contours, to which is added drainage features, gives a curiously effective impression of a three-dimensional surface. Then the cartographer is able to demonstrate his artistic skill, using these mechanically drawn inclined contours as a base, by putting in detailed shading to give an impression of a light-

[1] A. H. Robinson and N. J. W. Thrower, 'A New Method of Terrain Representation', *Geographical Review*, vol. 47, pp. 507–20 (New York, 1957).

source from the north-west, employing to enhance the detail such additional information as aerial photographs and field notes. Finally, any required names can be inserted, and the final draft is inked in. Students are recommended to consult the clear diagrams showing the several stages of the work and the finished map given in the reference cited, and to study the simple geometrical principles behind the method.[1]

(j) *Landform Type, 'Terrain Type' or 'Geomorphological Province' Maps*

Various methods of indicating broad and generalized landform types have been used by a number of American geographers.[2] These methods differ from E. Raisz's physiographic maps (although he sometimes uses the general word 'landform' for this type), which do not seek to delineate 'types' or categories. The earliest examples were drawn to illustrate the numerous texts written mainly by American geologists on North America as a whole.[3]

Preston E. James[4] has made extensive use of this type of map, where a physical background was needed for economic or ecological surveys. He divided the land surface of each continent into the following categories:

Plain	Hilly upland and plateau	Low mountain
High mountain	Hamada	Erg
Mountain and bolson	Intermont basin	Ice-covered area

[1] N. J. Thrower further develops the uses of orthogonal mapping, with some striking examples, in 'Extended Uses of the Method of Orthogonal Mapping of Traces of Parallel Inclined Planes with a Surface, especially Terrain', *International Yearbook of Cartography*, vol. 3, pp. 26–38 (London, 1963).

[2] A convenient brief summary is provided by J. E. Dornbach, 'An Approach to the Design of Terrain Representation', *Surveying and Mapping*, vol. 16, pp. 41–4 (Washington, D.C., 1956).

[3] To quote merely a few: (i) J. W. Powell, *The Physiography of the United States*, pp. 98–9 (New York, 1896); (ii) R. D. Salisbury, *Physiography*, p. 18 (Chicago, 1907); (iii) Isaiah Bowman, *Forest Physiography*, plate 4 (New York, 1911); (iv) N. M. Fenneman, 'Physiographic Divisions of the United States', *Annals of the Association of American Geographers*, vol. 18, end folding map (Lancaster, Pa., 1928). Now, as G. M. Lewis says (see footnote on p. 107), 'The delimitation of physiographic regions has ceased to be a popular pursuit in the United States. . . .'

[4] P. E. James, *A Geography of Man* (Boston, 1949). Note the supplement of Reference Maps, pp. 583–618. P. E. James also produced in a previous work, 'On the Treatment of Surface Features in Regional Studies', *Annals of the Association of American Geographers*, vol. 27, p. 213 (Lancaster, Pa., 1937), a surface configuration map of Kentucky, on which he distinguished four categories: (i) flat-topped interfluves; (ii) colluvial slopes; (iii) residual slopes; and (iv) alluvial bottom-lands.

To each category he ascribed a stipple or line-shading, and so produced a series of clear 'terrain type' maps of the continents.

Several other methods of drawing maps of landscape or terrain types based on a non-genetic classification have been devised. V. C. Finch and G. T. Trewartha[1] used, in earlier editions of their well-known text-book, maps on a world and continental scale, based on a simple four-fold division of plain, plateaux, hill-lands and mountains. Much the same basis was utilized by another famous team of H. M. Kendall, R. M. Glendinning and C. H. MacFadden.[2] E. H. Hammond made a 'geomorphic study' of part of southern California on a scale of 1 : 560,000.[3] E. H. Hammond has paid particular respect to the small scales required on world and continental maps,[4] so as to afford an empirical analysis of the configuration without any genetic implications. While some degree of quantitative definition, some reasonable approximation, was involved, in order to attain a certain objectivity and to make possible comparisons, this was to be only in general terms, involving nothing so elaborate as statistical analysis, for which, indeed, uniform material is not available. For North America he divided the continent into $7\frac{1}{2}$-minute rectangles, and used three main distinguishing categories: *local relief* (i.e. maximum difference in elevation), *slope* (in percentages), and *profile* (proportion of near-level land), subdividing these into 5, 9 and 4 classes respectively. Combinations based on this classification resulted in eight groupings or *terrain types* (op. cit., 1954, pp. 36–7), as follows:

I	Nearly flat plains	V	Hills
II	Rolling and irregular plains	VI	Low Mountains
III	Plains with widely spaced hills and mountains	VII	High Mountains
IV	Partially dissected tablelands	VIII	Ice-caps

[1] V. C. Finch and G. T. Trewartha, *Elements of Geography*, pp. 257, 344, 355, 370 and Plate V (New York, 1949).

[2] H. M. Kendall, R. M. Glendinning and C. H. MacFadden, *Introduction to Geography*, pp. 182–6, Plate IV (New York, 1951).

[3] E. H. Hammond, 'A Geomorphic Study of the Cape Region of Baja California', *University of California Publications in Geography*, vol. 10, p. 50 (Berkeley, 1954). The terms 'geomorphography' and 'geomorphic' have been used in America to indicate the surface configuration or 'solid geometry' of the land surface, without implying the genetic elements inherent in geomorphology. See also E. H. Hammond, 'Landform, Geography and Landform Description', *Californian Geographer*, vol. 3, pp. 71–2 (Berkeley, 1962); and J. E. Kesseli, 'Geomorphic Landscapes', *Yearbook of the Association of Pacific Coast Geographers*, vol. 12, pp. 3–10 (Cheney, Wash., 1950).

[4] E. H. Hammond, 'Small-scale Continental Landform Maps', *Annals of the Association of American Geographers*, vol. 44, pp. 33–42 (Lancaster, Pa., 1954).

To these he added two others: plains and tablelands having many lakes or swamps, and plains and tablelands above 8,000 feet. Using all available published maps, and making quantitative determinations for his carefully chosen $7\frac{1}{2}$-minute rectangles, he drew boundaries for these terrain types for North and South America (op. cit., pp. 39, 41), and added distinctive shading. Later he produced an end-paper world-map of these terrain types in the 1957 edition of Finch and Trewartha.[1]

In 1960, N. J. W. Thrower[2] contributed a landform study of the island of Cyprus in the form of a series of maps in colour, based on the methods used by E. H. Hammond. Using the G.S.G.S. map series on a scale of 1 : 50,000, he divided the island into 5-km squares, and constructed (i) a map showing the percentage of the land with a slope of less than 1 in $12\frac{1}{2}$, i.e. a 'gentle slope' (three grades, 0–50, 50–90 and 90–100 per cent); and (ii) a map showing relative relief, i.e. the difference in altitude between the highest and lowest points in each square (three grades, 0–100, 100–1,000 and 1,000–4,000 feet). By combining the categories on these two maps, four terrain types (mountains, hills, rolling and irregular plains, and nearly level plains) were distinguished and depicted on the final map.

A development of this 'terrain type' map was produced by G. M. Lewis[3] to show the west-central United States east of the Rocky Mountains; to this type he gave the name *choromorphographic*, '. . . because it delimits and classifies areas of land (Greek, *khora*) according to their surface configuration (Greek, *morphe*)' (op. cit., 1962, p. 88). He prefers this term to 'terrain map', because the latter often includes surface materials and even vegetation in addition to the 'solid geometry' of the land. He found the available data for this area to be remarkably variable in detail, quality and scale, but using much the same grouping of factors as Hammond he produced a map with twelve 'choromorphographic types' (Fig. 30).

A comprehensive effort has been made by G. M. Goldberg to determine geomorphological regions or provinces on an objective quantitative basis, handling a wide range of statistical information in a

[1] V. C. Finch, G. T. Trewartha, E. H. Hammond and A. H. Robinson, *Elements of Geography* (New York, 1957).
[2] N. J. W. Thrower, 'Cyprus: A Landform Study', Map Supplement No. 1, *Annals of the Association of American Geographers*, vol. 50 (Lawrence, Kansas, 1960).
[3] G. M. Lewis, 'Changing Emphases in the Description of the Natural Environment of the American Great Plains Area', *Transactions and Papers, 1962: Institute of British Geographers*, no. 30, 75–90 (London, 1962).

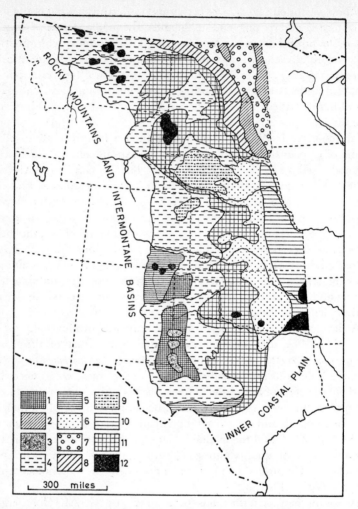

Figure 30. CHOROMORPHOGRAPHIC MAP OF THE WEST-CENTRAL UNITED
STATES EAST OF THE ROCKY MOUNTAINS

Based on G. M. Lewis, 'Changing Emphases in the Description of the Natural Environment of the American Great Plains Area', *Transactions and Papers, 1962: Institute of
British Geographers*, no. 30, p. 82 (London, 1962).

The numbers in the key are as follows: **1.** Extensive and flat high plains; **2.** Flat
valley-floor plains; **3.** Flat sand plains; **4.** Coarsely dissected high plains; **5.** Closely
and deeply dissected high plains; **6.** Closely dissected low plains; **7.** Rolling plains with
many lakes and swamps; **8.** Hills with many lakes and swamps; **9.** Sand hills; **10.** Low
rolling hills; **11.** High hills; **12.** Hill and mountain enclaves.

computer.[1] He selected eight 'terrain factors' for each proposed unit: maximum elevation, range of relief (the difference between the highest and lowest points in a grid square), the arithmetic mean elevation (a mean height of all grid-line intersections), the frequency of peaks (as indicated by closed contours) which exceeded this mean elevation, the 'relative peak relief' (i.e. the height to which a peak rises above the lowest valley within a certain radius), 'peak anomalies' (i.e. the percentage of peaks above the relative peak relief), the 'upland–lowland proportion' (derived from subtracting the lowest elevation from the mean elevation and dividing by the elevation range, thus giving a decimalized proportion, e.g. 0·239), and the average slope tangency (five east-west and five north-south traverses were made across each unit-region, the contour-crossings per line were counted and the slope tangents computed). All these data were systematically analysed and applied to the mountains of Central Germany. He also applied various scale coverages to the United States, using in this case five terrain factors: the 'grain-spacing' between major ridges and valleys, the frequency of peaks, average relief, slope direction changes and 'characteristic plan profiles'. From these data a series of geomorphological regions were derived. These examples are quoted to show how by means of computer techniques large masses of data can be readily, objectively and profitably handled.

(k) *Landscape Evaluation Maps*

Of wider implication than a relief map is a landscape map, which seeks to show the sum total of any area, rural or urban (the latter may be specifically termed a *townscape*). An effort has been made by K. D. Fines[2] to establish a series of landscape and townscape values as a contribution to planning development and conservation. The problem was to replace, as far as is possible, pure subjective judgement and inevitable personal bias with some scale of values, though of course personal observation (or 'view evaluation'), possibly supplemented by colour photographs, from a chosen series of viewpoints is required, so that the subjective element cannot be obviated. Fines suggested a continuous series of scale-values from 0 to 32, with six classes or categories: *unsightly*, 0–1; *undistinguished*, 1–2; *pleasant*, 2–4; *distinguished*,

[1] G. M. Goldberg, 'The Derivation of Quantitative Surface Data from Gross Sources', *Surveying and Mapping*, vol. 22, pp. 537–48 (Washington, D.C., 1962).
[2] K. D. Fines, 'Landscape Evaluation: a Research Project in East Sussex', *Regional Studies*, vol. 2, no. 1, pp. 41–55 (London, 1968).

4–8; *superb*, 8–16; *spectacular*, 16–32. An obvious disadvantage of such a scale is the compression of most landscapes into the lower ranges, leaving half the scale for the rare 'spectacular'; the highest value in Great Britain he adjudges as 18 (a view of the Black Cuillins of Skye across Loch Coruisk), while 12 (the prospect from Newlands Corner near Guildford across the Weald) is the highest in Lowland Britain. Using a system of twelve grades of shading, Fines produced a Landscale Evaluation Map of East Sussex, using a large number of *tracts* (units largely related to geological and physiographical features), the smallest being about 1 sq. km, and shaded on a choropleth basis.

(l) *Configuration Maps*

This name can be applied to a simple yet accurate outline map of surface features, such as is used to illustrate a textbook in physical or regional geography. The detail is abstracted by selectively tracing from a relevant scale topographical map, and is then conventionalized and symbolized as required.

A simple configuration map is shown in Fig. 31. It was required to produce an outline relief map as a locational basis for a regional study of the Lötschenthal, a valley in the Bernese Oberland drained by the river Lonza, a right-bank tributary of the Rhône. The outline of the snow-fields and glaciers was abstracted from the 1 : 50,000 *Siegfried* map of the area, and then the main crest-lines were defined by heavy black lines.

(m) *Morphological Maps*

It has long been necessary for geographers to depict on a map the surface *forms* of the earth, as distinct from the quantitative expression of its relief in terms of contours. These can be on all scales. Thus D. L. Linton compiled maps of the world (1 : 48 million) and of Europe (1 : 16 million) for the *Oxford Atlas*, in which the elements of structure were shown by colour and shading, indicating such features as 'flat-bedded', 'slightly folded' and 'strongly disturbed' newer sedimentaries, similar categories of older sedimentaries, metamorphic zones of younger fold mountains, undisturbed Tertiary volcanics, etc., with symbols of volcanoes, and various lines for scarp-forming faults and for escarpments themselves. In the same volume he included a more detailed map of Great Britain, including also axes of anticlinal flexures, fault-lines with an indication of direction of downthrow, and outcrops of thrust-planes.

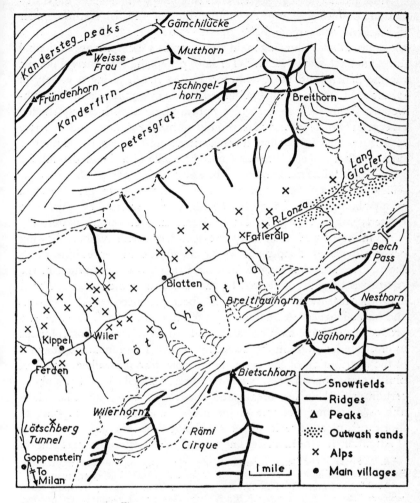

Figure 31. A CONFIGURATION MAP

Based on *Topographischer Atlas der Schweiz* (Atlas Siegfried).

The Lonza, rising in the snow-fields of the Bernese Oberland, flows through the Lötschenthal, and then joins the Rhône near Brigue. This map was drawn to emphasize the main features of the configuration of the valley.

Most geomorphological papers include maps which show the distribution, pattern and mutual relationships of selected features under discussion. For example, B. W. Sparks[1] produced a generalized map of the South Downs, showing main and secondary escarpments, anticlinal and synclinal axes, the mid-Tertiary peneplain, the early Pliocene marine bench, and the Goodwood Raised Beach, using a variety of symbols. M. M. Sweeting[2] published a most informative map of northwestern County Clare, on which appeared eight distinctively shaded categories of surfaces and flats, prominent breaks of slope, and a variety of features (caves, swallow-holes, springs, etc.) shown by distinctive point-symbols. Special mention may be made of a detailed symbolic geomorphological map drawn by B. Robitaille[3] of part of Canada on a scale of 1 : 72,000, deriving his material from the careful interpretation of air-photographs. Landform details were depicted by thirty-five located symbols, superficial deposits by shading and stipple patterns.

An interesting series of morphological maps, with symbols showing beaches, terraces, flood-plains, escarpments, closed depressions, sand-sheets, barchans and longitudinal sand-ridges (the orientation of the last two is carefully indicated by the direction of the linear symbols) and other features of desert relief has been produced by A. T. Groves and his colleagues.[4] A final example of this work is by J. Lewin,[5] who produced a series of most revealing and attractive morphological maps to illustrate his work on the Yorkshire Wolds. These included analyses of the distribution of the escarpments, cols, through-valleys, benches, bluffs, planation surfaces, valley-networks, valley-heads, slopes above certain critical angles, breaks and changes of slope, etc. The whole affords a striking example of the use of morphological mapping to illustrate a piece of regional geomorphology (see Fig. 32).

Attention has been increasingly paid to a large-scale mapping of

[1] B. W. Sparks, 'The Denudation Chronology of the Dip-slope of the South Downs', *Proceedings of the Geologists' Association*, vol. 60, facing p. 166 (Colchester, 1949).

[2] M. M. Sweeting, 'The Landforms of Northwest County Clare, Ireland', *Transactions and Papers, 1955: Institute of British Geographers*, no. 21, facing p. 37 (London, 1955).

[3] B. Robitaille, 'Presentation d'une Carte Géomorphologique de la Région de Mould Bay, Île du Prince-Patrick Territoires du Nord-Ouest', *Canadian Geographer*, vol. 4, no. 15, pp. 39–43 (Toronto, 1960).

[4] A. T. Grove and A. Warren, 'Quaternary Landforms and Climate on the South Side of the Sahara', *Geographical Journal*, vol. 134, pp. 194–208 (London, 1968); and A. T. Grove, 'Landforms and Climatic Change in the Kalahari and Ngamiland', *Geographical Journal*, vol. 135, pp. 191–212 (London, 1969).

[5] J. Lewin, *The Yorkshire Wolds: a Study in Geomorphology*, Occasional Papers in Geography, no. 11 (University of Hull, 1969).

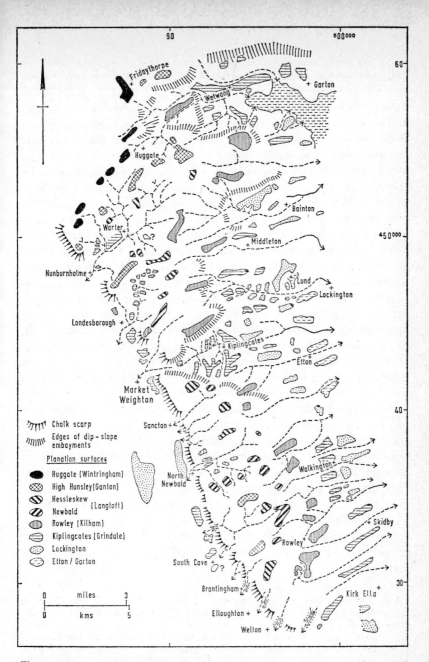

Figure 32. PLANATION SURFACES IN THE SOUTHERN YORKSHIRE WOLDS

Based on J. Lewin, *The Yorkshire Wolds, A Study in Geomorphology*, Occasional Papers in Geography, no. 11, p. 24 (University of Hull, 1969).

The figure is based on the Ordnance Survey Map with the sanction of the Controller of H.M. Stationery Office, Crown copyright reserved. The area includes the southern part of the Wolds, from Fridaythorpe to Kirkella, and eastward from the escarpment.

individual 'facets' of the land surface,[1] in the form of a purely empirical (i.e. not interpretative) survey in the field. On such a map the boundaries of morphological units are defined by various kinds of solid and broken lines, together with graduated stipples indicating degrees of slope, and a range of symbols and letters representing individual recognizable and plottable features which can be added if desired. The article quoted illustrates a suggested classification of these features, the relevant portion of the 6-inch O.S. map used, and the completed morphological map. The equipment used is simple: the 6-inch base map on a mapping board, a clinometer for angle measurement, ruler and field notebook. The method demands chiefly an aptitude for careful observation and accurate plotting. In the words of R. S. Waters (op. cit., p. 12), 'By virtue of its exacting demands on the student for careful and purposeful observation and clear and accurate delineation on a base-map of morphologically indivisible units of the land-surface, its value as a training method in field-work was quickly appreciated. . . .' An example as applied to beaches is shown on Fig. 36.

The most systematic survey of this kind in Britain, as yet, was organized (under the direction of D. L. Linton) to produce a complete morphological map of Great Britain on the occasion of the London meeting in 1964 of the International Geographical Union. The broad principle was the depiction of areas with certain morphological characteristics by means of applied tints, and of specific morphological feature by means of symbols. Large features appeared as individuals, accurately located, smaller grouped features by means of conventional representation. A team of contributors compiled their individual allocations on a ¼-inch topographical base, for reduction to the final map on a scale of 1 : 625,000. The areas and features to be depicted were classified into six groups on a broadly genetic basis, each distinguished by a particular specified colour, each broken down where required; in all, fifty-five carefully defined separate tints and symbols were used.

---

[1] R. S. Waters, 'Morphological Mapping,' *Geography*, vol. 43, pp. 10–17 (Sheffield, 1958). See also R. A. G. Savigear, 'A Technique of Morphological Mapping', *Annals of the Association of American Geographers*, vol. 55, pp. 514–38 (Lawrence, Kansas, 1965), which gives a full account of the recognition, measuring and mapping of breaks of slope, facets and segments, using various lines, symbols, arrows and figures, with a full bibliography and numerous examples; and E. M. Bridges and J. C. Doornkampf, 'Morphological Mapping and the Study of Soil Patterns', *Geography*, vol. 48, pp. 175–81 (London, 1963), affording an interesting correlation of morphological and soil maps.

It is interesting to note that the Royal Dutch–Shell group have developed, after many years of experience and trial, a series of symbols which are used in photo-geological work, to denote both morphology and geology.[1] Printed in five colours, the symbols differ according to whether the features are 'certain' or 'uncertain'.

A striking example of geomorphological mapping on a national scale is being carried out in Poland, in connection with the preparation of plans for the economic development of various regions of the country. One area which has received special attention is the Upper Silesian Industrial District.[2] The map is based on a detailed combined genetic and chronological classification, together with morphometric information such as degrees of slope by colour shading, detailed contours and spot-heights, and height and depth of minor forms by varying line thicknesses. As Klimaszewski says, '. . . the map reveals the appearance (*morphography*), the dimensions (*morphometry*), the origin (*morphogenesis*) and the age (*morphochronology*) of the landforms. . . .' The volume cited reproduces (though in black and white) a portion of the finished map, with a detailed legend of some sixty different symbols. The map is intended to have practical economic applications, and in fact was sponsored by the Committee for Affairs of the Upper Silesian District. Other maps have been prepared from it to show land favourable or unfavourable to specific types of economic development, particularly maps of potentiality for improvement ('*bonitative maps*').

A map on the same scale of 1 : 25,000 is being prepared to show the hydrography of Poland. On to a topographical base-map is plotted all possible information relating to water and its circulation, underground supplies, watersheds, springs, degrees of impurity, flood areas, marsh and bog, drainage works, etc.; on the map-fragment in the work cited, nearly eighty separate symbols are used. The area for which mapping is already completed covers broadly the Upper Silesian Industrial District, where the hydrographic problems are extremely complex.

[1] The Shell Petroleum Co. Ltd., St Helen's Court, Great St Helen's, London, E.C.1 have published (n.d.) a most attractive booklet listing the symbols in colour, together with two aerial photographs accompanied by maps showing the symbols as plotted.

[2] M. Klimaszewski, 'The Problems of the Geomorphological and Hydrographic Map on the Example of the Upper Silesian Industrial District', *Problems of Applied Geography*, pp. 73–81, Polish Academy of Sciences, Institute of Geography, Geographical Studies No. 25 (Warsaw, 1961).

MORPHOMETRIC ANALYSIS

In recent years increasing attention has been paid by geomorphologists to morphometric techniques, that is, the utilization in various cartographical and diagrammatic ways of the statistical information about the earth's surface provided by published topographical maps, or actually measured in the field. In the words of A. A. Miller,[1] 'Much can be learnt from inspection of the map, but when its possibilities by this method have been exhausted there still remains infinitely more that can be extracted by quantitative analysis and rearrangement of the data.' Most of these morphometric techniques are concerned with slope and altitude; to the patterns of distribution provided by the dimensions of length and breadth, the map affords specific information about height, in the form of contour-lines and spot-heights (see pp. 88–92). This information may be classified, interpreted and presented by various devices, the subject of the next 54 pages. As G. M. Goldberg says,[2] and illustrates so efficiently, 'The study of Geomorphometry . . . puts us on the threshold of an even more precise age than before.'

While, however, we may accept that 'geomorphologic science, at its present state of development, suffers often from a lack of quantitative exactness',[3] it is important to realize that morphometric methods alone are quite inadequate. Even G. H. Dury,[4] one of their leading and most skilful exponents, states quite categorically, 'In geomorphological research there can be no substitute for lengthy and painstaking fieldwork. A multitude of features of the highest significance will appear only on those maps drawn by the field geomorphologist.' Moreover, J. P. Bakker and A. N. Strahler say,[5] '. . . full use of the modern morphometric statistical method for slope development can only be made

[1] A. A. Miller, *The Skin of the Earth*, p. 43 (London, 1953).

[2] G. M. Goldberg, 'The Derivation of Quantitative Surface Data from Gross Sources', *Surveying and Mapping*, vol. 23, pp. 537–48 (Washington, D.C., 1963). See also J. K. Gregory and E. H. Brown, 'Data Processing and the Study of Land Form', *Zeitschrift für Geomorphologie*, vol. 10, pp. 237–63 (Berlin, 1966).

[3] J. P. Bakker and A. N. Strahler, 'Report on Quantitative Treatment of Slope-Recession Problems', *Premier Rapport de la Commission pour l'Etude des Versants (Union Géographique Internationale*, p. 30 (Amsterdam, 1956).

[4] G. H. Dury, 'Methods of Cartographical Analysis in Geomorphological Research', *Indian Geographical Society : Silver Jubilee Souvenir Volume*, p. 136 (Madras, 1952). A summary of the main methods of Morphometric Analysis is given by Dury in his *Map Interpretation*, chap. XV, pp. 167–79 (London, 1952), including a bibliography; he classes the main groups of techniques as: (i) geometric analysis; (ii) arithmetic analysis; (iii) volumetric analysis; and (iv) clinometric analysis.

[5] *Premier Rapport de la Commission pour l'Etude des Versants*, p. 32 (Amsterdam, 1956).

if it is based on field-work. . . .' Again, as H. Baulig wrote,[1] 'There remains to be proved that mathematics has ever revealed an actual relationship in geomorphology that has not been discovered without its aid.' S. W. Wooldridge,[2] while recognizing the illustrative value of morphometric work, sounded a repeated warning against '. . . the urge to dabble in elementary mathematics', or to adopt it '. . . for snobbish reasons – i.e. because it sounds impressive and it is fashionable to clothe our thought in the jargon of mathematics'.

Another claim sometimes made is that morphometric methods can yield precise, accurate and objective results, superior to general descriptive and deductive methods used in the field. A survey by J. I. Clarke and K. Orrell[3] assesses the reliability of a number of the morphometric methods and stresses their inherent limitations. On the one hand, they emphasize that the map itself is 'subjective, selective and conventional; and, therefore, students dissatisfied with the imperfections of map interpretation and distributional mapping will find no peace of mind in morphometric analysis'. On the other hand, they claim that the objectivity sometimes claimed is illusory; that elements of subjectivity enter into this work whether the exponents be experienced or not; and that the same exercise carried out conscientiously by a number of operators may produce results significantly different in detail, as their experiments show. Moreover, they conclude that although the various techniques and methods are useful supplementary demonstrations, the considerable labour and tedious expenditure of time involved may not always be justified. This same caution was sounded by L. O. Quam:[4] 'Efforts towards a more quantitative analysis of land-forms is highly commendable, and it is hoped that through investigation of this sort our science may become more objective. Is there not a danger, however, that the use of mathematical

[1] H. Baulig, 'William Morris Davis: Master of Method', *Annals of the Association of American Geographers*, vol. 40, p. 195 (Lancaster, Pa., 1950) (contribution to the W. M. Davis Centenary Symposium).

[2] S. W. Wooldridge, 'The Trend of Geomorphology', *Transactions and Papers, 1958, Institute of British Geographers*, no. 25, pp. 32–3 (London, 1958).

[3] J. I. Clarke and K. Orrell, 'An Assessment of some Morphometric Methods', *Department of Geography, Durham Colleges in the University of Durham, Occasional Papers Series*, no. 2 (Durham, 1958); this is a most valuable paper, the product of much tedious experimentation.

[4] Contribution to discussion, following the paper by A. N. Strahler, 'Davis's Concepts of Slope Development viewed in the Light of Recent Quantitative Investigations', *Annals of the Association of American Geographers*, vol. 40, p. 213 (Lancaster, Pa., 1950).

formulae and statistical analysis give a false impression of objectivity
to sample measurements? . . . Are we not in danger of giving an
impression of accuracy which is not warranted?'

With these limitations, qualifications and reservations, however,
there still remains much of value; the most useful methods will be
reviewed in turn.[1]

PROFILES

The drawing of a profile from a contour-map may be of very great
assistance in visualizing the relief, and in the description and explanation of the land-forms. A geomorphologist in particular, seeking to
analyse the nature of relief, is interested in surfaces with different
slopes, corresponding to periods of peneplanation and of aggradation,
but contour-maps often fail to bring out these significant surfaces.
'Since the contoured map, while an indispensable aid to the geomorphologist, is incomplete in its indications, we may consider certain
methods by which its testimony can be supplemented, rendered clearer,
or translated into other terms. Such methods are, of course, no substitute for work in the field. They may, however, usefully precede such
work and can also assist in portraying its results.' Various methods,
often involving the drawing of some form of profile, may therefore be
employed.[2]

[1] A valuable critical analysis of some morphometric techniques is J. I. Clarke,
'Morphometry from Maps', in *Essays in Geomorphology* (ed. G. H. Dury), pp. 235-74
(London, 1966). See also C. A. M. King, *Techniques in Geomorphology* (London, 1966).
[2] S. W. Wooldridge and R. S. Morgan, *The Physical Basis of Geography*, p. 259
(London, new impression, 1946). The valuable chapter in which this quotation appears
was omitted from the 1959 edition, entitled *An Outline of Geomorphology* (London,
1959). See also A. F. Pitty, 'Some Problems in the Location and Determination of
Slope Profiles', *Zeitschrift für Geomorphologie*, vol. 10, pp. 454-61 (Berlin, 1966); 'Some
Problems in Selecting a Ground-surface Length for Slope-angle Measurement',
Revue de Géomorphologie Dynamique, vol. 17, pp. 66-71 (Paris, 1967); 'Some Comments
on the Scope of Slope Analysis Based on Frequency Distributions', *Zeitschrift für
Geomorphologie*, vol. 12, pp. 350-5 (Berlin, 1968); 'A Simple Device for the Measure-
ment of Hill-slopes', *Journal of Geology*, vol. 76, pp. 717-20 (Chicago, 1968); and *A
Scheme for Hillslope Analysis*, University of Hull, Occasional Papers in Geography,
no. 9 (Hull, 1969), which discusses several important devices (notably subdivided
histograms and triangular graphs). Other important recent references to slope are
given in the following: A. Young, 'Some Field Observations of Slope Form and
Regolith, and Their Relation to Slope Development', *Transactions of the Institute of
British Geographers*, no. 32, pp. 1-29 (London, 1963); C. E. Everard, 'Contrasts in the
Form and Evolution of Hill-side Slopes in Central Cyprus', ibid., pp. 31-47; K. J.
Tinkler, 'Slope Profile and Scree in the Eglwyseg Valley, North Wales', *Geographical*

The term 'section' and 'profile' are used with little precision and much confusion. The literal meaning of a section is a cutting, or a surface exposed by such a cutting, and the term is correctly used only when the geological structure is shown. A profile, on the other hand, is the outline produced where the plane of a section cuts the surface of the ground. A profile of a river valley, for example, may be either longitudinal or transverse; in the former case it is the outline of the valley on the surface from source to mouth, in the latter case it is drawn across the valley at right-angles to its general direction.

A *soil-profile* is in fact really a section since it shows the successive layers or zones from the surface downwards. The construction of a soil-profile is of major importance in pedological studies. A column is drawn to vertical scale and the various layers are shaded and labelled. These profiles may be located along lines of transect.[1]

The longitudinal profile of a river may afford valuable evidence of its geomorphological history, particularly in a polycyclic drainage basin, with marked breaks of slope between separate graded reaches. R. J. Chorley[2] drew a longitudinal profile of the river Heddon in Devonshire, which reveals a marked break of slope at 520 feet. If the curve of the upper reach is extended it will reach zero slope at 460 feet above present O.D.

The Drawing of Profiles

The first step in the drawing of an accurate profile is to lay a straight-edge of paper along the chosen line on the map, then mark accurately with sharp clean ticks all contour intersections, spot-heights, rivers,

Journal, vol. 132, pp. 379–85 (London, 1966) (on to the profiles he adds an indication of the nature of their surfaces by symbols (bedrock, grass, scree and clitter – loose rock fragments at an angle determined by the underlying material)); G. Robinson, 'Some Residual Hillslopes in the Great Fish River Basin, South Africa', *Geographical Journal*, vol. 132, pp. 386–90 (London, 1966), uses annotated slope profiles.

[1] This was done very successfully by Joy Tivy, 'An Investigation of Certain Slope Deposits in the Lowther Hills, Southern Uplands of Scotland', *Transactions and Papers, 1962: Institute of British Geographers*, no. 30, pp. 59–73 (London, 1962). She chose lines of transect down selected spurs, and dug a large number of pits. The spurs were levelled, the sites of the pits plotted, and the profiles were located in series on slope-diagrams.

[2] R. J. Chorley, 'Aspects of the Morphometry of a "Poly-cyclic" Drainage Basin', *Geographical Journal*, vol. 124, pp. 370–3 (London, 1958). This valuable paper uses other methods of dealing with longitudinal profiles, notably by treating the heights along the valley as a statistical sample, and then making frequency distribution curves on both arithmetic and logarithmic scales, and also plotting the distributions in cumulative form on both arithmetic and logarithmic probability paper.

summits and other defined points. Draw the base-line of the profile
on a sheet of graph-paper, and transfer the ticks carefully to this.
Rule vertical lines at either end of the base-line, and mark off a
vertical scale, which should be carefully chosen, bearing in mind the
height range involved and the nature of the country. Allow 100 feet,
or some such exact figure, to each horizontal line on the graph-paper,
for ease of plotting. Number the vertical scale at suitable intervals,
avoiding strings of noughts. Unless the horizontal scale is large and
the range of altitudes considerable, the vertical scale must be considerably larger than the horizontal, otherwise the undulations along
the profile will hardly be perceptible. On the other hand, too large
a vertical scale will produce a ridiculously caricatured effect of the
land surface. This relation between the horizontal and vertical scales
is known as the *vertical exaggeration*. Thus if the horizontal scale is 1 inch
to 1 mile (i.e. R.F. 1 : 63,360) and the vertical scale is 1 inch to 1,000
feet (i.e. 1 : 12,000), the exaggeration will be 5·28 times. Always state
the exaggeration below the profile. No exaggeration should be used
for accurate geological sections (see p. 188), which would otherwise
give an inaccurate dip to the strata. Either by following the vertical
graph-lines, or by ruling perpendiculars from the ticks on the base-
line, mark the position of each point according to the vertical scale,
with a fine accurate cross. When all points are plotted, join them by
a smooth line, not by a series of straight lines. To interpret the detail
between two widely spaced contours, use any other indications of the
relief, such as spot-heights near the line of profile, the position of
streams and information from hachures. It may even be necessary, if
profiles are being used for advanced landform study, to level accurately
each line of profile in the field, especially when the contour-interval
of the map is considerable. When drawing in the profile, care must be
taken (*a*) to start from the exact height at either end; (*b*) to distinguish
between dips and rises where there are two successive contours numbered identically; and (*c*) to draw carefully the outlines of summits,
whether peaked or flattened.

Finish off the line-work in ink, add significant place-names, lettered
in sideways above the point on the section to which they refer, together
with a title, and give the grid reference of either end. Unless a location-
map is included, indicating the plan of the profile, specify also the
orientation. If the profile is originally drawn on graph-paper for convenience, it may be traced on to drawing-paper, omitting the horizontal graph-lines.

Serial Profiles

A series of profiles to illustrate the edge of a plateau,[1] the transverse shape of a valley from the source of a river to its mouth,[2] the character of a coastline, a series of projecting spurs,[3] or a series of planation surfaces,[4] may be drawn with effect. They can be arranged in a vertical column, representing for example a series from north to south, accompanied by a location map. They can also be constructed as serials in time (Fig. 34), if short-term changes are appreciable.[5] Equally they can be arranged by the time and place and movement of beach material, as in Fig. 38.

Longitudinal Profiles

The chief problem involved in drawing longitudinal profiles is the transference of points along a curved or winding road, railway or river to the straight-edge which forms the base of the profile. The distance between each contour, spot-height or other feature must be accurately determined by stepping it off with dividers, and then ticking it off along the straight-edge. Railway profiles are difficult to draw, since

[1] F. J. Monkhouse, *The Belgian Kempenland*, p. 24 (Liverpool, 1948). See also J. Lewin, *The Yorkshire Wolds: A Study in Geomorphology*, Occasional Papers in Geography, no. 11, p. 8 (University of Hull, 1969); he uses a series of profiles, with the geology in section, to illustrate the edge of the escarpment (Fig. 33).

[2] M. Jones, 'The Development of the Teifi Drainage System', *Geography*, vol. 34, pp. 136–45 (London, 1949); K. M. Clayton, 'The Denudation Chronology of part of the Middle Trent Basin', *Transactions and Papers, 1953: Institute of British Geographers*, no. 19, p. 30 (London, 1953); C. Embleton, 'Some Stages in the Drainage Evolution of Part of North-east Wales', *Transactions and Papers, 1957: Institute of British Geographers*, no. 23, pp. 19–35 (London, 1957); G. H. Dury, 'Tests of a General Theory of Misfit Streams', *Transactions and Papers, 1958: Institute of British Geographers*, no. 25, pp. 105–18 (London, 1958), makes use of a large number of located serial profiles to illustrate his argument. J. A. A. Jones, 'Morphology of the Lapworth Valley, Warwickshire', *Geographical Journal*, vol. 134, pp. 216–26 (London, 1968), adds to his long-profile of a thalweg the flats on either side by means of shaded segments, differentiating between those on both banks, and on right or left bank only.

[3] M. M. Sweeting, 'The Land-forms of North-West County Clare, Ireland', *Transactions and Papers, 1955: Institute of British Geographers*, no. 21, p. 39 (London, 1955). These were accurately levelled in the field.

[4] R. B. McConnell, 'Planation Surfaces in Guyana', *Geographical Journal*, vol. 134, pp. 506-20 (London, 1968).

[5] See, for example, profiles of the beach at Marsden Bay, Co. Durham, during 1950, surveyed and drawn by C. A. M. King, *Transactions and Papers, 1953: Institute of British Geographers*, no. 19, p. 20 (London, 1953). Extensive use of shoreline relation diagrams is made by F. M. Synge and N. Stephens, 'Late- and Post-Glacial Shorelines, and Ice Limits in Argyll and North-east Ulster', *Transactions of the Institute of British Geographers*, no. 39, pp. 101–25 (London, 1966).

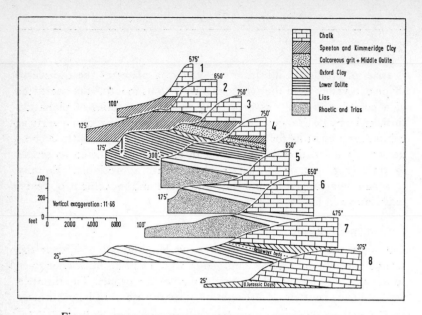

Figure 33. SCARP PROFILES OF THE YORKSHIRE WOLDS

Based on J. Lewin, *The Yorkshire Wolds, A Study in Geomorphology*, Occasional Papers in Geography, no. 11, p. 8 (University of Hull, 1969).

    The numbers refer to eight samples, extending from Ganton in the north to near North Ferriby in the south.

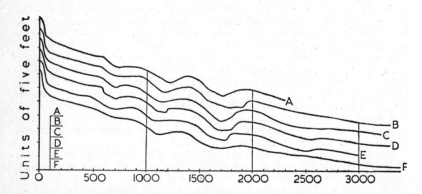

Figure 34. TRANSVERSE BEACH-PROFILES

Compiled by R. Kay Gresswell, from data personally surveyed on Southport beach. The datum for each profile was fixed at 10 feet below O.D. The letters refer to successive dates, viz. A. 21 October 1935; B. 3 January 1936; C. 5 March 1936; D. 13 May 1936; E. 25 July 1936; and F. 28 October 1936. The profiles have been placed in series, each with a different base, in order to avoid confusion, and yet to facilitate ready comparison.

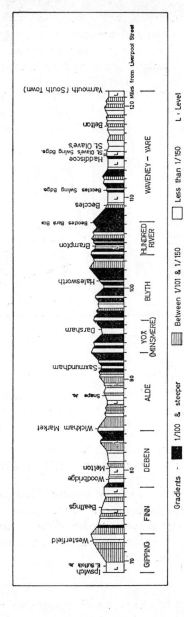

Figure 35. GRADIENT PROFILE OF THE RAILWAY FROM IPSWICH TO GREAT YARMOUTH
Based on J. H. Appleton, *The Geography of Communications in Great Britain*, Fig. 10 (Oxford, 1962).

'cutting and filling', i.e. the grading of the track by cuttings and embankments, may have removed or smoothed out minor changes of slope, and contours often end abruptly at the edge of the cutting symbol. Railway profiles as a rule can be accurately drawn when a large-scale map is available, showing numerous spot-heights along

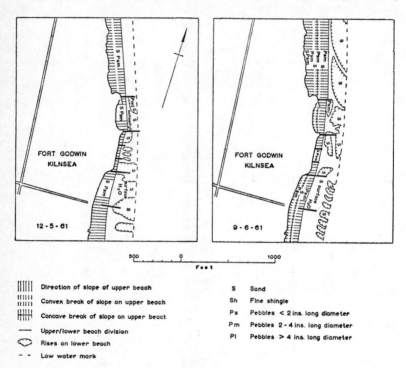

Figure 36. RECONNAISSANCE MAP OF BEACH–FORMS
Based on a manuscript map compiled by Dr A. Phillips.

the track itself. Information can be obtained from official track descriptions, or the actual gradients can be read from the gradient posts along the track. Fig. 35 shows a profile of the railway from Ipswich to Great Yarmouth, on which the four main gradient categories are emphasized by distinctive shading.

Superimposed Profiles

It is a useful practice to compare and to correlate profiles spaced at regular intervals across a piece of country, and then to plot them on a single frame.[1] Each individual line should be carefully numbered and located on an accompanying map. These are known as superimposed profiles (Fig. 37, *top*). Unless, however, the landforms have

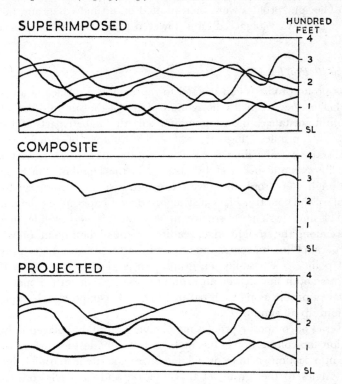

Figure 37. PROFILES

Superimposed, composite and projected profiles have been drawn of the same piece of country.

some morphological unity, such as an erosion platform, which will be shown up by the general uniformity of level of the various profiles, the result is apt to look somewhat muddled, and the several profiles should then be placed separately. E. H. Brown plotted the profiles of

[1] There are numerous examples of superimposed profiles as applied to geomorphological problems. J. W. Wright, 'The White Nile Flood Plain and the Effect of Proposed Control Schemes', *Geographical Journal*, vol. 114, pp. 173–90 (London, 1949), has two series of superimposed bank-profiles; C. A. M. King and W. W. Williams, 'The Formation and Movement of Sand Bars by Wave Action', *Geographical Journal*, vol. 113, pp. 70–85 (London, 1949), have two series of superimposed beach-profiles, representing the same beach at successive dates.

a series of valley-spurs, in order to bring out breaks of slope which might not be apparent in the valley-sides. He drew a profile along the crest-line of each spur, 'pulling-out' each profile into a straight-line. The spur-profiles were then plotted on a single diagram, but for clarity each was 'staggered', and lowered in relation to the preceding one.[1]

Composite Profiles

A composite profile is constructed to represent the surface of any area of relief, as viewed in the horizontal planes of the summit-levels from an infinite distance, and so including only the highest parts of a series of parallel profiles (Fig. 37, *middle*). A number of closely spaced equidistant parallel lines is ruled across the area; the orientation of these lines should be carefully chosen; in the Highlands of Scotland they might run from north-west to south-east, i.e. transverse to the 'grain' of the country. Place a straight-edge of paper along the outside parallel line, with a set-square at right-angles to it. Slide the set-square along the straight edge, reading off the highest point, on which-ever parallel line it occurs, and noting it on the straight-edge. Transfer these points to a base-line on graph-paper, plot them and join them with a smooth line, as for an ordinary profile. A lengthier method is to draw a profile along each parallel line, super-impose each in pencil, and join up all summit lines.

A series of composite profiles may be very profitably placed in juxta-position for comparison. One method is to divide the surface of the map into sections a mile or a kilometre wide along parallel lines at right-angles to the 'grain' of the country, and for each construct a composite profile. These profiles can be cut out of stiff cardboard and mounted in wooden slots, thus forming in effect a model from which minor complications of the landscape have been removed.[2]

Projected Profiles

It is possible to plot on a single diagram a series of profiles, including only those features not obscured by higher intervening forms. This will give a panoramic effect, with a distant sky-line, a middle-ground

[1] E. H. Brown, 'Erosion Surfaces in North Cardiganshire', *Transactions and Papers 1950: The Institute of British Geographers*, no. 18, pp. 51–66 (London, 1952).

[2] This method has been used by P. R. Shaffer, 'Correlation of the Erosion Surfaces of the Southern Appalachians', *Journal of Geology*, vol. 55, pp. 343–52 (New York, 1947); and by E. H. Brown, op. cit., pp. 51–66 (1952).

and a foreground; it is, in fact, an outline landscape drawing showing only summit detail (Fig. 37, *bottom*). The profiles should be spaced at equal intervals, but it is possible to add selected lines, running along, for example, a crest-line.[1]

A device akin to the projected profile has been termed a *compressed profile* by P. R. Schaffer and E. H. Brown (op. cit.). The series of composite profiles referred to above is compressed together, and when viewed at right-angles only those features not obscured by higher ones in the foreground are visible. E. H. Brown (op. cit., p. 60, 1950) produced a compressed profile of northern Cardiganshire viewed from both north and south.

Reconstructed Profiles

It is sometimes useful in the elucidation of geomorphological problems to reconstruct a pre-existing profile. This is especially important in the examination of the effects of rejuvenation of a drainage system on the present form of a valley, when it is necessary to reconstruct the original profile below the knickpoint clearly visible on the present profile. The latter is first drawn; obviously it is tempting to prolong by eye the curve of the upper course, but this would be most unsatisfactory. Various formulae have been devised, involving the height of the source, length of stream, the distance of any point from the mouth, and

[1] Five examples from the many available will suffice:

(*a*) J. Barrell, 'The Piedmont Terraces of the Northern Appalachians', *American Journal of Science* (4th series), vol. 49 (New Haven, Conn., 1920).

(*b*) A. A. Miller, 'The Entrenched Meanders of the Herefordshire Wye', *Geographical Journal*, vol. 85, pp. 160–78 (London, 1935).

(*c*) D. L. Linton, 'Some Scottish River-Captures Re-examined', *Scottish Geographical Magazine*, vol. 67, p. 36 (Edinburgh, 1951), uses a double projected profile of the Tilt through-valley in the Grampians in a most effective manner. He projected the heights of the bordering hills and valley walls as depicted on the One-inch map 'from each side on to a composite line made up of straight lengths following the centre of the valley', and drew two projected profiles, one on either side of a central line on which distances in miles and true bearings in degrees were marked.

(*d*) K. M. Clayton, 'The Denudation Chronology of Part of the Middle Trent Basin', in *Transactions and Papers, 1953, Institute of British Geographers*, no. 19, p. 30 (London, 1953).

(*e*) R. Common, 'A Suggestion for Terrain Profiles, based upon the Thalweg and Watershed Lines', *Irish Geography*, vol. 4, no. 3, pp. 209–12, offers a method of profile construction which demonstrates erosion surface remnants, produces suggestive and natural-looking profiles, yet is less demanding of tedious effort. It is analogous to the outline of the shadow cast by horizontal light rays falling upon a watershed, and so projected on to a vertical plane.

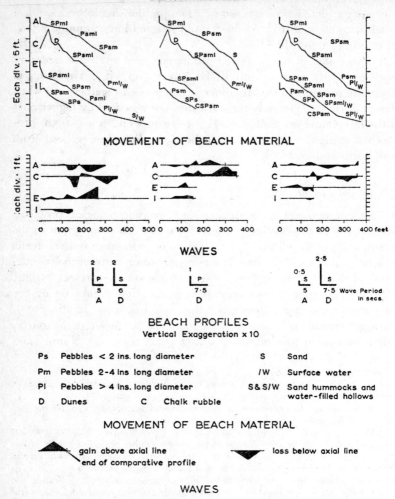

BEACH PROFILES

MOVEMENT OF BEACH MATERIAL

WAVES

BEACH PROFILES
Vertical Exaggeration x 10

Ps Pebbles < 2 ins. long diameter   S   Sand

Pm Pebbles 2-4 ins. long diameter   /W   Surface water

Pl   Pebbles > 4 ins. long diameter   S & S/W Sand hummocks and
                                                       water-filled hollows
D   Dunes   C   Chalk rubble

MOVEMENT OF BEACH MATERIAL

gain above axial line   loss below axial line
end of comparative profile

WAVES

S   Spilling   P   Plunging

Figure 38. CONSOLIDATED RECORD OF BEACH CHANGE AT SPURN HEAD,
APRIL 1960

Based on a manuscript diagram compiled by Dr A. Phillips.

The beach profiles **A, C, E** and **I** and the wave diagrams **A** and **D** refer to observation points on the seaward side of the spit at Spurn Head. Movement of beach material has been plotted about a central axial line which represents the profile of the previous date of observation in a straightened form.

certain constants determined by trial.[1] Using such a formula, a number
of points can be plotted and the lower portion below the knickpoint
reconstructed.

GRADIENT AND SLOPE

The Significance of Slope Determination

The topographical map presents much precise information, 'but it
has had the effect of elevating simple height to an unmerited eminence.'[2]
The geographer is concerned with many other features of the landscape,
particularly the slope[3] (its amount and changes), the mean height,
surface-levels and platforms, 'breaks' and 'edges', and the like. The
exact height of a point above sea level is for some purposes much less
important than the relation of its altitude to that of surrounding
areas. Some of these considerations have been discussed already when
dealing with profiles.

It must be remembered that while many methods of morphometric
analysis depend on statistical information extracted from a topographical map, it is often necessary for a geomorphologist to carry out
his own slope measurements in the field. Various methods may be used,
conveniently summarized by A. F. Pitty.[4] These include *precise surveying*

[1] O. T. Jones, 'The Upper Towey Drainage System', *Quarterly Journal of the Geological Society*, vol. 80, pp. 568–609 (London, 1924).

J. F. N. Green, 'The River Mole', *Proceedings of the Geologists' Association*, vol. 45,
pp. 35–69 (London, 1934).

A. A. Miller, 'Attainable Standards of Accuracy in the Determination of Pre-
Glacial Sea-Levels by Physiographic Methods', *Journal of Geomorphology*, vol. 4, pp.
95–115 (New York, 1941).

R. F. Peel, 'The North Tyne Valley', *Geographical Journal*, vol. 98, pp. 5–19 (London,
1941).

[2] A. A. Miller, 'The Dissection and Analysis of Maps', Presidential Address, 1948,
to the *Institute of British Geographers*, published in *Transactions of the Institute*, Publication
No. 14, p. 3 (London, 1949).

[3] D. R. Macgregor, 'Some Observations on the Geographical Significance of
Slopes', *Geography*, vol. 42, pp. 167–73 (London, 1957). A detailed example of the
geomorphological study of slopes is afforded by R. A. G. Savigear, 'Some Observations
on Slope Development in South Wales', *Transactions and Papers, 1952, Institute of
British Geographers*, no. 18, pp. 31–51 (London, 1952). He uses a large number of
profiles, located on a key-map, on which angles of slope derived from precise survey
are marked, in some cases with additional information such as bed-rock, a cover of
talus or grass, etc. See also Savigear's contribution, 'Technique and Terminology in
the Investigation of Slope Forms', *Premier Rapport de la Commission pour l'Etude des
Versants (Union Géographique Internationale)*, pp. 66–75 (Amsterdam, 1956).

[4] A. F. Pitty, *A Scheme for Hillslope Analysis*, Occasional Papers in Geography, no. 9,
pp. 27–30 (University of Hull, 1969).

with theodolite and chain, which is of course highly accurate though slow; *tacheometry*,[1] using a type of theodolite adapted for the measurement of distance, with which both the horizontal and vertical position of a point can be established by instrumental observations, thus eliminating the slowness of chaining; *barometric levelling*,[2] using a small pocket aneroid; *Abney levelling*, using an Abney level, a spirit-level mounted above a sighting-tube, which affords a rapid and convenient field-survey method when an accuracy of $\pm$ a half-degree is acceptable;[3] the technique of a board laid on a slope and its inclination measured, useful for a very small section of slope;[4] and A. F. Pitty's *slope pantometer*, consisting of two uprights with two linked cross-pieces, fitted with a protractor scale and a spirit level.

The analysis of slope and its representation on a map has been the subject of much research, particularly by American geomorphologists. The calculation of average gradient, either along the steepest slope (i.e. at right-angles to the contours) or along a road, is a simple matter. But to work out some representation of *average* slope, particularly in an area of complicated relief, and to express this on a map to provide a clear picture which may help the geomorphologist to make important deductions, is a much more complicated affair.[5] The analysis of average slope may be quite objective and arbitrary, using methods of random sampling[6] or of uniform grids as bases for calculations of each unit area.

A geomorphologist examining a specific problem, such as the slope of an erosion platform, may well choose to omit valleys produced in this platform by subsequent erosion, which may have steep sides and so would profoundly affect the figure for the whole area computed by

---

[1] J. Tivy, 'An Investigation of Certain Slope Deposits in the Lowther Hills, Southern Uplands of Scotland', *Transactions of the Institute of British Geographers*, no. 30, pp. 59–73 (London, 1962).

[2] M. M. Sweeting, 'The Karstlands of Jamaica', *Geographical Journal*, vol. 124, pp. 184–99 (London, 1958); and B. W. Sparks, 'Effects of Weather on the Determination of Heights by Aneroid Barometer in Great Britain', *Geographical Journal*, vol. 119, pp. 73–80 (London, 1953).

[3] R. A. G. Savigear, 'Some Observations on Slope Development in South Wales', *Transactions of the Institute of British Geographers*, no. 18, pp. 31–51 (London, 1952); and C. S. Carter and R. J. Chorley, 'Early Slope Development in an Expanding Stream System', *Geological Magazine*, vol. 98, pp. 117–30 (London, 1961).

[4] A. N. Strahler, 'Quantitative Slope Analysis', *Bulletin of the Geological Society of America*, vol. 67, p. 574 (New York, 1956).

[5] An immense amount of information is contained in *Premier Rapport de la Commission pour l'Etudes des Versants* (*Union Géographique Internationale*) prepared for the Rio de Janeiro Congress (Amsterdam, 1956). This contains 155 pp., and a large number of articles, maps and bibliographies on the subject of slope.

[6] See Appendix, p. 473.

an arbitrary method. He wishes to find the slope of the original platform and so omits the irrelevant slopes; this is analogous to the generalized contours discussed on p. 90. Moreover, the difference in value between the arbitrary and the selective figures is in itself an index of the amount of subsequent dissection, which can in fact be used as a comparison with other morphological areas.

The Calculation of Gradient

It is sometimes essential to be able to express exactly the steepness of a uniform slope. If two points on a hillside are projected on to a horizontal plane, as they are on a map, the distance between them is known as the *Horizontal Equivalent* (H.E.). The difference in vertical height between the two points is known as the *Vertical Interval* (V.I.). The gradient is expressed as a proportion, V.I./H.E., with the V.I. reduced to unity. Thus if the H.E. is 500 yards, and the V.I. is 150 feet, the gradient will be 150/1,500, or one in ten.

The gradient may be expressed as an angular measurement between the horizontal plane and the line of slope. This can be given approximately by multiplying the gradient, expressed as a fraction, by 60, which is reasonably correct to a slope of about 7 degrees.

Thus, a slope of 1 degree = gradient of 1 in 60 (actually 57·14)
a slope of 2 degrees = gradient of 1 in 30 (actually 28·65)
a slope of 3 degrees = gradient of 1 in 20 (actually 19·08)

The slope in degrees can be accurately computed by looking up the angle in a table of cotangents, for the tangent of the angle of slope equals V.I./H.E. This can obviously be used, given an angle of slope (obtained, for example, by a clinometer) to find the H.E. for a given contour interval, or to find the V.I. for any given H.E.

To obtain maximum slopes on a hillside, the H.E. is measured as far as possible at right-angles to the contours. If an area has a uniform slope in a more or less constant direction, a series of equidistant lines drawn down the steepest slope will be more or less parallel. The slope is calculated along each line and the mean of the results obtained; this will give an indication of the average slope of the area as a whole. Where the land has slopes of differing degrees of steepness in various directions, it will be necessary to divide the map into areas of broadly similar slope, or 'facets', by inspection, and these units will be used as bases for calculation. Facets with certain critical slopes may be distinguished and mapped; this would be necessary in, for example, a study of soil erosion, 'sheetwash', flood and run-off.[1]

If road gradients are required, the H.E. must be carefully measured along the bends (see p. 80). Spot-heights should be used rather than contour intersections, as far as possible. These road gradients can be very misleading, since on a small-scale map, at any rate, minor dips and rises are masked. The same caution should be noted when calculating gradients of railway tracks, as when drawing track-profiles (see p. 124). However, such calculations can be of practical value if a road-profile in hilly country is to be constructed; any striking gradients can be expressed precisely in figures on the profile, and thus give much practical information to a motorist or cyclist. This is the method employed by compilers of Road-books.

Scales of Slopes

A useful graphical exercise in connection with contour-maps is the construction of some form of line-scale of slopes. Examine the horizontal scale and the contour-interval of the map in question. If the V.I. is 50 feet, then the H.E. will be 2,857 feet on the ground for a slope of 1 degree (as worked out from tangent tables), 1,435 feet on the ground for a slope of 2 degrees, 954 feet for a slope of 3 degrees, and, for example, 187 feet for 15 degrees. Draw a horizontal line, and mark off along it these and other horizontal equivalents according to the scale of the map. Thus, on a 6-inch map, when the contour-interval is 50 feet, a length of 3·24 inches between two successive contours will represent a slope of 1 degree, and a length of 0·21 inch will represent a slope of 30 degrees. Slopes can be determined immediately by stepping off the distance on the map between two successive contours, transferring this to the line-scale of slopes, and reading off the degrees of slope indicated.

Another simple way of constructing a scale of gradients is shown in Fig. 39. It may be explained as follows:

For a vertical interval of 250 feet and a gradient of 1 in 20,

$$\frac{V.I.}{H.E.} = \frac{1}{20} = \frac{250}{x} \quad \ldots \quad x = 5000 \, feet.$$

---

[1] This was done by A. T. A. Learmonth, 'The Floods of 12th August, 1948, in South-east Scotland' (circulated in manuscript form, 1951). He produced a map of the Lammermuir Hills, on which facets with gradients of 1 in 2 or over and those with gradients of from 1 in 2 to 1 in 3 were differentiated.

On a map of scale actual distance $= 1:63,360$ (1 *inch* $= 1$ *mile*),

$$\frac{5000}{5280} = 0\cdot947$$

Draw a horizontal line *AB* any convenient length, and divide it into twenty equal parts. Draw a line *BC*, 0·947 inch long, and join

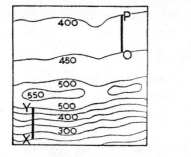

Gradient from X to Y

I in 15

Gradient from P to O

I in 85

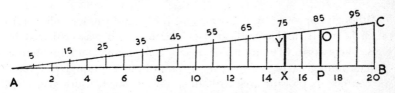

Figure 39. A SCALE OF GRADIENTS

The gradient of slope **XY** can be read off immediately from the lower scale-line **AB** as 1 in 15; the gradient of **OP** can be read off from the upper scale-line **AC** as 1 in 85.

AC. Draw parallels from the upper line to pass through each point on the lower line; the lengths of these parallels serve as a scale of horizontal equivalents. Slopes between 1 in 1 and 1 in 20 are read from the base-line, using a contour-interval of 250 feet; those between 1 in 20 and 1 in 100, are read from the upper line, using a contour-interval of 50 feet.

For slopes steeper than 1 in 20, lay a straight-edge of paper as far as possible at right-angles across the contours of the slope to be measured, and mark the horizontal distance between any contour-interval of 250 feet; alternatively, dividers may be used to step off this distance. Move the straight-edge or the dividers along the base-line of the scale until the distance coincides with a parallel between *AB* and *AC*. The corresponding gradient can then be read off from the *AB* scale-line.

For slopes gentler than 1 in 20, lay a straight-edge of paper at right-angles across the contours of the slope to be measured, mark the horizontal distance between a contour-interval of 50 feet. In the same way, determine the corresponding perpendicular, and read off the gradient along the *AC* scale-line. Obviously, the distance *BC* represents a V.I. of 50 feet instead of 250 feet, as in the previous case; the intervals along *AC* therefore represent gradients proportionally gentler (i.e. five times as gentle, as the V.I. is five times smaller).

Methods of Average Slope Determination

A method of average slope determination was used as long ago as 1890 by S. Finsterwalder and also (separately) by K. Peucker. This involved measuring the total length of all contours with an opisometer, and applying the formula:

$$\frac{Sine\ of\ degrees}{of\ average\ slope} = \frac{total\ length\ of\ contours \times contour\ interval}{total\ area}$$

Although approved by J. Tricart and J. Muslin,[1] this formula entails much laborious and complicated calculation, and was criticized by C. K. Wentworth.[2] He in turn devised a 'general and random' method of determining average slope over an area from a map. He covered the contour-map of the area with an east–west, north–south grid, then counted all contour crossings, and tabulated them, so determining the average number of contour crossings per mile. The procedure was repeated using an oblique grid over the same area, and the two results were averaged. He then applied the following formula:[3]

$$\frac{Average\ number\ of\ contour\ crossings\ per\ mile \times contour\ interval}{3361\ (constant)}$$

The result gave the average slope in terms of the tangent of the average

[1] J. Tricart and J. Muslin, 'L'Etude statistique des Versants', *Revue de Géomorphologie Dynamique*, vol. 2, pp. 173–81 (Paris, 1951).

[2] C. K. Wentworth, 'A Simplified Method of Determining the Average Slope of Land Surfaces', *American Journal of Science*, series 5, vol. 20 (New Haven, Conn., 1930). Besides explaining clearly his method and formulae, he describes the earlier methods, with references, of A. Penck, S. Finsterwalder and J. L. Rich, and discusses their limitations.

[3] The constant figure 3361 is derived from a formula which is explained fully by Wentworth; it is 5280×0.6366, which figure is the mean of all possible values of $\sin \theta$, where θ is the angle between the grid-lines and the contours.

angle of slope, which can be converted into degrees with tangent tables.

G.-H. Smith's Method of Slope Analysis

It is often important to relate the altitude of the highest and lowest points in any particular area, that is, to ascertain *the amplitude of available relief*.[1] This problem has been examined by G.-H. Smith,[2] who used the term 'relative relief' or 'local relief'. This type of relief analysis was developed mainly in Germany by such workers as N. Krebs, H. Schrepfer, V. Paschinger, H. Kallner and others, and was applied by K. H. Huggins to an analysis of the Highlands of Scotland.[3] G.-H. Smith used the method to make an analysis of the surface of the state of Ohio; his paper may be summarized briefly to illustrate the principles involved. He took a contour-map of Ohio on a scale of 1 : 600,000 and divided it into rectangles of 5 minutes each by longitude and latitude, representing approximately 4·40 by 5·75 miles on the ground, but, of course, varying slightly between the north and south of the state. He then calculated the difference in height between the highest and lowest points in each rectangle, obtaining about 2,000 values, which were plotted in the centre of each square on the base-map. Isopleths to indicate areas having the same amplitude in absolute altitude (i.e. the same 'local relief') were interpolated for each 100 feet of difference. The map was then shaded in eight tints, to indicate areas with the same 'local relief'. Further information was obtained by measuring the area of each 'relief province' (i.e. from 0 to 100 feet, from 100 to 200 feet and so on), and then expressing each area as a percentage of the total land area of Ohio (41,263 square miles).

The resulting map brings out most strikingly the areas of high relative relief in the south and east (the outlines of the Allegheny

---

[1] Available relief is defined by W. S. Glock, in 'Available Relief as a Factor of Control in the Profile of a Land-Form', *Journal of Geology*, vol. 40, p. 74 (Chicago, 1932), as 'the vertical distance from an original fairly flat upland down to the initial grade of the streams'.

[2] G.-H. Smith, 'The Relative Relief of Ohio', *Geographical Review*, vol. 25, pp. 272–84 (New York, 1935). Smith gives a full bibliography of German, Polish and other workers who developed the method. Apart from the interesting technique he describes, it is useful to read his analysis of vegetation, land utilization and settlement in relation to 'local relief'.

[3] K. H. Huggins, 'The Scottish Highlands: A Regional Study', op. cit. (1935). In an effort to delimit the Highlands, he used the grid of 2 mile squares which is superimposed on the One-inch O.S. Series (Popular Edition), and blackened in all squares in which the difference in height between the highest and lowest squares exceeded 700 feet.

Plateau), and the areas of low relative relief in the north-west (the Maumee Plain and the area south of Lake Erie). Smith's method has been applied to an analysis of the relative relief of the Dorking area (Fig. 40).

One obvious shortcoming is that the map presents *amplitudes* of maximum relief distances, which may be either between two points on opposite ends of a diagonal of any square, or in the case of a perpendicular cliff, may have no horizontal equivalent at all. Smith suggested that squares with extreme points far apart should be subdivided, but this would destroy much of the symmetry of the map. Miller[1] put forward the elaboration that the difference in height between maximum and minimum points in each square might be divided by the respective horizontal equivalent, the values plotted, and isopleths drawn.

The Raisz and Henry Method of Average Slope Determination

Raisz and Henry[2] applied Smith's method of analysing 'relative relief' to New England, but concluded that the results were not satisfactory in this particular case. Here the narrow valleys cut deeply into the peneplain and so give a big relief amplitude, as do isolated knolls projecting steeply from level surfaces. The relative relief figures for most of the area, therefore, are high, and mask important features. Raisz and Henry concluded that '. . . the method applied by Smith is good only on maturely dissected plateaux of horizontal sedimentary rock structure with uniform slopes and a simple physiographic history. In geologically complex regions, different methods must be applied.' They then tried to bring out the detailed differences which were lost in Smith's large 5-minute rectangles by covering the state with a grid of mile squares, but the result was 'a complex patchwork', unreproducible on a small-scale map of the whole state.

Their next refinement was to divide the topographical map, not into equal rectangles, but into irregular areas with some physiographic identity – monadnocks, incised valleys and so on. But the unequal areas proved unsatisfactory; thus an extensive though gently sloping plain might have a higher relative relief value than a monadnock of small extent.

Finally, Raisz and Henry divided the large-scale topographical map

[1] A. A. Miller, op. cit., p. 8 (1949).
[2] E. Raisz and J. Henry, 'An Average Slope Map of Southern New England', *Geographical Review*, vol. 27, pp. 467–72 (New York, 1937).

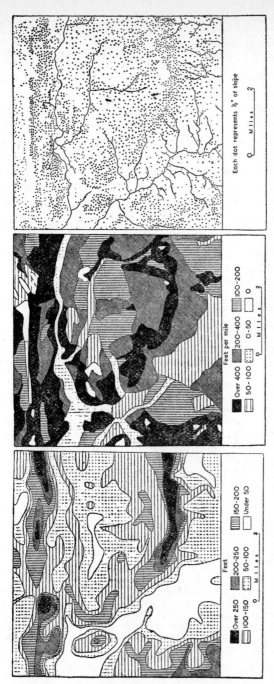

Figures 40–2. THREE METHODS OF SLOPE REPRESENTATION

Each map was compiled from data extracted from sheet 51/14, 1 : 25,000 series, Ordnance Survey. In the north is the line of the North Downs, in the west the Holmbury valley, and in the south-centre is Leith Hill.

Figure 40 (*left*) was drawn using G.-H. Smith's method, Fig. 41 (*centre*) using the method devised by E. Raisz and J. Henry, and Fig. 42 (*right*) using A. H. Robinson's method.

into small regions, within each of which the contour-lines had the same standard spacing, i.e. the same number of contour-lines per mile of horizontal equivalent. The chosen categories were seven in number, representing slopes of under 50 feet per mile, 50–100 feet, at each intermediate 100-foot interval to 500 feet, and above that figure. A horizontal scale of standard contour-spacings was drawn, so that the number of contours per mile on the map-scale could be checked off with dividers, and the slope category ascertained by careful inspection. When the six categories had been demarcated on the large-scale maps, their boundaries were transferred to the small-scale state map, and tinted in seven shades. The southern part of the map is reproduced in the reference cited above.[1]

An average slope map of Illinois was constructed by W. Calef and R. Newcomb,[2] following in general principles the Raisz and Henry method, though using the Wentworth equation (see p. 134) instead of delimiting areas of standard contour-spacing. Four slope classes (less

[1] Two applications of this method may be cited:

(a) R. B. Batchelder applied the methods both of Smith and of Raisz and Henry to an identical area, comparing the results, in 'Application of Two Relative Relief Techniques to an area of Diverse Landform: A Comparative Study', *Surveying and Mapping*, vol. 10, pp. 110–18 (Washington, D.C., 1950). The area chosen extended from Puget Sound eastward across the Cascade Range to the Columbia Plateau. He claimed that for this particular area Smith's method emphasized major landforms – their extent, continuity and inter-relationships. The method devised by Raisz and Henry represented secondary features in greater detail, but tended to obscure major relationships.

(b) G. B. Cressey, 'The Land Forms of Chekiang, China', *Annals of the Association of American Geographers*, vol. 28, pp. 259–76 (Lancaster, Pa., 1938), analysed from the 118 maps of the Chinese General Staff series, 1 : 100,000, with the aid of 2,500 miles of field reconnaissance and a quantitative study of surface configuration in terms of slopes. He distinguished the following five categories:

	Spacing of Contours at 50 metres interval	Slope
(i) Coastal Flatlands	None	Flat. 0°–2°
(ii) Interior Lowlands	More than 5 mm apart	1°–5° (1 in 100 to 1 in 10)
(iii) Rolling Hills	2·6 mm apart	4°–10° (1 in 12 to 1 in 4)
(iv) Mountainlands	1·2 mm apart	10°–30° (1 in 4 to 1 in 2)
(v) Steeplands	1 mm or less	30° and over (1 in 2 and steeper)

The resultant data were plotted on a base map on a scale of 1 : 400,000 and reproduced as a folded map in the *Annals*, with five grades of shading.

[2] W. Calef and R. Newcomb, 'An Average Slope Map of Illinois', *Annals of the Association of American Geographers*, vol. 43, pp. 304–16 (Lancaster, Pa., 1953). A valuable feature of this paper is the careful appraisal by the authors of the method and result, in respect of its objectivity, efficacy and usefulness.

than 1, 1–5, 5–9 and over 9 per cent were delimited and tinted distinctively.

The student should try this method, using a One-inch or a 1 : 25,000 Ordnance Survey map of an area of not too diversified relief. This has been done in Fig. 41, using the same area (the neighbourhood of Dorking) as was selected for Fig. 40. The map was first carefully examined, and areas which seemed to have more or less the same density of contours were demarcated on superimposed tracing-paper. Then the map was worked over in more detail, and the boundaries of the areas were modified where necessary. Finally, the slope categories were chosen, a system of density shading was selected, and the various regions were tinted accordingly. Obviously, the arduous and critical part of this method was the delimitation of the areas, and much depended upon individual judgement.

Coefficient of Land Slope. Raisz developed a method of establishing an exact 'coefficient of land slope', a process which, however, is laborious

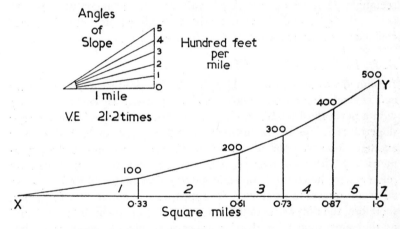

Figure 43. THE COEFFICIENT OF LAND SLOPE
The area of the plane figure **XYZ** provides a coefficient of slope for that portion of the map (arbitrarily selected as a rectangle) for which it has been drawn. The upper diagram provides a convenient scale of angles of slope corresponding to 100 feet, and each succeeding 100 feet, up to 500 feet per mile.

in the extreme.[1] The map is covered with rectangles of some reasonable size; the more uniform the slopes the larger they can be. Each rectangle is divided into areas within each of which the slope is uni-

[1] F. E. Raisz, *General Cartography*, p. 278 (2nd edition, New York, 1948).

form (under 100 feet per mile, 100–200 feet per mile and so on) by counting contours as described above for average slope-maps. Then the area of each category is measured with a planimeter. A line is drawn proportional in length to the total area of the rectangle, and the areas plotted to scale for each category as points along this line, with the zero on the left. Next a series of lines is drawn, each at the angle of slope corresponding to a vertical rise of 100 feet, 200 feet and so on, to the maximum per mile; on Fig. 43 the maximum slope is 500 feet per mile. Each change of slope occurs perpendicularly above the point on the horizontal scale representing that particular contour. The area of the plane figure XYZ provides a coefficient of slope for the rectangle for which it has been drawn. This is calculated by finding through simple geometry the area of each plane trapezium enclosed by the two adjacent perpendiculars, the horizontal line and the line of the angle of slope, each of which is measured graphically, and then by adding the five results. Obviously, completely flat land will have a coefficient of zero. When the coefficient has been established for every rectangle on the map, plot each value at the geometrical centre of each rectangle, and draw isopleths through points with the same coefficient, or interpolate between them in the usual way.

A. H. Robinson's Method of Slope Analysis

A method was devised by A. H. Robinson[1] to produce a quantitatively accurate relief map from areal slope data (Fig. 42). He covered his map with a network of squares each 0·01 square miles in area. The average slope of each square was estimated, and one dot for each degree of average slope was placed within it. The dots were not placed symmetrically within each square, but their positions were determined by reference to the contours on the topographical map and to the dots in adjacent squares to produce some appearance of continuity. In theory, therefore, the dots are countable, so as to give precise information, and at the same time their density appearance gives a good visual impression. The size of the dot must be carefully chosen, in order to produce effective visual contrasts of light and shade. The chief problem is, of course, to estimate the average slope of each square. Obviously, as Robinson points out, the most accurate method is by levelling in the field, following a traverse across the line of average slope. He quotes a complicated formula produced by J. A. Barnes, from which a table

[1] A. H. Robinson, 'A Method for Producing Shaded Relief from Areal Slope Data', *Surveying and Mapping*, vol. 8, pp. 157–60 (Washington, 1948).

was prepared, enabling the computer to count the number of contours in each square, read off an average angle of slope, and therefore determine the number of dots.

The student should try the effect of this dot method, without necessarily going into details of slope calculation by elaborate formulae. Examine each square, estimate by an inspection of the contours the average gradient of the square, and convert this into degrees of slope (see p. 131).

This method has been carried out (Fig. 42), using the same region as for the two previous methods. The scale of the base-map was 1 : 25,000, and one dot was used for each half degree.

A. N. Strahler's Methods of Slope Analysis

For a number of years A. N. Strahler of Columbia University has pursued investigations into the problems of slope analysis,[1] in order, in his own words, 'to substitute quantitative statement for Davis's qualitative statements'. In 1956 he published two new types of slope-map,[2] using isopleths, in contrast to Raisz and Henry's choropleths. Basically, a large number of slope-values are plotted, derived either from using an Abney Level in the field, or by computing them from a map of sufficient scale. The tangent and sine functions are then plotted, and a series of isopleths (*isotangents* and *isosines* respectively) are interpolated. A further development was to measure the areas between successive isotangents and isosines by means of a planimeter, and so produce a percentage frequency distribution from which a histogram may be drawn.

A summary of this method, with its application to a district in Upper Teesdale, is provided by J. I. Clarke and K. Orrell.[3] This involved contouring by levelling in spot-heights at 20-foot intervals on a grid, and interpolating contours at 2-foot intervals in the field. Measurement

---

[1] Examples of his work include: (i) 'Davis's Concepts of Slope Development viewed in the Light of Quantitative Investigations', *Annals of the Association of American Geographers*, vol. 41, p. 209 (Ann Arbor, 1950–1); (ii) 'Dimensional Analysis in Geomorphology', *Abstract, Bulletin of the Geological Society of America*, vol. 64, pp. 1479–80 (New York, 1953); (iii) 'Equilibrium Theory of Erosional Slopes approached by Frequency Distribution Analysis', *American Journal of Science*, vol. 248, pp. 673–96, 800–14 (New Haven, Conn., 1950); (iv) 'Statistical Analysis in Geomorphic Research', *Journal of Geology*, vol. 62, pp. 1–25 (Chicago, 1954).

[2] A. N. Strahler, 'Quantitative Slope Analysis', *Bulletin of the Geological Society of America*, vol. 67, pp. 571–96 (New York, 1956).

[3] J. I. Clarke and K. Orrell, op. cit. (1958).

of angles of slope were made with an Abney Level, first on a 20-foot grid, then at significant breaks of slope. They stress, however, that 'this technique is not likely to produce more reliable isosine and isotangent maps than those constructed in the laboratory from large-scale contour maps. The latter method also takes much less time.' Four maps were published, one of surveyed contours, one derived from the Raisz and Henry method of slope analysis (see p. 136), and one each of isosines and isotangents by Strahler's method. Clarke and Orrell conclude, 'In the writers' opinion, the Raisz and Henry method, given reliable contours and intimate knowledge of the terrain, permits a more satisfactory analysis of slope. Much however depends upon this personal knowledge of the terrain under investigation, hence accuracy is directly related to a purely subjective factor.'

Slope-zone Maps

O. M. Miller has been concerned with working out 'some method preferable to hypsometric tinting or oblique hill-shading for presenting the pattern of relief on small-scale aeronautical charts used for contact flying at high altitudes and high speeds'.[1] He first suggested this slope-zone method in 1951,[2] and developed it in the article cited (1960). His aim was to emphasize slope rather than mere elevation, to divide the surface into a series of zones with successive degrees of slope, and indicate these zones by clear though subdued tinting. He adopted the concept of A. Wood's fourfold slope elements in the development of a hill-side:[3] his 'waxing slope', 'free face', 'constant slope' and 'waning slope'. He examined various mathematical functions of slope and calculated the ranges of $(1 - cos\ \alpha)$, α, $sin\ \alpha$ and $\sqrt{sin\ \alpha}$, where α is the angle of slope; the values for the $\sqrt{sin\ \alpha}$ function, when divided into four equal parts, are $3°\ 35'$, $14°\ 24'$, $34°\ 14'$ and $90°\ 00'$, thus giving four zones ($0°$ to $3°\ 35'$, $3°\ 35'$ to $14°\ 24'$, etc.), which agree well with several important aspects of Wood's slope elements.

Contour-maps on a scale of 1 : 62,500 of the area to be depicted were then laid down, and covered with a transparent overlay. Using guide-

[1] O. M. Miller and C. H. Summerson, 'Slope-Zone Maps', *Geographical Review*, vol. 50, p. 196 (New York, 1960).

[2] O. M. Miller, 'Relief on Maps and Models: Some Conclusions and a Proposal', *The Ohio State University Research Foundation, Mapping and Charting Research Laboratory, Technical Paper No. 151* (Columbus, Ohio, 1951).

[3] A. Wood, 'The Development of Hillside Slopes', *Proceedings of the Geologists' Association*, vol. 53, pp. 128–40 (London, 1942).

samples of the contour-spacing at the various slope boundaries, he drew the boundaries of the slope-zones. The compilation sheets were assembled, reduced photographically to a scale of 1 : 250,000, and four areal grades of flat greyish-brown tints were applied. On the final map (published in the 1960 volume of the *Geographical Review*) of the Williamsport area, Pennsylvania, the drainage pattern was added in blue, roads and towns in purple, and airfields in brown. Another map, of part of New York, New Jersey and Pennsylvania, was compiled from the same original material, but further reduced to a scale of 1 : 2 million. Each has a very clear and effective appearance.

With regard to this reduction of scale, Miller makes the point that the areal patterns can be reduced without generalization and loss of detail, whereas reduction of a contour map usually means an increased vertical interval and some generalization, with resulting loss of relief detail.[1]

Divided Slope Histograms

In the analysis of a slope, the frequency distribution of the angles comprising that slope may be presented in the form of a histogram. A valuable description of this technique, with illustrations, is provided by A. F. Pitty.[2] After noting the greatest and least angles within the slope, he then made subdivisions into classes (a procedure which involved the complex problem of where to place the class boundaries), counted the number of observations falling within each of these classes, and drew proportionally the heights of each rectangle of the histogram. He presents an illuminating diagram revealing how the various alterations in the class-groupings of the slope-angles can materially modify the visual impact of the histograms (Fig. 44). An elaboration of this type of histogram is provided by the subdivision of the columns according to the frequency of angles within the basal, midslope and summit portions of the slope (Fig. 45).

Other Methods of Slope Analysis

Further refinements concerned with the slope of the land can be developed. Any particular slope can be calculated for each square on

---

[1] See also J. B. P. Angwin, 'A Small-scale Portrayal of Relief Patterns by Slope Zones', *The Ohio State University Research Foundation, Mapping and Charting Research Laboratory, Technical Paper No. 204* (Columbus, Ohio, 1956).

[2] A. F. Pitty, *A Scheme for Hillslope Analysis*, Occasional Papers in Geography, no. 9, pp. 31–8 (University of Hull, 1969).

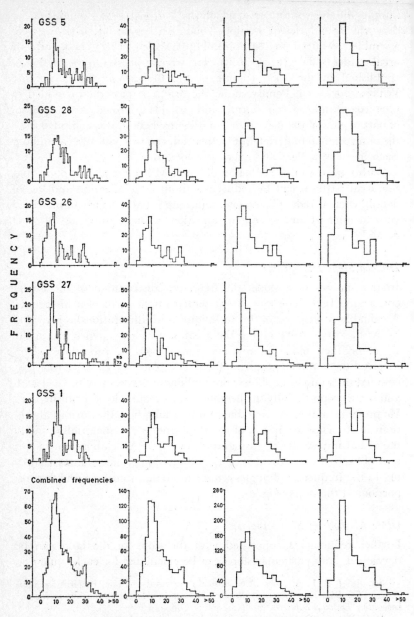

ANGLE IN DEGREES OF SLOPE UNITS FIVE FEET LONG

a gridded map and isopleths drawn. For example, what Raisz[1] calls *flatland-ratio* maps can be constructed by calculating the percentage of each square occupied by land *below* any given crucial slope, plotting these values, and then drawing isopleths. If a ratio map is thought to be unnecessary, an *absolute flatland map* can easily be drawn by outlining all areas with slopes below any critical figure, and then shading them.

There is another elaborate method in which pictorial symbols of hills in profile are used; their heights are made proportional to the relative relief figure, their bases to the average slope. These symbols are distributed over the map, and in the case of a country with a strongly marked relief they give quite a vivid impression.

Slopes may also be indicated simply on land-use maps by pecked lines (marking change of slope) and arrows (giving direction of slope). R. M. Glendinning,[2] for example, superimposes these slope symbols on a map, on which are also shown by tinting the effects of soil erosion and by index figures the grades of land use. A figure representing the slope in degrees can be placed beside the arrow; this is analogous to information concerning the dip of the strata on a geological map.

In a case where slope-categories have been determined, further useful information can be derived by measuring the area of each category, and then expressing it as a percentage of the total area. This can be done for areas with some physical unity, such as a river basin. If a series of graphs is drawn, the results may produce some striking data to assist in an analysis of catchment areas, run-off, flood regimes and the like (Fig. 46).

Combined Techniques. Various methods may be used in combination to illustrate some specific relationship of elements in the surface configura-

---

[1] E. Raisz, *General Cartography* (2nd edition, New York, 1948).
[2] R. M. Glendinning, 'The Slope and Slope-Direction Map', *Michigan Papers in Geography*, vol. 7 (Ann Arbor, Mich., 1937).

Figure 44. SLOPE-ANGLE HISTOGRAMS

Based on A. F. Pitty, *A Scheme for Hillslope Analysis: 1. Initial Considerations and Calculations*, Occasional Papers in Geography, no. 9 (University of Hull, 1969).

The histograms illustrate the visual effect of progressively wider slope-angle class groupings; some degree of skewness to the right of the distribution is shown. The letters and numbers refer to an index code for identification of slope profiles.

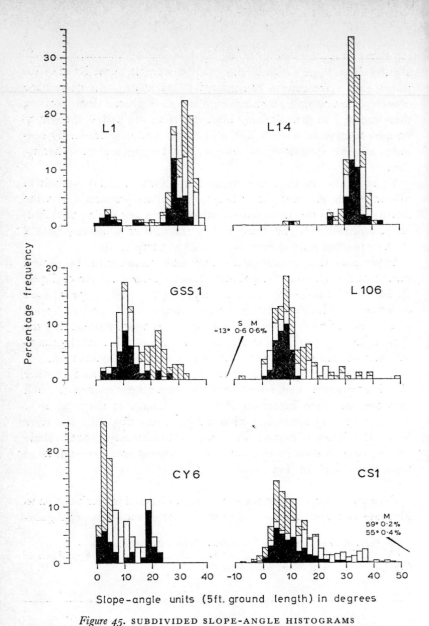

Figure 45. SUBDIVIDED SLOPE-ANGLE HISTOGRAMS

Based on A. F. Pitty, *A Scheme for Hillslope Analysis: 1. Initial Considerations and Calculations*, Occasional Papers in Geography, no. 9, p. 34 (University of Hull, 1969).

The letters and numbers (**L1**, etc.) refer to Dr Pitty's slope-profile index code. The black subdivision of the columns represents the basal third of the slope-profile, the shaded subdivision the summit (**S**) third and the unshaded portion the midslope (**M**) third.

tion. F. E. Elliott[1] developed a technique of presenting slope and relative relief on one map. He put a grid of $2\frac{1}{2}$ minutes square on a map quadrangle, and determined the relative relief figure for each (see p. 135); he then drew isopleths to show amplitude of relative relief, following in general G.-H. Smith's method, but introducing some degree of subjectivity by departing from it where obvious breaks of slope occurred. He then shaded four categories in horizontal ruling. Secondly, using the same contour-map base, he distinguished categories of slope, using the Raisz–Henry method of measuring contour-frequency spacings with dividers, and shaded three categories in vertical lines. When the two maps were superimposed, the combinations of the two sets of categories produced twelve composite slope and relative relief regions, obviously distinguished by degrees of cross-shading; these were outlined by pecked lines.

Area–Height Diagrams

Various types of diagram have been used to indicate the relationship between area and altitude, since this is obviously of fundamental geomorphological interest. However, a fair summary of these is afforded by Clarke and Orrell:[2] '. . . statistical methods to show these two aspects have never proved entirely satisfactory and are often misinterpreted'.

The area–height curve shows the actual area of land between two adjacent contours, depending on the contour-interval selected. The usual method is to measure the actual area between the two contours by means of a planimeter (see p. 79), repeated usually two or three times, so that a mean value can be used; this is extremely laborious.[3]

One form of area–height curve uses a vertical scale of heights (in feet), and a horizontal scale of areas (in square miles). Columns proportional in length to the area between each successive pair of contours are drawn opposite the relevant vertical height-scale. Alternatively and more usually, instead of in terms of absolute areas, the horizontal scale may be expressed as a percentage of the total area occupied by the area between each contour-pair. The points are plotted against

---

[1] F. E. Elliott, 'A Technique of presenting Slope and Relative Relief on one Map', *Surveying and Mapping*, vol. 13, pp. 473–8 (Washington, D.C., 1953).

[2] J. I. Clarke and K. Orrell, op. cit. (1958).

[3] A possible alternative (though F. J. M. and H. R. W. have admittedly never tried it, nor are likely to) is said to be practicable, by which the map is cut up along the significant contours, and then the groups of similar interval-strips are weighed in a delicate torsion-balance, and the areas thereby calculated.

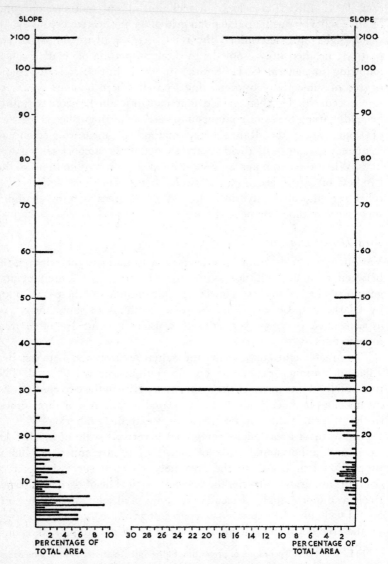

Figure 46. COLUMNAR DIAGRAM OF SLOPES

Compiled by A. T. A. Learmonth, in 'The Floods of 12th August, 1948, in South-east Scotland' (circulated in manuscript form, 1951).

The bars represent the percentage of the total catchment area of the Monynut and Blackadder (tributaries of the Tweed) at a certain slope (1 in 10, 1 in 20, etc., as shown on the vertical scale), and so enable the configuration of the two river basins to be compared at a glance.

the vertical height-scale (for example, in the case of the 400–500-foot area against the 450-foot point) opposite the relevant percentage on the horizontal scale, and are joined by straight lines to form a continuous curve.

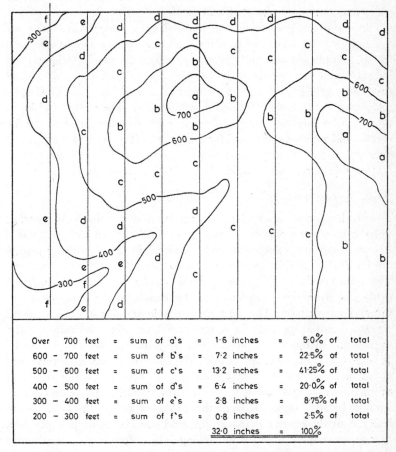

Over	700	feet	=	sum	of	a's	=	1·6	inches	=	5·0% of	total
600 – 700		feet	=	sum	of	b's	=	7·2	inches	=	22·5% of	total
500 – 600		feet	=	sum	of	c's	=	13·2	inches	=	41·25% of	total
400 – 500		feet	=	sum	of	d's	=	6·4	inches	=	20·0% of	total
300 – 400		feet	=	sum	of	e's	=	2·8	inches	=	8·75% of	total
200 – 300		feet	=	sum	of	f's	=	0·8	inches	=	2·5% of	total
								32·0	inches	=	100%	

Figure 47. ESTIMATION OF AREA BY LENGTH OF INTERCEPT
Based on A. A. Miller, *The Skin of the Earth*, pp. 65–70 (London, 1953).

Another method, not involving planimetric measurement of area, is described by A. A. Miller.[1] This is by means of quasi-random sampling. A system of parallel lines is drawn across the area under consideration.

[1] A. A. Miller, *The Skin of the Earth*, pp. 65–70 (London, 1953).

Taking each line in turn, the length of the intercepts between each pair of selected contours is measured, the lengths noted in columns, and the totals between each pair of contours for all lines are obtained by addition (Fig. 47). The sum of the intercepts between each pair may be taken as proportional to the area covered, and these totals can be plotted against the vertical altitude scale.

Obviously the position of the parallel lines is significant. The indefatigable J. I. Clarke and K. Orrell constructed area–height diagrams for the islands of Gozo and Guernsey,[1] using the planimetric method for each, and then repeated the exercise no less than four times, using the 1-km grid-lines for Gozo, 1,000-yard grid-lines for Guernsey, measuring the intercepts for both north–south and east–west grid-lines, and then for intermediate grid-lines in each direction. The results for each island were superimposed for comparison. The general correspondence was good, although (in their own words) '. . . closer examination will reveal discrepancies which should curb rash conclusions'.

Trend Surface Analysis

This technique of analysing a distribution of continuously varying point values over a particular area under consideration is of great value in examining the possible existence of a now dissected and fragmented erosion surface and in reconstructing its former character and nature. The data consist of spot-heights defined relative to a superimposed grid, and from them isopleths can be drawn in the form of three-dimensional surfaces. The very considerable amount of data involved renders essential the use of a computer.[2] A valuable illustration of the application of the technique is given by J. B. Thornes and D. K. C. Jones;[3] this contains maps of isopleths showing linear, quadratic and cubic surfaces, and interprets their significance. The input consisted of maximum height values for each National Grid square and actual summit heights, obtained from the 1 : 25,000 series; the values were referred to the central points of the grid squares, using interpolation by eye where spot-heights were lacking. The output of the co-ordinates

[1] Op. cit. (1958), pp. 9–13.

[2] The general principles of trend surface analysis are dealt with succinctly by J. P. Cole and C. A. M. King, *Quantitative Geography*, pp. 375–9 (London, 1968); and by C. Board, in R. J. Chorley and P. Haggett, *Models in Geography*, pp. 716–19 (London 1967). See also R. J. Chorley and P. Haggett, 'Trend-surface Mapping in Geographical Research', *Transactions of the Institute of British Geographers*, no. 37, pp. 47–65 (London 1965), which has a bibliography of seventy-seven items.

[3] J. B. Thornes and D. K. C. Jones, 'Regional and Local Components in the Physiography of the Sussex Weald', *Area*, no. 2, pp. 13–21 (London, 1969).

and the actual, computed and residual values were programmed for and processed on a computer, and plotted in symbolic form.[1] Similarly, the value of trend surface analysis has been demonstrated by J. Lewin[2] (Fig. 48) in his study of the Yorkshire Wolds, and by D. E. Smith, J. B. Sissons and R. A. Cullingford,[3] who used accurate levelling based on Ordnance Survey bench-marks to obtain altitudes on either side of the Firth of Forth in order to compute linear trend-surfaces for the Main Perth Raised Shoreline. The large number of points were expressed to the nearest 10 metres in terms of the National Grid, and then they used three different programmes (Fortran II, Fortran IV and a modification by P. Kahn of the latter) to produce maps of the surfaces. Finally, a full and well-illustrated exemplification of the method is given by C. A. M. King, in her study of the erosion surfaces of the central Pennines.[4]

Hypsometric Curves

Akin to the area–height curves previously described, a hypsometric (sometimes known as a hypsographic) curve is used to indicate the proportion of the area of the surface at various elevations above or depths below a given datum. If, for example, a hypsometric curve is to be drawn of an island, calculate the total area, the area enclosed within the 100-foot contour, and then those areas within successively higher contours, to the highest point. A less accurate method is to cover the map with parallel lines drawn at equal distances and then to find the sum of the intercepts between each of the successive pairs of contours, which are proportional to the areas.

Choose a convenient horizontal scale to represent the areas in square miles, putting zero on the left. Construct a vertical scale on the left side of the base-line to represent height above sea level, at appropriate intervals. Plot each area against its corresponding contour interval,

[1] See also H. Svensson, 'Method for Exact Characterising of Denudation Surfaces, especially Peneplains, as to the Position of Space', *Lund Studies in Geography*, ser. A, no. 82 (Lund, 1956); and C. A. M. King, 'An Introduction to Trend Surface Analysis', *Nottingham Bulletin of Quantitative Data*, no. 12 (Nottingham, 1967).

[2] J. Lewin, *The Yorkshire Wolds: A Study in Geomorphology*, Occasional Papers in Geography, no. 11, pp. 25–8 and Fig. 6 (University of Hull, 1969).

[3] D. E. Smith, J. B. Sissons and R. A. Cullingford, 'Isobases for the Main Perth Raised Shoreline in South-east Scotland as determined by Trend-surface Analysis', *Transactions of the Institute of British Geographers*, no. 46, pp. 45–52 (London, 1969).

[4] C. A. M. King, 'Trend-Surface Analysis of Central Pennine Erosion Surfaces', *Transactions of the Institute of British Geographers*, no. 47, pp. 47–59 (London, 1969).

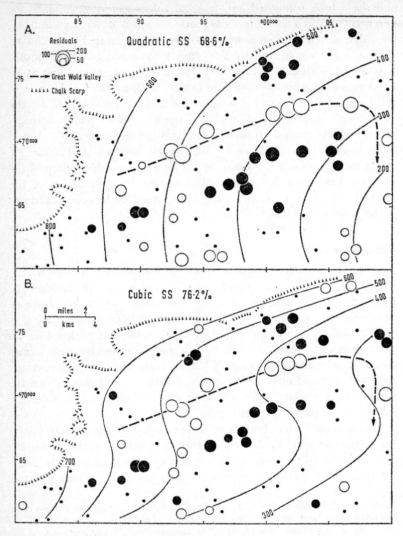

Figure 48. TREND SURFACE ANALYSIS IN THE NORTHERN
YORKSHIRE WOLDS

Based on J. Lewin, *The Yorkshire Wolds, A Study in Geomorphology*, Occasional Papers in
Geography, no. 11, p. 22 (University of Hull, 1969).

Residuals of more than 50 feet are shown by proportional circles, positive by solid
and negative by open ones. Other control points are shown as dots, and the percentage
sum of squares accounted for by the surfaces is given at the top of each diagram.

mark its position with a cross, and join up each point with a smooth curve.

The hypsometric curve should be used for an area with some physical unity, such as an island (which will bring out the relative area of coastal plain, plateau and mountain range) or an erosion platform. For example, S. W. Wooldridge drew a series of hypsometric curves for various parts of south-eastern England, calculated from the Ordnance Survey Half-inch series, which show clearly the 200-foot 'platform' or planation surface,[1] and K. M. Clayton a series for the

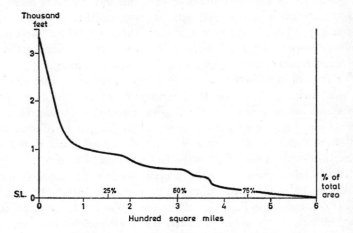

Figure 49. A HYPSOMETRIC CURVE

middle Trent basin.[2] Hypsometric curves were also used by E. de Martonne[3] to compare young fold mountains (the Alps and the Pyrenees) with ancient massifs (Armorica and the Central Massif).

Percentage Hypsometric Curves

An alternative method is to express the measured area between each particular pair of contours as a percentage of the total area of the region in question, and draw a horizontal percentage scale, as on Fig. 49.

[1] S. W. Wooldridge, 'The 200-foot Platform in the London Basin', *Proceedings of the Geologists' Association*, vol. 39, pp. 1–26 (London, 1928).

[2] K. M. Clayton, 'The Denudation Chronology of part of the Middle Trent Basin', *Transactions and Papers, 1953, Institute of British Geographers*, no. 19, p. 27 (London, 1953).

[3] E. de Martonne, 'Morphometrie et Morphologie', *Annales de Géographie*, vol. 50, pp. 24–54 (Paris, 1941).

Similarly, on the vertical (ordinate) scale is plotted as a percentage the ratio of the height of any contour above the base-level of the area on the map to the total height. A curve is drawn through the plotted points.[1]

Clinographic Curves[2]

The clinographic curve seeks to illustrate the average gradient between any two contours, and to express a series of these averages in a single curve. Its chief value, therefore, is that it indicates both sudden changes and breaks in the general relief of any region, and moreover it emphasizes uniform areas such as plateaux. It gives at the same time average gradients, the percentage extent of each average gradient, and exact breaks of slope. It is much more sensitive to small changes than a hypsometric curve, and in some cases it is less misleading.

The area of land between two successive contours is measured. If the area enclosed by the sea-level contour is represented by x square inches and the area between the sea-level and the 100-foot contours is y square inches, these areas can be represented by circles of equivalent area, with radii R and r, for $x = \pi R^2$ and $y = \pi r^2$. Find R and r in inches, and convert these values by the scale-factor into feet. If a right-angled triangle is drawn, with BC proportional to the contour-interval (h) and the base AB representing the difference in length between the radii ($R-r$), the angle BAC (i.e. the angle of slope) can be calculated, using the tangent formula.[3] Similarly the angle of slope between each succeeding pair of contours is calculated. A clinographic curve can then be drawn, using the contour-intervals as vertical components and inserting each section of average slope between each two contours with a protractor, starting at the top (Fig. 50). If the clinographic curve is of gentle slopes, a vertical exaggeration can be introduced, by increasing each contour interval by a constant factor.

It is useful to draw the hypsometric and clinographic curves on

[1] This method was used by A. N. Strahler in studying small drainage basins, 'Hypsometric (Area–Altitude) Analysis of Erosional Topography', *Bulletin of the Geological Society of America*, vol. 63, pp. 1117–42 (New York, 1952).

[2] A detailed comparative study of various types of clinographic curve is given by J. I. Clarke, 'Morphometry from Maps', *Essays in Geomorphology* (ed. G. H. Dury), pp. 235–74 (London, 1966).

[3] J. Ross Mackay, 'Arithmetic–Square Root Graph Paper', *Professional Geographer*, vol. 6, no. 1, pp. 15–16 (New York, 1954), indicates a method by which this process can be speeded; if this type of graph-paper is used, with the Y-axis arithmetically divided and the X-axis of square roots, the square root value will readily give the radii of circles with area equal to the total areas enclosed by each successive pair of contours.

the same diagram for comparison; the former will show the area of land involved, the latter the actual slope, between each successive pair of contours. Hanson-Lowe illustrates this by superimposing the clinographic curve of the island of Jersey (with an exaggeration of ten times) and the hypsographic curve on the same diagram.[1] A very

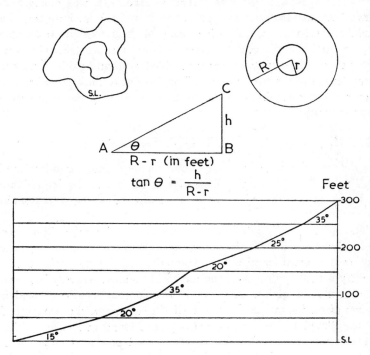

Figure 50. A CLINOGRAPHIC CURVE
The contour-interval in this case is 50 feet.

striking application of the value of comparative hypsometric and clinographic curves is shown by W. G. V. Balchin, who determined them for four sub-regions of north Cornwall and for the area as a whole.[2]

A different method was suggested by F. Debenham.[3] He measured the actual length of each contour, using an opisometer, and plotted this

[1] J. Hanson-Lowe, 'The Clinographic Curve', *Geological Magazine*, vol. 72, pp. 180–4 (London, 1935).
[2] W. G. V. Balchin, 'The Erosion Surfaces of North Cornwall', *Geographical Journal*, vol. 90, p. 59 (London, 1937).
[3] F. Debenham, *Exercises in Cartography*, p. 60 (London, 1937).

on a horizontal scale against a vertical scale representing the height of each contour.

A. N. Strahler's Mean Slope Curve

A method was proposed by A. N. Strahler[1] of obtaining a curve of mean slopes, plotted in the same way as Hanson-Lowe's clinographic curve. The area between each pair of adjacent contours is measured by a planimeter. Then the length of each of these contours is obtained with an opisometer, and their mean lengths computed. The mean width of each intercontour area is obtained by dividing this area by the mean length of the two contours. The tangent of the angle of slope is then derived from dividing the contour interval by the mean intercontour width:

$$tan\ \alpha = \frac{contour\ interval}{mean\ intercontour\ width}$$

For the area above the highest contour, the mean intercontour width is replaced by half the length of the highest contour as the denominator in the formula; this gives what is considered to be a reasonable approximation.

Though extremely laborious, this method is of greater significance and value than the Hanson-Lowe formula (above), since the latter ignores the actual lengths of the contours, and therefore if an area of country is acutely dissected, the method becomes markedly less accurate. Strahler's method takes the higher contour lengths implied by greater dissection into account. J. I. Clarke[2] compared the two methods for the islands of Gozo and Guernsey, and found appreciable differences in the results; Hanson-Lowe's method produced longer intercontour widths and therefore lower angles than did Strahler's, the latter obtaining slopes from $1\frac{1}{4}$ to 5 times as steep, with no consistent differences.

De Smet's Curve

R. de Smet developed a simple method of depicting a slope by graphing the average width between each contour.[3] This is found by dividing

---

[1] A. N. Strahler, 'Hypsometric (Area–Altitude) Analysis of Erosional Topography,' *Bulletin of the Geological Society of America*, vol. 63, pp. 1117–42 (New York, 1952).

[2] J. I. Clarke, 'Morphometry from Maps', in *Essays in Geomorphology* (ed. G. H. Dury), pp. 235–74 (London, 1966).

[3] R. de Smet, 'Courbe Hypsographique et Profil Moyen de l'Ardenne', *Bulletin de la Société Belge d'Etudes Géographiques*, vol. 23, pp. 143–67 (Brussels, 1954).

the area between each two contours by their mean length, plotted on a horizontal axis against a vertical axis of heights.

F. Moseley's Slope Maps

F. Moseley made a detailed cartographic analysis of Bowland Forest as part of his study of its erosion surfaces.[1] The area covers about 140 square miles, which he divided into fifty-two rectangles, each 2 miles north–south, by 1½ miles east–west. Using maps of 6-inch scale, he then computed the angle of slope at horizontal intervals of 1,000 feet around

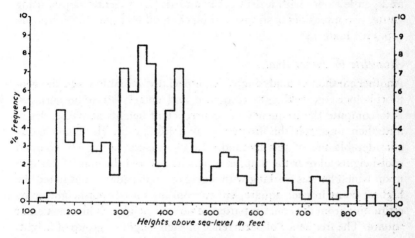

Figure 51. AN ALTIMETRIC FREQUENCY HISTOGRAM

Compiled from data extracted from sheet 51/14, 1 : 25,000 series, Ordnance Survey, representing the district south of Dorking, including the North Downs and Leith Hill. (Compare Figs. 40–2, which cover the same area.) Instead of drawing vertical columns to each plotted height, these may be joined by a single line to give a continuous graph.

each contour, at vertical intervals of 25 feet, using a calibrated scale laid at right-angles to the contours. As a piece of masterly under-statement, he says 'the method of map preparation was undoubtedly laborious', and he recommended the plotting of heights on a rectangular grid, to be programmed for a computer. The total number of measurements in fact exceeded 26,000. From these data a variety of maps and diagrams was constructed. He first plotted all the calculated degrees of slope, and interpolated isopleths of equal slope (*isoclines*) at degree

[1] F. Moseley, 'Erosion Surfaces in the Forest of Bowland', *Proceedings of the Yorkshire Geological Society*, vol. 33, pt. 2, no. 9, pp. 173–96 (Hull, 1961).

intervals; this he termed his first derivative map. He then plotted a second derivative map, showing the rate of change of slope in degrees per 1,000 feet, thus indicating breaks of slope, and interpolated isopleths of rate of change at 2 degree intervals. Finally, he combined these two maps with the original contour map, and used the breaks of slope shown on the second derivative map to distinguish categories of relative slope. Then adding further information from field observations and areas of drift from the geological map, he obtained four categories of (i) steep slopes, (ii) moderate slopes with no drift cover, (iii) moderate and gentle slopes with a drift cover and (iv) very gentle slopes, using white, two intensities of stipple and black to differentiate the categories (op. cit., Plate 14).

Altimetric Frequency Analysis

Another method of analysing relief, especially valuable when the geomorphologist is seeking to recognize and correlate erosion surfaces, is to compute the frequency of occurrence of heights above sea level, and then to graph this frequency as a histogram. H. Baulig made considerable use of this method.[1] In some cases he merely counted spot-heights all over the map, in others he covered the map of the area upon which he was working with a grid of small squares and noted the highest point in each square, either from an actual spot-height or by estimation from the contour pattern if no spot-height fell in a particular square. The frequency of occurrence of each height or group of heights (e.g. from 20 to 30 metres) was tabulated, and these frequencies were graphed on a vertical scale against the actual altitudes represented on a horizontal scale. S. E. Hollingworth used the same principle with minor modifications.[2] Instead of using all spot-heights, he was concerned

---

[1] H. Baulig, 'Sur une Méthode Altimétrique d'Analyse Morphologique Appliquée à la Bretagne Péninsulaire', *Bulletin de l'Association de Géographes Français*, no. 10 (Paris, 1926); in this paper he included altimetric frequency curves for the Armorican Massif as a whole and for the north and south zones, as well as for Léon, La Vendée and the Paris Basin.

Baulig's other works in which he includes examples of this technique are 'Les hauts Niveaux d'Erosion eustatique dans le Bassin de Paris', *Annales de Géographie*, vol. 37, pp. 288–385 (Paris, 1938); *Le Plateau Central de la France et sa Bordure méditerranéenne* (Paris, 1928); and 'The Changing Sea-Level', *Publication No. 3 of the Institute of British Geographers* (London, 1933).

[2] S. E. Hollingworth, 'The Recognition and Correlation of High-Level Erosion Surfaces in Britain: A Statistical Study', *Quarterly Journal of the Geological Society*, vol. 94, pp. 55–84 (London, 1938). He reproduces altimetric frequency curves for Devon and Cornwall, the Lake District, South-west Scotland, the Cheviots, North-west Wales, and South-west Wales. On Plate XI he puts all these curves on a single chart,

only with actual summit-levels, and he tabulated for each of the areas with which he was concerned the number of summit-level spot-heights occurring within 10, 20, 40 and 50-foot intervals. Obviously the broader groupings smoothed out minor inequalities, but emphasized the major features.

A number of experiments by various workers has been made in order to determine the effectiveness of the several methods of computing the frequency of occurrence of the heights to be plotted, on the basis, firstly, of fully random sampling using only spot-heights, and secondly, of quasi-random sampling using a grid. In the case of the first, if all spot-heights are used, the result may depend on the number of spot-heights inserted on the map by the surveyors and cartographers. J. I. Clarke and K. Orrell,[1] in drawing a series of these curves for four islands (Jersey, Anglesey, Arran and Man), note the marked differences in the provision of spot-heights on the respective topographical maps used. The question also arises whether all spot-heights, all summit spot-heights, all spot-heights plus unmarked summits or summit spot-heights plus unmarked summits should be used. S. E. Hollingworth, in the work referred to above, used only summits, since he was concerned only with high-level erosion surfaces, where the significance of the summits is striking. The main difficulty of utilizing *all* spot-heights is that the *number* of spot-heights is what matters, and thus a dissected area may count more. As A. A. Miller[2] aptly comments, '. . . it is a weakness that the method emphasizes number and minimizes area'.

Where a grid is used, the problem of the size of the grid-mesh is important. Clarke and Orrell constructed three altimetric frequency curves for Jersey, using $\frac{1}{2}$-, 1- and $1\frac{1}{2}$-inch grids, and super-imposed the results; as they say, 'the accordance of the curves between 200 and 350 feet is not entirely satisfactory'. They also plotted curves for the highest point in each square mile and in each square kilometre. All these various curves are superimposed by Clarke and Orrell (op. cit. (1958), p. 15) for comparison. Other refinements include plotting values derived from the mean of the highest and lowest points in each

---

graphing the number of actual summit-heights falling within each fifty-foot group. Especially valuable is his Fig. 3 (p. 62), on which he compares two altimetric frequency curves of Cornwall and Devon, one representing the number of summits in a 40-foot-grouping by a line-curve, the other the number of summits in each 20-foot group represented by a histogram. In addition, he included a series of projected profiles of the same region drawn on the same vertical scale.

[1] J. I. Clarke and K. Orrell, op. cit. (1958), pp. 14–17.

[2] A. A. Miller, *The Skin of the Earth*, p. 72 (London, 1953).

square, and from the formula suggested by Dr M. H. P. Bott, as follows:

$$\frac{Height\ of\ each\ corner + 4\ (height\ of\ centre)}{8}$$

Instead of using all spot-heights, they may be grouped in altitude class intervals, as in the case of Fig. 51, with 20-foot groups, using a grid. This covers part of the Dorking area of Surrey, covering exactly the same area for which the three Figs. 40–2 were produced. The result of the grouping depends very largely on the group-interval. Clarke and Orrell (op. cit. (1958), p. 17) produced a series of both simple and running sum class intervals for various values between 5 and 50 feet, and concluded 'Such is the diversity arising from the use of different class intervals that interpretations are likely to vary greatly.' All these methods are, to say the least of it, tedious and laborious.

Another form of altimetric frequency analysis involves the construction of a histogram showing the frequency at successive altitudes of *areas* of summits, wherein the area of each summit within the highest closed contour is calculated.[1] Again, it is possible to compute and depict as a histogram the altimetric frequency of 'bench-units'; a bench was defined as a unit-area of 16 hectares by P. Macar[2] in 1938, as a unit of 4 hectares by him in 1957, with a slope of 2·5 per cent or less, obtained from a detailed analysis of large-scale (1 : 20,000) topographical maps, with a contour-interval of 5 metres.

Finally, W. F. Geyl[3] used a method of frequency analysis, by which the areas of summits and the lengths of shoulders and cols were tabulated and plotted as histograms, counting one point for each specific area of summit and one point for each length of shoulder or col. After careful experimentation, working on a map-scale of 1 : 63,360, he awarded one point for a shoulder with an intercontour distance of 5 mm (⅛ inch), and an extra point for each further 5 mm or part thereof; one point for any summit, however small, denoted by a closed contour, with an extra point for each 25 sq. mm or part thereof above the first 25 mm; and one point for each 5 mm of col, with an extra point for each further 5 mm or part thereof. He used a transparent ruler for the measurements. The map area was gridded, the points counted

[1] H. D. Thompson, 'Topographic Analysis of the Monterey, Staunton and Harrisonburg Quadrangle', *Journal of Geology*, vol. 49, pp. 521–49 (Chicago, 1941).

[2] P. Macar, 'Contribution à l'Etude Géomorphologique de l'Ardenne', *Annales de la Société Géologique de Belgique*, vol. 61, pp. 824–37 (Liége, 1937–8).

[3] W. F. Geyl, 'Morphometric Analysis and the World-wide Occurrence of Stepped Erosion Surfaces', *Journal of Geology*, vol. 69, pp. 388–416 (Chicago, 1961).

and a histogram drawn. It is not clear why areas and lengths should be correlated; the same category of dimension would seem more consistent. This article includes a large number of histograms, a contour map on which points awarded for these significant features are located and plotted, and a remarkable composite diagram summarizing the occurrence of stepped erosion surfaces in five areas of the world (continental Europe, Britain, the United States, Africa and Australia), derived from the work of a large number of geomorphologists.

F. Moseley's Slope–Height Curve

As described on p. 157, F. Moseley obtained 26,000 values of slope in the Forest of Bowland upland. He grouped his fifty-two rectangles into twelve major regions and constructed a slope–height curve for each. He determined the mean slope value between each pair of contours for each of the regions, and plotted these values along a horizontal axis against a vertical axis in feet. Thus uniform slopes are straight lines parallel to the vertical axis, a concave slope with a uniform rate of change is a straight line inclined to the right, a convex slope a straight line inclined to the left. Another pair of slope-height curves was made for the overall highland and lowland portions of Bowland, and using the same vertical scale clinographic curves were constructed for the two areas. From these (and other diagrams, including a number of slope frequency distribution histograms), he obtained cartographic and diagrammatic confirmation of the presence of two major sub-aerial erosion surfaces, the Bowland Surface (undulating fell-tops, at 1,200–1,700 feet) and the Wyresdale Surface (425–1,000 feet, old valley floors).

This method was employed for slope values measured on maps, but it can also be utilized for values directly measured in the field, or for mean slopes for inter-contour areas obtained by either the Hanson-Lowe (p. 155) or the Strahler (pp. 156) method.

Height–Range Diagrams

Various attempts have been made to devise effective cartographic or diagrammatic representations of flats or levels, representing fragments of former erosion surfaces or of old valley floors. The height–range diagram was devised by B. W. Sparks[1] in connection with the plotting and interpretation of marine erosion surfaces in the South Downs.

[1] B. W. Sparks, 'The Denudation Chronology of the Dip-slope of the South Downs', *Proceedings of the Geologists' Association*, vol. 60, pp. 165–215 (Colchester, 1949).

He identified and surveyed in the field no less than 283 of these flats, including also summits and cols, tabulating them according to their location, origin, grid reference, elevation and length, and to the geological zone in which they are developed. He then plotted the height–range of each flat (i.e. the difference between the upper and lower height of each surface) as ordinates on a vertical height-scale against a horizontal scale proportional to the greatest length of the flat, thus forming a series of black rectangles. Further refinements, where required, included angular indentations for back-eroded flats and angular extensions for front-eroded flats, and he used the symbol **T** for an isolated hill, and **⊥** for a col. The symbols were grouped on the diagram in relation to the three main river valleys (Arun, Adur and Ouse). In a subsequent paper,[1] Sparks identified all fragments of erosion surfaces with an index mark, plotted on an accompanying map and entered against each rectangle on the height–range diagram. Essentially the same method was used by C. E. Everard,[2] who also linked all the upper surfaces and all the lower surfaces with pecked lines, in order to emphasize the vertical spacing of the platforms.

A number of variations and modifications of this method have been published.[3] Thus both A. F. Coleman and E. H. Brown plotted the extension of the flats from measurements made throughout in the same constant direction, rather than use Sparks' maximum extension, and gave accurate scale values to the spaces between the flats. Each used

---

[1] B. W. Sparks, 'Stages in the Physical Evolution of the Weymouth Lowland', *Transactions and Papers, 1952: Institute of British Geographers*, no. 18, p. 21 (London, 1952).

[2] C. E. Everard, 'Erosion Platforms on the Borders of the Hampshire Basin', *Transactions and Papers, 1956: Institute of British Geographers*, no. 22, p. 35 (London, 1956).

[3] See, for example, (i) E. H. Brown, 'Erosion Surfaces in North Cardiganshire', *Transactions and Papers, 1950: Institute of British Geographers*, no. 16, pp. 51–66 (London, 1950); this was of rather a different nature, using a vertical line for the height–range of each flat, and a symbol ⊙ for a horizontal one. See especially p. 54, his interesting diagram illustrating the height–range of surfaces in that area, classified into marine and sub-aerial, from Ramsay in 1846, via W. M. Davis, *et. al.*, to Brown himself; (ii) A. F. Coleman, 'Some Aspects of the Development of the Lower Stour, Kent', *Proceedings of the Geologists' Association*, vol. 63, pp. 63–86 (London, 1952); (iii) E. H. Brown, 'The River Ystwyth (Cards.)', *Proceedings of the Geologists' Association*, vol. 63, pp. 244–69 (London, 1952); (iv) A. F. Coleman, 'The Use of the Height–Range Diagram in Morphological Analysis', *Geographical Studies*, vol. 1, no. 1, pp. 19–26 (London, 1954); (v) M. M. Sweeting, 'The Land-Forms of North-west County Clare', *Transactions and Papers, 1955: Institute of British Geographers*, no. 21, pp. 33–49 (London, 1955); and (vi) C. Embleton, 'The Planation Surfaces of Arfon and Adjacent Parts of Anglesey', *Transactions of the Institute of British Geographers*, no. 35, pp. 17–35 (London, 1964).

irregular quadrilaterals, trapezia and even triangles, drawn to scale and recording accurately the height–range but also indicating the degree of erosion of the flat. A. F. Coleman (op. cit. (1954), p. 20) indicates clearly the importance of the vertical plane of projection, the horizontal location of the flats relative to the map and field-sketch, and the relation of the shapes of the quadrilaterals to those of the flats. E. H. Brown uses broken lines to outline related flats above and below. M. M. Sweeting used triangles to indicate summits, rectangles to indicate flats, projected and correctly spaced on a section taken from west to east across the area under consideration; further, those occurring on shales and flagstones were filled in solid black, those on limestone were left open.

Obviously this type of diagram has very great value. One should bear in mind, however, certain cautions and criticisms expressed by A. F. Coleman, op. cit. (1954), and also very pertinently by A. A. Miller,[1] who stresses 'its dangers of misrepresentation and risks of misuse', and 'its potentialities for self-deception'. Perhaps the most important thing to bear in mind is that such a diagram can have no possible claim to objectivity, since it represents surfaces established in the field by aneroid survey, accurate levelling, and personal judgement by the operator of where the flat begins and ends.

Triangular Graphs for Slope Analysis

In his analysis of hillslopes, A. F. Pitty used, among other devices, a triangular graph.[2] Having divided a slope into three portions (summit, basal and midslope), he analysed the frequency of occurrence of each slope-angle observation in each of these categories, and calculated the percentage of the total observations falling in each slope-angle class. On the triangular graph (Fig. 52) the bottom left-hand corner is the point at which all slope-angles (i.e. 100 per cent) occur in the summit-portion of the slope-profile; the top corner is the point at which all slope-angles are in the midslope portion; and the bottom right is the point at which all are in the basal position. A single point on the graph will represent the three percentages, and when all the slope-angle classes have been plotted, their positions on the graph will reveal significant features; e.g. a clustering in the bottom right-hand corner

[1] A. A. Miller, 'Notes on the use of the Height–Range Diagram', *Geographical Studies*, vol. 2, no. 2, pp. 111–15 (London, 1955).
[2] A. F. Pitty, *A Scheme for Hillslope Analysis*, Occasional Papers in Geography, no. 9, pp. 62–9 (University of Hull, 1969).

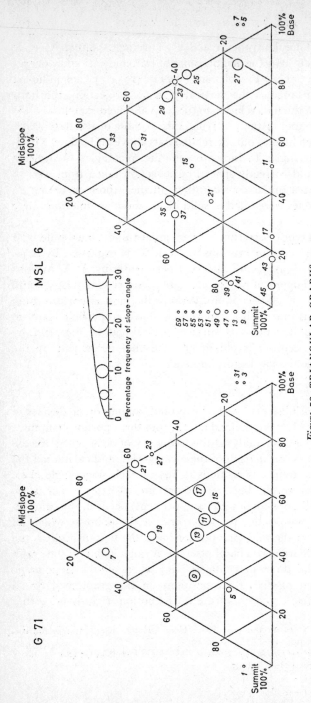

Figure 52. TRIANGULAR GRAPHS

Based on A. F. Pitty, *A Scheme for Hillslope Analysis: 1. Initial Considerations and Calculations*, Occasional Papers in Geography, no. 9 (University of Hull, 1969).

The letters and numbers refer to an index code for identification of slope profiles. Angles which occur exclusively in the summit third or in the base third of the slope respectively are shown outside the triangle.

will indicate a relative frequency of angles in the basal portion. The frequency of each slope-angle class can be indicated by placing a proportional circle at the point which represents the percentages of the three profile divisions, so distinguishing between slope-angle classes containing many or few observations (Fig. 52).

Intervisibility Exercises from Contour-maps

Gradients are also involved in the determination either of inter-visibility between two points, or of the whole area of 'dead-ground' from any particular position. There is a number of methods of solving intervisibility between two points, three of which may be briefly summarized (Fig. 53).

In the first method, draw a cross-section and rule a line of sight from the point of observation to the other point; if the line clears all intervening rises, the points are visible from each other. It is not necessary to draw a complete section; perpendiculars to scale representing any possible intervening eminences are sufficient. It must be remembered that minor irregularities not revealed by the contour-interval, as well as woods and hedgerows, may have to be allowed for.

A second method is to calculate the overall gradient of the line of sight from the observer to the second point, and also the gradients from any possible points of interference of view to the second point. Clearly, if the gradient of the line of sight to the far point is gentler than the gradients to any intervening points, then the two stations are intervisible.

A third and more accurate method of determining intervisibility employs the principle of similar triangles. Estimate the altitude above Ordnance Datum of the two points, using the nearest contour or spot-height, and by subtraction find the difference in height; thus, if A is 740 feet and B is 650 feet (Fig. 53), then the difference is 90 feet. Draw a line parallel to the line of view, a convenient distance away, nine units in length, and mark off as in the diagram. Join each end of the line of sight to the opposite end of the parallel line; these lines will cross at X. From any point P along the line of sight which would seem to interrupt visibility, draw a line through X and produce it to meet the parallel line at P_1, then read off the height indicated on this line, which actually gives the height of the line of sight at the

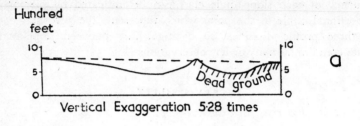

Vertical Exaggeration 5·28 times

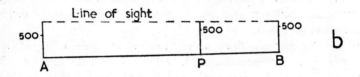

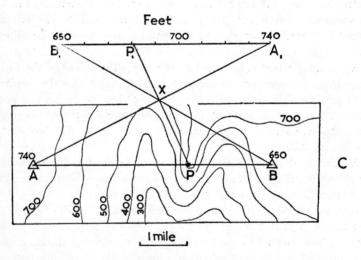

Figure 53. INTERVISIBILITY

The methods illustrated use **a** a line of sight and a section, **b** a line of sight and perpendiculars from each possible interruption and **c** the principle of similar triangles.

particular point. Obviously, if this height is greater than the height of *P*, the line of sight will clear *P*, and the points will be intervisible.

'Dead-ground'

Dead-ground may be plotted on a relief map with a fair degree of accuracy, but the reservations already noted as to the interruption of visibility by minor obstacles apply with equal force. Examine the map, noting all irregularities. Obviously, every point higher than the view-point will cause the area beyond this to be dead-ground. In other cases, draw a profile and rule in a line of sight; this will show where the line reaches the ground beyond the intervening obstacle, and therefore the extent of dead-ground. Skirting profiles are necessary to find the side limits of dead-ground round each obstacle. Plot the edge of the dead-ground, and shade in the area thus delimited.

A useful method is to draw a series of radiating lines from the view-point. Examine the points at which the lines of view are obviously interrupted, and estimate their heights. By comparison with the height of the point of observation, it can be seen at what rate the line of view declines in altitude. Thus the height of the ray can be calculated at any point along it, and in fact the heights of the ray can be scaled along it. Where the ground-level is shown by the contours is below the height of the ray, the ground is 'dead'.[1]

REGRESSION CURVES AND SCATTER DIAGRAMS

In the study of geomorphological and hydrological processes increasingly frequent use is being made of regression analyses. For example, the measurements of rates at which sediment is removed from river catchments and the evaluation of the relative significance of the several contributing factors is of fundamental importance in the study of erosion and denudation processes. Rates of denudation of dissolved and suspended sediment expressed in volumes per unit per year can

[1] An application and elaboration of the general principle of plotting 'dead-ground' maps was devised by Dr A. Garnett, 'Insolation, Topography and Settlement in the Alps', *Geographical Review*, vol. 25, pp. 601–17 (New York, 1935), and also in 'Insolation and Relief', Publication no. 5 of the *Institute of British Geographers* (London, 1937). She produced a series of maps showing (*a*) winter noonday shadow areas, (*b*) the periods of potential sunshine experienced at the equinoxes and (*c*) the intensity of insolation at noon at the equinoxes. The last maps bear 'iso-intensity lines', which show the percentages of intensity of insolation of the maximum possible. The article describes fully how these maps were plotted. The maps are closely related to land utilization and settlement in a number of Alpine valleys.

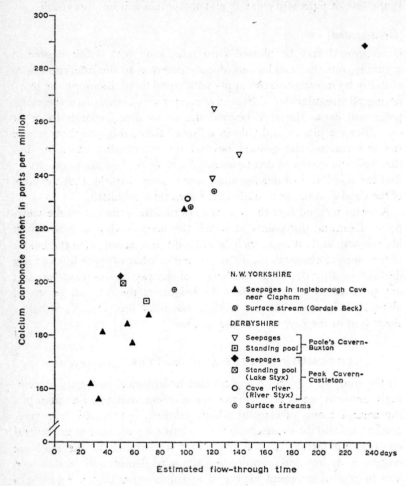

Figure 54. A SCATTER DIAGRAM

Based on A. F. Pitty, 'Calcium Carbonate Content of Karst Water in Relation to Flow-through Time', *Nature*, vol. 217, no. 5132, pp. 939–40 (London, 1968).

The diagram shows the relationship between estimated 'flow-through' time and mean calcium carbonate content of karst waters from the southern and central Pennines.

be determined from sampling river water. Variations in these rates can then be tested against a number of factors, including (as in Fig. 55) variation in stream flow as measured by discharge at the stations sampled. The fitting of regression curves then enables different catchments to be compared.[1]

The assumption of a straight-line relationship between sets of given variables may not always be justified. The relationship, however, can be inferred from a scatter diagram. As an example, A. F. Pitty's work[2] on calcium carbonate content of karst water in relation to flow-through time may be cited (Fig. 54).

LANDSCAPE DRAWING AND FIELD-SKETCHING

With the development of photography and of methods of reproducing photographs by half-tone blocks, landscape drawing has seemed to have lost much of its former importance. In the past, such geographers as G. K. Gilbert and W. M. Davis used simple yet most effective sketches to illustrate their concept of landforms.[3] Recently, however, both in Britain and America, there has been an increasing tendency of geomorphologists to make use of clear bold sketches to help to clarify geomorphological problems.[4] These drawings show economy of line, and enable a geographer to select and emphasize salient features by omitting minor foreground and other irrelevant details.[5]

A geographer should practise sketching in the field, to amplify and

[1] For further discussion, see Ian Douglas, 'Erosion in the Sungei Gombak Catchment, Selangor, Malaysia', *Journal of Tropical Geography*, vol. 26, pp. 1–16 (Singapore, 1968).

[2] A. F. Pitty, *An Approach to the Study of Karst Water*, Occasional Papers in Geography, no. 5 (University of Hull, 1966). See also *Nature*, vol. 217, no. 5132, pp. 939–40 (London, 1968).

[3] G. K. Gilbert, *Report on the Geology of the Henry Mountains (Utah)* (Washington, 1877), used field-sketches, closely associated with block-diagrams and geological sections, in a most graphic and revealing manner, as did W. M. Davis, *Geographical Essays* (Boston, 1910); an unabridged republication of this classic work appeared in 1954 (New York).

[4] See D. L. Linton, 'Watershed Breaching by Ice in Scotland', *Transactions and Papers, 1949*, the Institute of British Geographers, no. 15, pp. 1–16 (London, 1951); and R. F. Peel, 'A Study of Two Northumbrian Spillways', ibid., pp. 75–89; the latter uses some half-dozen sketches of remarkable clarity.

[5] An outstanding survey of this technique is G. E. Hutchings, *Landscape Drawing* (1960). Apart from 100 illustrations of the author's beautiful work, it contains information in detail about the art of observation, interpretation and reproduction of a landscape, the value of tone, light and shade, principles of perspective, the representation of rocks, trees and houses, etc.

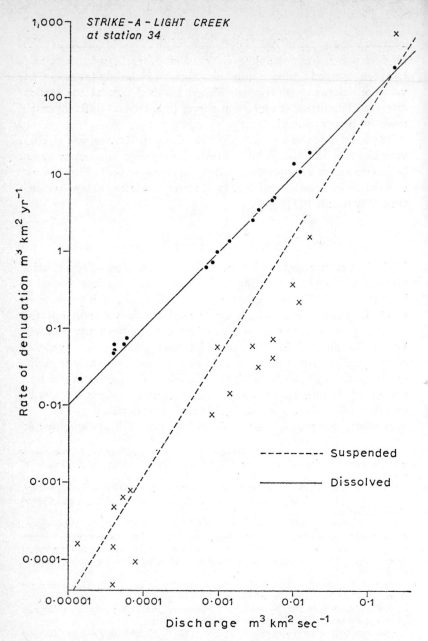

Figure 55. REGRESSIONS OF RATES OF DENUDATION OF SUSPENDED AND DISSOLVED SEDIMENT LOAD AGAINST DISCHARGE

Regression lines fitted to measurements of suspended and dissolved load in the case of a small Australian catchment. Both rates of denudation and discharge are on log scales and comparatively good fits are obtained. Based on unpublished data on rates of denudation in selected small catchments in eastern Australia by I. Douglas.

illustrate his maps and text. This will help the development of an eye for country, which is of first-rate importance, and will stimulate his powers of observation and appreciation. In addition, a student can illustrate essays concerned with landforms by means of rapidly drawn sketches. Figures 56–9 show some examples of how simple sketches can be used to illustrate various topographical features.[1]

The teacher can also draw simple landscape sketches on the board, built up during the lesson as a 'synthetic regional landscape'.[2] E. Raisz is a supreme exponent of this art. Field-sketching may also be used very effectively in urban studies, so producing 'skyscapes' and urban profiles, giving in the first case a silhouette panoramic outline of the horizon, in the second a full urban sketch in depth, with variations in density between foreground and background.

Sketches from Contour-maps

It is a useful exercise in the interpretation of a contour-map to make a simple sketch of one's concept of the landscape as it would appear to an observer at some specific point. If the sketch is then compared with a photograph taken from the same point, or better still with the actual country, the results are sometimes very chastening. There are various geometrical methods available, one of which is described by D. Sylvester.[3] It is based on the projection of suitable points on a contour-map drawn within a cone, on to a plane surface placed at right-angles to the median line of the cone. These points must be chosen to delimit foreground, middle-ground and background hill profiles, and in addition 'water-lines' such as the coast and rivers, and any other essential detail. These points are then joined to form a line-drawing.

Another method is clearly demonstrated by Raisz.[4] Two lines are drawn radiating from the point of observation, enclosing an angle of about 45 degrees, to represent the field of view; these can either be

[1] The geometrical relationship between the landscape, a photograph, a map, and a block-diagram or relief model is very clearly illustrated by E. Imhof, *Gelände und Karte*, p. 156 (Zürich, 1950). This example is reproduced in *Geographica Helvetica*, vol. 5, p. 301 (Bern, 1950).

[2] P. J. M. Bailey, 'Teaching Regional Geography. The Synthetic Landscape Method', *Geography*, vol. 48, pp. 285–92 (London, 1963). See also J. S. Crichton and G. Rae, *Blackboard Drawing for Geography* (London, 1958).

[3] D. Sylvester, 'A Method of Panorama Construction from Contoured Maps', *Geography*, vol. 28, pp. 12–18 (London, 1943). This article fully explains the procedure, with examples.

[4] E. Raisz, *General Cartography* (New York, 1948).

ruled lightly in pencil on the actual contour-map, or on a sheet of
superimposed tracing-paper. A horizontal line is then ruled across the
map at right-angles to the line of sight of the observer. This line will
be placed according to the size of the panorama required; obviously

Figures 56–9. LANDSCAPE SKETCHES

Figure 56 depicts a corrie, with tarn, screes and crags (Dow Crag and Goats Water,
near Coniston); Fig. 57 is of the Grand Canyon of the Colorado; Fig. 58 is a U-shaped,
glaciated valley, with benches, cascades and alluvial cones (Lauterbrunnen); and
Fig. 59 shows a cove formed by marine erosion (Lulworth Cove, in Dorset).

the nearer the line is placed to the viewpoint, the smaller will be the
area of the landscape shown. Then a number of significant points,
such as summit spot-heights, are projected on to this horizontal line
by lines of sight from the point of observation drawn through the particular point on the map. This will locate the horizontal component
of each point relative to the observer. The vertical component, which

will enable each point to be fixed on the panorama, is then obtained, either by estimating a vertical scale which appears to give a reasonable impression of height (usually involving some exaggeration), or by means of a hyperbolic vertical scale. On such a scale, the baseline represents the height of the observer's horizon (i.e. it is at the same height above sea level as the point of observation), while the vertical scale decreases in inverse proportion to the distance from the point of observation.

Sketching in the Field

It is not necessary to have any artistic ability to be able to sketch in the field, for it is possible to employ various 'rule of thumb' principles. Use a sketching-block with stiff cardboard backing, an H pencil, and a ruler or Service protractor. From the viewpoint, examine carefully the country to be depicted, note the sky-line profile, then the middle-ground detail, and finally the foreground. Decide on the extent of the panorama to be included; a horizon arc of about 30 degrees is usually sufficient.[1] Note any prominent point on the horizon about the centre of the area to be sketched, and draw a vertical line on the paper through it. Rule also a horizontal line across the middle of the paper, following if possible some definite line on the country. The sketch can then be drawn by eye alone; estimate the vertical and horizontal positions relative to the guide-lines of main features such as hilltops, plot them on the paper, and then work from these to minor features. To obtain greater precision, fix the positions of important features relative to the guide-lines with a ruler, and mark them proportionally on the drawing. For greater exactness, but with more tedium, cut a hole in a piece of card the size of the intended sketch, covered with a piece of celluloid ruled with a grid.[2] This is held at a fixed distance from the eye, and detail is transferred, square by square, to the paper.

Often a disappointingly flat impression of country is given by a photograph; a landscape drawing must introduce a vertical exaggeration to appear natural. A twofold exaggeration is suitable, i.e. the height

[1] *The War Office Manual of Map-Reading, Photo-Reading and Field-Sketching* (H.M.S.O., 1929) states (p. 75) that if a Service protractor is held 11 inches from the eye it will subtend an angle of 25 degrees, i.e. it will obscure the country with a horizon arc of 25 degrees. The protractor can be kept at a uniform distance by a piece of string 11 inches long, one end fastened to the protractor, the other held in the observer's teeth.

[2] E. Imhof, *Gelände und Karte*, p. 200 (Zürich, 1950), describes and illustrates an instrument known as a *Sitometer*, by which the angular distance relative to the observer of points in a panorama can be accurately measured.

of a point from the bottom of the sketch should be increased twice compared with the horizontal measurement. This is especially necessary for a sketch of country viewed from a hilltop.

Once the framework of the sketch has been built up, detail can be easily added. The rules of perspective should be obeyed, to convey the impression of distance, and the scales of known objects such as trees and houses must be carefully observed. Detail should be simplified and to some extent symbolized: woods by a wavy outline, lightly stippled; clumps of trees in outline; rivers by a double line in perspective with light longitudinal shading; and villages by a few rectangles, with such elaboration as a church-tower and chimneys. Draw in the foreground detail by heavier lines, more distant detail by fine or even lightly pecked lines.

Annotation can add valuable information to a sketch.[1] Names and spot-heights can be placed among hills, and names of villages, lakes and so on lettered beside those particular features. Land utilization and anything else clearly visible may be indicated by capital letters or symbols. The sketch should be orientated by a note in the caption, as 'view from X, looking south-east', and an indication of scale by a statement: 'The distance between A and B is 3 miles.' Ink in the finished sketch, or make a new tracing in ink from the field-sketch.

Sketching from a Photograph

Often the only source of illustration for some theme is a poor photograph, indistinctly showing the desired feature; this will appear even worse on reproduction by means of a half-tone block. It may be far better to redraw the scene in ink for reproduction as a line-block. Place the photograph on a well-lighted tracing table, with a piece of tracing-paper over it, and outline the main features in pencil, picking out the essential elements and omitting minor foreground detail, as is done in field-sketching. A small photograph can be covered with a grid, and the detail then transferred on a larger scale in the same manner as in map-enlarging (see p. 64). Ink in when the pencil sketch is completed, adding any desired annotation. Obviously no ver-

[1] Griffith Taylor,'British Columbia: A Study in Topographical Control', *Geographical Review*, vol. 32, pp. 372–402 (New York, 1942), includes a number of simple yet exceedingly clear and informative annotated sketches.

M. Hardy and A. Geddes, 'Lewis', *Scottish Geographical Magazine*, vol. 52, facing p. 224 (Edinburgh, 1936), have an annotated 'diagrammatic view' in colour across the Isle of Lewis from south-west to north-east. Land utilization on the diagram was shown in colour, to correspond to an accompanying land utilization map.

tical exaggeration is involved, as mentioned above, and in some cases the result may appear flat. It is possible to shorten these processes by drawing in waterproof Indian ink directly on to the surface of a bromide print; details can be accurately followed. Then bleach away the photograph in a suitable medium,[1] and re-photograph the line-work.

BLOCK-DIAGRAMS

Block-diagrams[2] are widely and effectively used to illustrate the features of the geomorphology of a particular area, or of specific land-forms. They are in effect sketches of relief models, rather than actual pictures of sections of the earth's crust. They have the added advantage that geological sections can be appended to the sides of the block (see pp. 188), thus enabling correlations to be made between structure and surface (Figs. 67, 68). A series of such diagrams enable the stages of landscape development to be illustrated.

Annotation may be added very effectively to block-diagrams, the value of which in teaching Geography is stressed by N. Proctor.[3] Effective examples are provided in two articles discussing the shorelines of parts of eastern Scotland.[4]

There are three main categories of block-diagram. These are, first, sketch block-diagrams of imaginary landforms, drawn merely by eye or with the guidance of an outline of a simple geometrical figure; second, sketch block-diagrams drawn from a contour-map, not strictly in perspective, but giving the general appearance of a relief model viewed obliquely;[5] and third, block-diagrams, drawn from a contour-map, either with a one-point perspective, or with a two-point, or true, perspective.[6]

[1] See *The British Journal Photographic Almanac, 1950*, p. 318 (London, 1950), for technical details.

[2] See A. Schou, *The Construction and Drawing of Block Diagrams* (London, 1961).

[3] N. Proctor, 'Using Block-Diagrams in Teaching Geography', *Geography*, vol. 48, pp. 393–8 (London, 1963).

[4] J. S. Smith, 'Morainic Limits and their Relationship to Raised Shorelines in the East Scotland Highlands', *Transactions of the Institute of British Geographers*, no. 39, pp. 61–4 (London, 1966); and R. A. Cullingford and D. E. Smith, 'Late Glacial Shore-lines in Eastern Fife', *Transactions of the Institute of British Geographers*, no. 39, pp. 31–51 (London, 1966).

[5] F. Debenham, *Exercises in Cartography*, p. 78 (London, 1937), calls this 'a pseudo-isometric projection of the contoured map'.

[6] Most geomorphological textbooks contain a wide variety of block-diagrams. The classic work is A. K. Lobeck, *Block Diagrams* (New York, 1924). G. K. Gilbert, in *Report on the Geology of the Henry Mountains (Utah)* (U.S. Geographical and Geological

Sketch Block-diagrams of Imaginary Land-forms

For these block-diagrams it is best to begin with simple geometrical figures, such as cylinders or cones, as a basis; draw a vertical plan, next sketch a plan in one- or two-point perspective, then the solid figure based on the ground-plan; this will afford accurate guide-lines for the finished drawing. Thus a cone can be used as the basis for volcanoes, fans and mountains generally, or for a series of diagrams to show the growth of atolls by subsidence. Pyramids can be used for mountain peaks and mesas; portions of spheres or spheroids for cirques, drumlins and acid volcanoes; cylinders for fold mountains, glacial valleys and fjords; and prismoidal blocks for cliffs, escarpments and ridges. A minimum of construction lines should be used. With experience, complex landforms can be built-up by adding to or cutting out portions from one or more geometrical blocks. Isometric graph-paper can be used quickly and effectively to produce simple block-diagrams (Fig. 60).

All this construction work is of course carried out in pencil. Ink in the edges of the block and the outlines of the relief, add light shading as in a perspective sketch to bring out slopes and minor details, and rule in vertical or horizontal shading to one or more sides of the block to give the impression of depth.

Survey of the Rocky Mountain Region, Washington, 1877), was probably the pioneer of their use. W. M. Davis used series of simple but extraordinarily effective block-diagrams in his many papers; see his *Geographical Essays* (Boston, 1910), and *Atlas for Practical Exercises in Physical Geography* (Boston, 1908). Other exponents are C. R. Longwell, A. Knopf and R. F. Flint, *Outlines of Physical Geology* (New York, 1947); C. A. Cotton, *Geomorphology of New Zealand* (Wellington, 1926), and *Landscapes as Developed by Normal Processes of Erosion* (Wellington, 1948); C. Barrington Brown and F. Debenham, *Structure and Surface* (London, 1929); and E. de Martonne, *Traité de Géographie Physique* (Paris, 1926).

S. W. Wooldridge and R. S. Morgan make abundant use of simple sketch block-diagrams in *The Physical Basis of Geography* (latest edition, London, 1946).

F. Debenham, *Exercises in Cartography* (London, 1937), gives ten examples of the way in which block-diagrams of particular landforms can be based on simple geometrical figures.

Block-diagrams, drawn accurately from large-scale contour-maps, can be used effectively for detailed regional studies. Two contrasting examples are the block-diagram of the Upper Teme Valley, in *The Land of Britain*, part 68, *Worcestershire*, Fig. 25, by K. M. Buchanan (London, 1944), and an annotated block-diagram, conveying a vast amount of information, of the San José Valley in California, compiled by E. N. Torbert, in 'The Specialized Commercial Agriculture of the Northern Santa Clara Valley', *Geographical Review*, vol. 26, pp. 247–63 (New York, 1936).

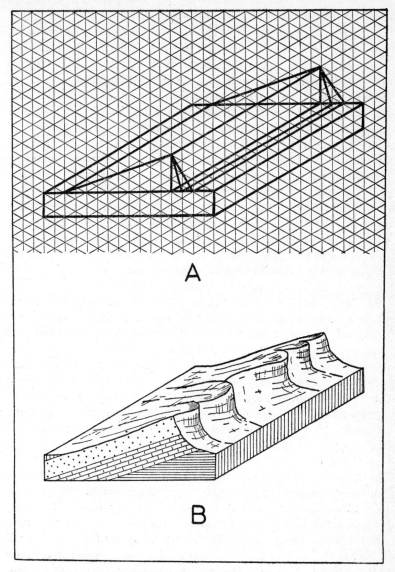

A

B

Figure 60. AN ISOMETRIC BLOCK-DIAGRAM

Isometric graph-paper provides ready guide lines for drawing geometrical figures
upon which pseudo-perspective block-diagrams can be based.

Block-diagrams drawn from a Contour-map

Trace a plan-view square from the contour-map of the desired area, showing only selected contours and streams, and cover the plan with a network of small squares. Examine carefully the contour-map, in order to determine the most satisfactory position from which the observer may be assumed to be looking. Obviously a range of high hills along one side will be placed in the background, a U-shaped valley will be viewed more or less longitudinally, a cirque from a point opposite its mouth. When the orientation of the block-diagram has been decided, project the square and its grid into a rhombus;[1] the sides of the square and of the grid square will remain their true lengths (Fig. 62). The angle between the base and side clearly determines the obliquity of the block. This angle should generally lie between 30 and 45 degrees; a higher angle will give the impression of a lofty viewpoint, a lower angle the converse. The relief details can now be transferred from the square to the rhombus, either by the multiple-section method or by the layer method.

The Multiple-section Method (Fig. 61). Mark off contour intersections, spot-heights and summits, rivers and valleys, as for an ordinary section, by placing a series of straight-edges along each side of the block and along each line of the square grid in turn. Transfer these points to the projected rhombus. Choose a vertical scale; the exaggeration compared with the horizontal scale should not exceed ten times for a piece of level land, and in mountainous country no exaggeration may be necessary. Draw sections along the four edges of the block and along the horizontal grid lines. It is not necessary to draw every section; some may obviously lie in dead-ground, while others may be replaced by occasional perpendiculars It may happen that certain conspicuous features are not adequately covered by the series of sections, so extra sections and perpendiculars may be inserted as required. Rivers, crest-lines, edges of escarpments and the like may be transferred directly from the ground-plan to the rhombus, and sections drawn along them.

Shade in the topography, using the sections, streams and other detail as guides, and carefully refer to the original map in order that minor

[1] W. A. White, 'Topographic Sketches from Contour Maps', *Surveying and Mapping*, vol. 3, no. 4 (Washington, 1943), describes two methods of foreshortening contours from a contour-map before starting to construct a block-diagram. One method involves the use of photography, the other involves parallel projection, using a powerful beam of light as the projecting agent.

details not shown by the sections will be accurately represented on the block. Take particular care over dead-ground. The representation of such features as lakes and tarns is difficult, for it is easy to make their surfaces appear sloping.

When the pencil draft is complete, ink in the edges of the block, the

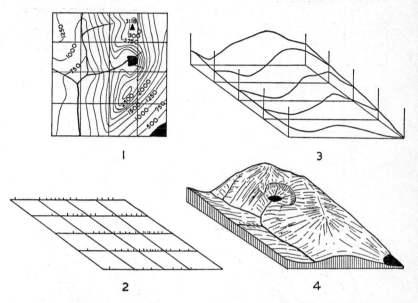

Figure 61. THE CONSTRUCTION OF A BLOCK-DIAGRAM BY THE MULTIPLE SECTION METHOD

1. Contour tracing, with grid; **2.** projection into a rhombus, and plotting of contour intersections; **3.** construction of sections along the horizontal grid-lines; **4.** completion of diagram with shading.

sky-line, the shading and other detail, and rub out construction lines. It may be easier and cleaner to retrace directly in ink, putting in only the required lines. Complete the block by shading the edge to give depth, add a title, a horizontal and a vertical scale, selected spot-heights, and any names of hills and streams. Put as much lettering as possible outside the block-diagram itself to avoid confusing the line-work. Rivers, lakes and their names can be inked in blue, spot-heights and hill-names in brown, railway lines in red and woods in green.

The Layer Method (Fig. 62). When the square and its grid have been projected into a rhombus, transfer on to it the contours and streams from the map, using the grid-lines as guides; this will produce a pseudo-perspective view of the contour-map. Choose a vertical scale, with some exaggeration if necessary. Draw on tracing-paper an outline of

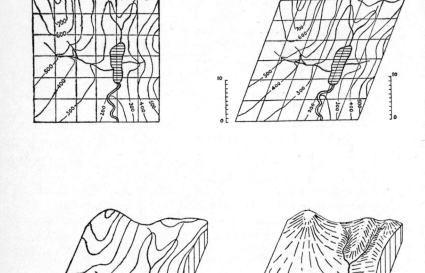

Figure 62. THE CONSTRUCTION OF A BLOCK-DIAGRAM BY THE LAYER METHOD

1. Contour tracing, with grid; 2. projection into a rhombus, with vertical scale-lines (in hundreds of feet); 3. each contour is drawn in at its correct scale-altitude; 4. completion of diagram with shading.

the rhombus, and drop perpendiculars at each corner of the base at the chosen scale, say, 0·1 inch to 100 feet. Place this tracing over the pseudo-perspective plan, fitting exactly at the corners, then slide it up the vertical scale until the highest contour lies in the same plane as its height on the vertical scale, and trace in this contour. Slide down the tracing until the next highest contour lies opposite the same figure on the vertical scale and trace that in, and so on. If a lower contour cuts a higher there is no need to continue it, since it would obviously be out of sight.[1] Draw in streams and lakes at their correct height. Complete the edges of the block, join the contour ends by lines between each of the four corners, i.e. by sections, taking care not to insert any dead-ground. Then, using the contours, projected to their scaled heights, and other detail such as streams as guides, shade and finish off the drawing as described in the multiple-section method. An example of 'sequential block-diagrams', illustrating the major stages in the evolution of the relief of part of southern Uganda, is given by J. C. Doornkampf and P. H. Temple.[2]

Perspective Block-diagrams

The block-diagrams produced by the methods already described will give a good general effect akin to an oblique view of a relief model. But it will be appreciated that the horizontal and vertical scales are maintained uniformly over the diagram, since the sides of the square and rhombus, and therefore of the grid, are kept identical. The diagram, in other words, is not drawn in perspective. If desired, however, block-diagrams can be constructed in either one- or two-point perspective, with considerably more labour.[3]

One-point Perspective. In this case the front of the diagram is a horizontal line, parallel to the back edge, viewed square-on to the observer. The sides of the diagram, and all horizontal lines on the plan, converge

[1] P. T. Dufour, 'Les Perspectives-Reliefs', in *Revue Géographie Annuelle*, vol. 8 (Paris, 1917), describes a device, a form of pantograph, consisting of a long rod, one end of which slides in a groove, the other has a tracer-point, with a pencil, the position of which can be varied, near the centre of the rod. The pencil is placed over the drawing-paper, the tracer over the map, and the contour-outlines are then followed round with the tracer. The pencil reproduces the contours, at any degree of tilt. The paper is moved for each successive contour, as in the ordinary layer method, along the chosen vertical scale. This is a rapid method for simple maps, but is difficult to use with involved contours. The shading and other detail are inserted as for any other method.

[2] J. C. Doornkampf and P. H. Temple, 'Surface, Drainage and Tectonic Instability in Part of Southern Uganda', *Geographical Journal*, vol. 132, p. 242 (London, 1966).

[3] See also Axel Schou, *The Construction and Drawing of Block Diagrams* (London, etc., 1962) with some beautiful illustrations. An example of the application of various perspectives to produce seven different block-diagrams of the Porta Westfalica near Minden in West Germany is given by F. Hölzel, 'Perspektivische Karten', *International Yearbook of Cartography*, vol. 3, p. 108 (London, 1963).

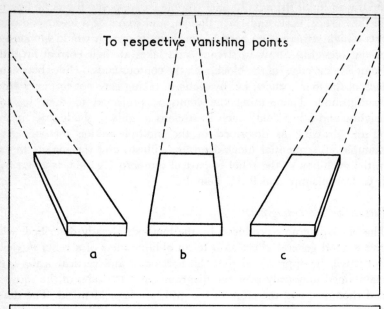

To respective vanishing points

a   b   c

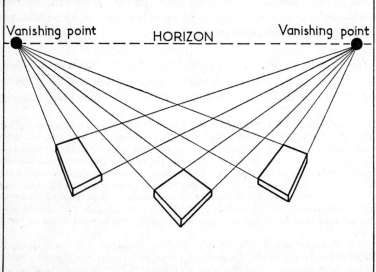

Vanishing point   HORIZON   Vanishing point

Figures 63, 64. ONE- AND TWO-POINT PERSPECTIVE BLOCK-DIAGRAMS

towards a distant vanishing point on the observer's horizon. The diagram can be placed in several ways to the observer; he may view it face on and see only the front edge, or from one side, either right or left, and see the front edge and a side (Fig. 63). In the latter case, the base angle opposite the observer should be about 45 degrees.

The base-plan, derived from the contour-map and its grid, has to be projected into perspective. Fix the vanishing point some distance away along a table with a pin. Draw the front edge its true size, and,

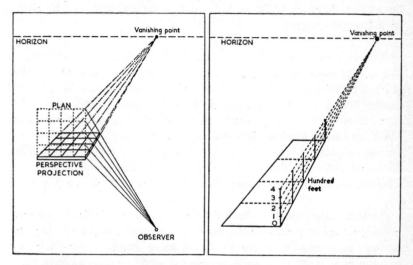

Figures 65, 66. A ONE-POINT PERSPECTIVE BLOCK-DIAGRAM
Figure 65 illustrates the method by which the base-plan (shown by pecked lines) is graphically projected into perspective (solid lines).
Figure 66 indicates how the vertical scale is determined for each perpendicular.

using a long ruler, mark the side edges, the vertical grid-lines, and the depth-line to converge at this vanishing point (Fig. 65). To project the back edge of the block, draw lines of view back from the corners of the base-plan to the observer's viewpoint. Clearly the horizontal scale will decrease from the front to the back of the block; this is naturally involved in the principle of perspective. What is not always appreciated is that the vertical scale also decreases. To find this, rule a perpendicular at the front and back corners and also at each intersection of a horizontal grid-line with the edge. Mark off the contour interval on the front perpendicular according to the chosen vertical scale and draw lines from each height to the vanishing point; the correct vertical scale is thus projected upon each perpendicular (Fig. 66). Plot each horizontal section line in the usual way from the base-plan, and transfer the ticks to the projected horizontal. Draw sections according to the particular vertical scale. When the sections have been completed, shading and other detail can be added in the usual way.

No scale should be stated, except in terms of the front edge. More conveniently, under the block can be written: 'The distance between X and Y (two prominent points) is 5 miles', which is quite sufficient for a block-diagram, while an indication of vertical scale can be given by a few spot-heights.

Two-point Perspective. In this case, the block-diagrams will present one corner towards an observer and two edges inclined away. The pairs of parallel sides when projected will meet at two vanishing points on the horizon, to the left and right of the observer (Fig. 64). The network of squares on the ground-plan is transferred to the perspective block by ruling lines to these two vanishing points, and then transferring contour intersections with the help of these guide-lines. The changing vertical scales are obtained as for the one-point perspective. This block-diagram allows the geological sections to be shown (Fig. 67).

Perspective Approximations. Both these mechanical methods of attaining perspective are somewhat laborious, and on a small-scale block-diagram perhaps unnecessarily elaborate. When using simple sketch-diagrams of imaginary landforms, an impression of one-point perspective may be attained by making the back edge slightly shorter than the front and so sloping the sides to it. An impression of two-point perspective can be obtained by making the angle at which the two back edges of the block meet slightly larger than the angle at the front.

Automatic Drawing of Block-diagrams

Machines of varying degrees of complexity have been developed in recent years which can be programmed to draw block-diagrams from contour maps. The student should be aware of their existence and potentialities, since various models have been adopted in some University and Government departments. An example is the *Perspektomat P-40*,[1] which comprises an elaborate linkage system by which contours on a map are followed by a tracer, which are transferred to another sheet on which the block diagram is being drawn.

[1] Made by F. Forster, 8200 Schaffhausen, Switzerland.

Three-dimensional Diagrams

There is a useful discussion of the use of three-dimensional diagrams in Geography by J. P. Cole and C. A. M. King.[1] They refer for example to the use of such a diagram to plot features of a till fabric using an equidistant zeniphal projection. The till fabric analysis includes orientation shown clockwise relative to north, while the dip is given by concentric circles with angle of dip from north to 90 degrees.

GEOLOGICAL MAPS

Geological maps are of very great value to the geographer. From the various geological maps available, the physical geographer can obtain much essential data to help his study of landforms.

The detailed interpretation of geological maps cannot be considered here. The geography student should examine samples of maps published by the British Geological Survey, and consider the detailed schemes of colour, stipple, symbol and lettering used, and the arrangement of the very detailed legends. He should familiarize himself with the method of lettering and numbering of the geological series in order of age. The geographer who wishes to study geological maps in some detail should read the references cited below.[2] In this brief section, it is merely necessary to consider ways in which geological information can be added to relief maps and diagrams.

Reproductions and Tracings

An outline geological map usually accompanies as a matter of course the relief maps illustrating the physical basis of a regional monograph. The main problem is how much detail to show and how to draw and reproduce it. Obviously all the details of a large-scale printed map in colours cannot be included, nor would it be desirable. The geographer chooses what he needs, and extracts it; examples include maps to show

[1] J. P. Cole and C. A. M. King, *Quantitative Geography, Techniques and Theories in Geography*, chap. 9 (London, 1969).
[2] (*a*) C. Barrington Brown and F. Debenham, *Structure and Surface* (London, 1929); (*b*) R. M. Chalmers, *Geological Maps; the Determination of Structural Detail* (London, 1926); (*c*) A. R. Dwerryhouse, *Geological Maps, Their Interpretation and Use* (London, 1924); (*d*) K. W. Earle, *The Geological Map* (London, 1936); (*e*) G. E. Elles, *The Study of Geological Maps* (Cambridge, 1921); (*f*) A. A. Miller, 'Geological Structures', being Sections I and II of *The Skin of the Earth*, pp. 11–27 (London, 1953), which describes determination of strike and dip, thickness of strata, nature of outcrop, folding, faulting, etc.; (*g*) B. Simpson, *Geological Maps* (Oxford, 1968).

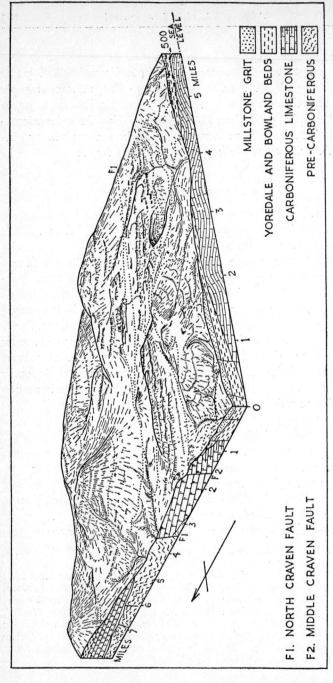

MILLSTONE GRIT

YOREDALE AND BOWLAND BEDS

CARBONIFEROUS LIMESTONE

PRE-CARBONIFEROUS

F1. NORTH CRAVEN FAULT

F2. MIDDLE CRAVEN FAULT

Figure 67. A TWO-POINT PERSPECTIVE BLOCK-DIAGRAM

The original diagram was compiled by P. Evans, with the assistance of R. G. S. Hudson; original drawing by R. A. Watling. This block-diagram is of the Craven district of the West Riding. In the right foreground lie Malham Cove and Gordale Scar, behind which are Malham Tarn and Fountains Fell. In the background is Pen-y-Ghent. The dotted lines on the surface of the block indicate outcrops.

the chalklands of England, the Palaeozoic rocks of the Lakeland Dome, the granite massifs of the Southwest Peninsula, the loess of Central Europe, the limestone (karst) of Jugoslavia and the volcanic areas of the Central Massif of France. These examples can be multiplied indefinitely, and few geography textbooks appear without a geological map, of however simplified a nature. The detail must nevertheless be abstracted accurately.

A dissertation map can be hand-coloured, but for a line-block black and white symbols (some of which are depicted on Fig. 19) must be

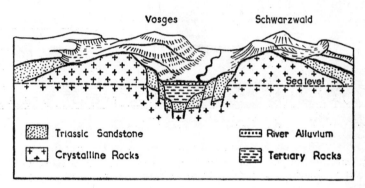

Figure 68. A BLOCK-DIAGRAM, WITH DIAGRAMMATIC GEOLOGICAL
SECTION, OF THE RHINE RIFT-VALLEY
Based on A. Cholley, *La France*, p. 110 (Paris, n.d.).

carefully chosen. It should be noted that some symbols for geological series have become standardized by usage. A few guides to location, such as rivers and one or two place-names, can be added, but it is difficult to include much detail over close back and white symbols.

One useful device is to superimpose the outline geology on a detailed relief map of the same scale, the former on transparent paper. For example, trace an outline map of part of the Craven Uplands to the north-west of Malham Tarn on to drawing-paper; extract contours at 250-foot intervals, add streams, mark the sink-holes, and name the main peaks. Then trace an outline map of the solid geology of exactly the same area, maintaining a careful register, from Geological Survey maps; the main outcrops can be lightly tinted. The relationships between relief and geology are most striking when the maps are superimposed; the high fells (the monadnocks of Pen-y-Ghent, Fountains Fell and Ingleborough), the limestone plateau above the mid-Craven fault, the relationship of surface streams and water-sinks to the bound-ary between the Yoredale Series and the limestone, the position of Malham Tarn, partly on the drift, partly on the exposed Silurian basement of the Pennines, are all emphasized most graphically.

Geological Sections

It is often practicable to add details of the underlying strata to a topographical profile or to the edges of a block-diagram, with very great effect (Fig. 67).[1] The information can be obtained from the

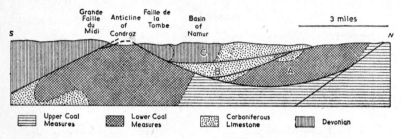

Figure 69. A GENERALIZED GEOLOGICAL SECTION

Based on A. Briant, 'Etude sur la Structure du Bassin Houiller de Hainaut dans le District du Centre', in *Annales de la Société Géologique de Belge*, vol. 21, p. 125 (Liége, 1894).

This section illustrates, in simplified form, the geological complexity of part of the Sambre-Meuse coalfield of Belgium. The Grande Faille du Midi is a great thrust-fault, along which some of the Devonian rocks of the Ardennes to the south have been driven over the Coal Measures. The letters **A, B, C** represent three immense over-thrust wedges, so that shafts have had to be sunk through these overlying Devonian and Lower Carboniferous rocks to reach the younger coal-bearing Upper Coal Measures below.

geological map itself, or often more conveniently from sections given in the margins of the geological map, from memoirs, or, for detailed large-scale work, from actual observation in the field. The various geological sections include diagrammatic, semi-diagrammatic and accurate sections.

Diagrammatic Sections. These are usually small-scale, generalized and with a considerable vertical exaggeration, but can be used very effec-tively to illustrate general concepts of physical geography. The geology may be applied either to a profile or to the edge of a block-diagram.

[1] A very attractive series of block-diagrams, with the geological detail added to the edges, is given in T. Hagen, 'Wissenschaftliche Luftbild-Interpretation', *Geographica Helvetica*, vol. 5, pp. 209–76 (Bern, 1950).

Familiar examples include sections of the Weald, the scarplands of England, the Pennines, the Rhine Rift-Valley (Fig. 68), and the Hercynian folding of central Europe (Fig. 69). Various landform diagrams (e.g. volcanic cones, sills, laccoliths, unconformities and fault-lines) are made much more clear and graphic when some idea of the solid geology is presented.

Semi-diagrammatic Sections. A topographical cross-section is first drawn accurately, using as small a vertical exaggeration as possible. Plot on this section the exact points of outcrop from the geological map; at these points the dip of the strata is drawn in by parallel lines. This approximation can only be used for a simple scarpland area, or where the strata have a known dip. The One-inch Geological Survey maps indicate inclined strata by arrows with numbers, showing the direction and amount of dip. If the same stratum outcrops twice, as on opposite sides of a valley, a line through these two points will give an indication of the dip. The aim of such a section in scarpland country is to illustrate the superposition of the strata, and the dip can only be approximate. In any case, if there is a vertical exaggeration involved, the dip cannot be shown accurately.

To illustrate this method, for example, a profile can be drawn using the One-inch Ordnance Survey series, from Lower Seagry to the Kennet–Og confluence at Marlborough. Then lay a straight edge along the same line of profile transferred to the corresponding sheet of the One-inch Geological Survey series, and mark carefully the outcrops. There is a long succession of these, from the Oxford Clay to the Upper Chalk, with complications of drift (which should be represented by the thinnest possible superficial layer), and it is quite exacting to fit the outcrops to the relief profile. But the result shows very clearly the vales to the west, the various scarp faces, and the long south-eastern back-slope. To complete the section, ink in the line-work, colour the strata to correspond to a careful legend, add topographical names above the section, and finish off the usual details of horizontal and vertical scales, vertical exaggeration, orientation and title.

Accurate Sections. These are used mainly by mining- or water-engineers, and are constructed usually from direct information obtained in the field. They must be drawn without any vertical exaggeration, or the dip will be falsified. The actual drawing of accurate sections involves the determination of *strike* (the line of intersection of a bedding surface with a horizontal plane), of *true dip*, which is at right-angles to the strike, and of *apparent dip*, which is a direction of slope at some angle between the strike and the true dip. Once the direction and amount of the dip has been established, the topographical section is drawn, the points of outcrop are plotted, and the strata are drawn in at the correct angle. In the case of anticlines or synclines there will be two opposite directions of true dip with the axis of the fold between. Faults introduce further complications.

The drawing of exact sections is somewhat outside the province of the geographer, and it is impossible to go into this here. The student is referred to the works cited on p. 185.

3

Climatic Maps and Diagrams

DATA

The chief *primitive elements* of climate include radiation from sun and sky, temperature, wind direction and speed, humidity, evaporation, cloud and sunshine, precipitation, snow-cover and atmospheric pressure. Each element is capable of further subdivision, for example, precipitation comprises hail, snow and rain; all may be measured by direct observation. *Combined elements* are those which are usually calculated by means of combining two or more elements, although some, such as humidity, may be directly observed. Examples include equivalent temperature (temperature and water-vapour pressure), drying power (saturation pressure, water-vapour pressure and wind velocity), cooling power (temperature, wind velocity and humidity,) and continentality (solar radiation and temperature). Finally, there are other elements which may be referred to as *derived elements*.[1] They include the variabilities, ranges, frequencies, probabilities and intensities of the various elements.

Published climatological data may include primitive, combined or derived elements; classification of data is therefore complex and variable.[2] An infinite variety of information may be available for some parts of the world, and comparatively little for others.[3] From the point of view of the geographer, primarily interested in climate rather than in weather, mean values for the various elements are of greater significance than actual measurements. Mean values are generally understood to reflect 'average weather', i.e. climate, but abnormal weather conditions cannot be overlooked. As W. G. Kendrew states: 'The

[1] V. Conrad, *Methods in Climatology* (Harvard, 1944).
[2] C. E. P. Brooks and N. Carruthers, *Handbook of Statistical Methods in Meteorology* (London, H.M.S.O., 1953).
[3] See *World Weather Records*, Smithsonian Miscellaneous Collections, vols, 79, 90, 105 (Washington, various dates), and International Meteorological Organization, *Réseau Mondiale, 1911–32* (published by British Meteorological Office, 1917–39).

abnormal weather conditions are very important, for the possibility of abnormally long or severe frost, or of a prolonged drought in a region which is usually well-watered, is the final consideration which may override for practical affairs of life the value of the mean conditions.[1] Tables of mean values therefore must be supplemented by reference to actual weather conditions, information about which is conveniently provided in synoptic charts. Not only do these charts bring together the elements which the climatologist is apt to separate one from another, but they also bring reality to his somewhat abstract conception of climate.[2]

Temperature

Daily Means. Where thermograph records of temperature are available, *true daily means* can be calculated. The mean is determined of twenty-four hourly observations; an example is given in the table below.

Hours	Observations in degrees Centigrade												True daily mean
	1	2	3	4	5	6	7	8	9	10	11	12	
A.M.	16·1	15·8	15·4	15·1	14·8	15·0	16·1	17·4	18·5	19·4	20·3	21·1	18·6
P.M.	21·9	22·6	22·7	22·5	21·9	21·0	19·8	18·8	18·1	17·5	17·0	16·6	

The term *daily mean* is used where only a limited number of observations is available. In the International Meteorological Code this is assumed to be the mean of daily observations taken at 07, 1400 and 2100 hours, and the use of the following formula is advocated:

$$Daily\ mean\ (^\circ C) = \frac{Readings\ at\ 07 + 1400 + 2100 + 2100\ hours}{4}$$

Thus using the observations given in the table above,

$$Daily\ mean\ (^\circ C) = \frac{16·1 + 22·6 + 18·1 + 18·1}{4} = 18·7$$

Where only the maximum and minimum temperatures for the day

[1] W. G. Kendrew, *The Climate of the Continents*, p. 6 (Oxford, 1927).
[2] A wide range of climatic maps and diagrams is given in R. G. Barry and R. J. Chorley, *Atmosphere, Weather and Climate* (London, 1968); and R. G. Barry, 'Models in Meteorology and Climatology', being pp. 97–144 in R. J. Chorley and P. Haggett (eds.), *Models in Geography* (London, 1967); the latter contains a bibliography of 192 items.

are available, the *daily average temperature* can be found by averaging the two, for example:

$$\text{Daily average } (°C) = \frac{\text{Maximum temperature } + \text{ Minimum temperature}}{2}$$

$$= \frac{22\cdot7 + 14\cdot8}{2} = 18\cdot8$$

Monthly Means. The *true monthly mean* is a weighted and corrected average of the temperature of each observation hour throughout the month, but the *monthly mean* is obtained simply by averaging the daily means. For climatological purposes the monthly means of specific years are less important than averages for each month over a long period, usually thirty-five years or more. Monthly means relating to long periods constitute the chief data of the climatologist in his analysis of temperature. But in addition to monthly means he must have details of mean monthly maxima, mean monthly minima, the absolute monthly maximum and the absolute monthly minimum temperatures.[1]

Annual Means. *True annual mean* temperature is a weighted and corrected average of true daily means for the year, but the *annual mean* is obtained by averaging the monthly mean temperatures. A slight inaccuracy results because each month is not of the same length. Annual mean temperatures have a general comparative value, but they are used chiefly in the determination of long-term temperature trends.

Significant Means, Extremes and Ranges. Mean and extreme diurnal ranges in temperature can be obtained from a comparison of maxima and minima figures. In addition to these basic data, some tables contain information about significant temperatures and their average or median dates of occurrence. For example, the highest and lowest temperatures ever recorded and their dates of occurrence, the average dates of the first and last killing frosts, the average dates of the rising or the falling of mean daily temperatures through 0° C and through 6° C (the temperature above which most plants begin to grow). Such information is particularly useful to the agricultural geographer.

Maps which attempt to show the classification of climates depend to a large extent on the selection of significant criteria. A good example

[1] See *Averages of Temperature for the British Isles, for periods ending 1935* (M.O. 407, 1936).

is the use of a 'temperateness index' by H. P. Bailey[1] in an attempt to
define and show the distribution of temperate climate throughout the
world. The index is based on the choice of a temperature of 14° C or
57·2° F as a critical point but also involves a spectrum of temperature
around this particular value. See in particular the chart used to plot
a temperateness index on the basis of average annual temperatures
against average annual range of temperature.

Rainfall

Precipitation may be measured by a standard rain-gauge which is
read daily, or, less usually, by a self-recording instrument. From these
measurements a number of records are compiled which are of interest
to the geographer.

Monthly Means. For general purposes the monthly means of rainfall
are simply arithmetic means of the total rainfall for each month over
a period of years.[2] Sometimes, however, it is useful to reduce all months
to an equal length in order to avoid false deductions, as, for example,
if a comparison has to be made between February and August. This
reduction may be effected in two ways: (*a*) The amounts for the 31-day
months are reduced by 3·2 per cent and the amount for February is
increased by 7·2 per cent. Hence each month is made equivalent to
30 days. (*b*) All months are altered to become months which are one-
twelfth of a year (i.e. 30·438 days). Therefore, the amount for February
has to be multiplied by 1·087 (1·049 in leap years), for 30-day months
by 1·015, and for 31-day months by 0·982.

Monthly means of rainfall are useful for illustrating the mean average
distribution of rainfall throughout the year at any one station. Means,
however, give no indication of the nature of the rainfall, whether heavy
or light, whether reliable from one year to another, or whether effective
in connection with plant growth. A great deal of additional information is therefore necessary to a proper understanding of this element
of climate.

Monthly and Yearly Totals. In many countries of the world details of
the rainfall are published year by year. *British Rainfall*, for example,

[1] Harry P. Bailey, 'Towards a unified concept of the temperate climate', *Geographical
Review*, vol. 54, p. 527 (New York, 1964).
[2] See Meteorological Office, *Average Monthly Rainfall of the British Isles, 1881–1915*,
by M. de C. S. Salter (Reprint, *British Rainfall*, London, 1920).

published by the Meteorological Office of the Air Ministry, incorporates information about total rainfall for each year for a number of stations in the British Isles. Yearly summaries of the total monthly rainfalls are also published.

The Number of Rain-days and Wet Days. British Rainfall also incorporates data concerning the number of *rain-days* in each year, i.e. days having more than 0·25 mm of rainfall, and the number of *wet days*, i.e. days having over 1·0 mm of rainfall. It should be noted that in some other countries a 'rain day' is considered to be a day with over 2·5 mm.

Rain Spells, Wet Spells, Droughts and Dry Spells. In *British Rainfall* there are tables giving information about *rain-spells*, i.e. durations of at least fifteen consecutive rain-days, *wet spells*, i.e. durations of at least fifteen consecutive wet days, *absolute droughts*, i.e. durations of fifteen or more consecutive days none of which has 0·25 mm of rain or more, *partial droughts*, i.e. durations of twenty-nine or more consecutive days, the mean rainfall of which does not exceed 0·25 mm, and *dry spells*, i.e. durations of fifteen consecutive days or more, none of which has 1·0 mm of rainfall or over.

Intensity of Rainfall. The rate at which rain falls is obviously related to problems of run-off, soil percolation, evaporation, soil erosion and flood control. Information about intensity of rainfall is therefore just as vital to an understanding of rainfall regimes as are mean values of total rainfall. Unfortunately, data about rainfall intensities are not very satisfactory, because there are few stations equipped with a reliable pluviograph or with sufficient personnel to make eye observations during rain-spells.

Most climatic tables do, however, give some information about the hourly duration of rain and about the total rainfall for a limited number of stations. Where such information is available, mean hourly or daily intensity of rainfall can be calculated using the formula:

$$I = \frac{A}{n}$$

where A = total rainfall over a given period and
*   n = total number of hours of rain, or number of rain-days.*

Using this formula, the intensity of rain at Boston, Mass., is, for example, 0·36 compared with 4·17 at Cherrapunji. Hourly intensities give better comparative results than do daily intensities.

Duration of Rainfall is compiled from pluviographic records. Tables of duration give the annual and monthly values in hours of rainfall for individual stations.

Deviations from Mean Rainfall are usually tabulated in the form of excess or deficiency of the total for any one month or year over the means; these data are of use in the analysis of rainfall probability (see p. 233).

Effectiveness of Rainfall[1] is usually taken as the actual total rainfall minus the total possible evaporation. There are numerous ways of measuring possible evaporation, none of which is entirely satisfactory. The Piche evaporimeter is commonly used, in which water in a tube is allowed to evaporate from a piece of porous paper, and the loss in a given time is measured on a scale graduated on the tube. Some evaporation statistics are calculated from measurements taken of the level of water in large open tanks or pans; the United States Weather Bureau makes extensive use of pan records. Various experiments have been made to measure evapo-transpiration, i.e. loss from evaporation plus moisture lost by plant transpiration since knowledge of water loss from vegetation surfaces is of vital importance for agriculture. Use of water by the plants can be calculated by repeated weighing of the soil-block contained in a tank sunk in the ground (or *weighing lysimeter*). A simpler approach, not requiring an expensive installation, is to determine the potential evapo-transpiration, i.e. the water loss from a vegetation surface which always has a sufficient supply of water from the soil. Potential evapo-transpiration rates are used in the calculation of moisture requirements of the soil, and in the determination of relative regional aridity[2] (see p. 214).

Snowfall and Hail are for record purposes usually measured by melting the solid precipitation and including it in the rainfall total as if it were rain. In addition, records are kept of the number of days in the year on which snow and on which hail fall, the number of days per year with snow lying, the duration of continuous snow-cover, mean and greatest snow-depth, and intensity of snow or hail-fall.[3] In the

[1] P. R. Crowe, 'The Effectiveness of Precipitation: A Graphical Analysis of Thornthwaite's Climatic Classifications', *Geographical Studies*, vol. 1, no. 1, pp. 44–61 (London, 1954).

[2] R. C. Ward, *Principles of Hydrology*, chs. 4, 5 (London, 1967).

[3] See tables in Meteorological Office, *The Baltic Sea* (London, 1947).

British Isles snowfall is so comparatively rare that records can be kept of individual falls of particular intensity.

Wind

Wind force is measured at certain fixed hours by direct observation of such an instrument as an air-meter or Robinson Cup Anemometer; at the same time its direction is recorded from a well-exposed weather-vane. The Dines Pressure-tube Anemometer produces an anemograph yielding continuous records of both direction and force. Wind direction and wind force are related and wind data can be analysed more efficiently by a consideration of both aspects at the same time rather than independently. Also duration of wind force and direction varies considerably during the day and the most valuable statistics indicate the time of observation. Because of all the measurements involved, wind tables take up quite a lot of space and are, perhaps, the most difficult of all weather data to express cartographically. Until 1949, when most countries of the world agreed to adopt the recommendations of the World Meteorological Organization, a variety of codes was in use whereby wind direction and force were measured. In British returns, for example, direction was usually expressed in terms of a number of cardinal points of the compass, and force in terms of the Beaufort Scale. Revised codes of force have now been drawn up, and returns of wind directions are made as true bearings in tens of degrees.[1] Most geographers, however, will find themselves working with average data assembled previous to 1949, of which the more important are summarized below.

Monthly Means of Wind Force and Direction. The number of occasions out of a hundred on which particular winds may be expected at different times of day, in different months and from different directions, are tabulated according to four, five, or more scales of wind force. Unfortunately, such information is not available for many places over a long period of years, but where it is available, it provides the bulk of the evidence about local winds.

Gust Levels. It is necessary to realize that statistics concerning wind velocity are usually expressed as a mean value for the period of observation, and 'gustiness' cannot be estimated from such figures. The *Gustiness Factor* (G) published for various stations provides some means of

[1] See Meteorological Office, *International Meteorological Code* (London, 1948).

calculating local gustiness if mean wind velocity is known. For example, if G is 0·5 for a mean wind velocity of 40 miles an hour, then gusts as high as 50 or as low as 30 miles an hour may be expected; if G is 1·0, then gusts of 60 miles an hour may occur.

Mean Diurnal Variations of Wind Direction. Changes in wind direction between day and night are particularly important in certain coastal regions, where land- and sea-breezes are experienced.

Resultant Wind Direction and Force. The average wind force from a given direction in miles per hour multiplied by the number of hours it blows from that direction in any given period is known as the *mean of wind.* If these data are available for any station it is possible to calculate from them the average resultant wind velocity, the average resultant wind direction, and the average 'steadiness' of wind at that particular station.[1] Occasionally these data are available in tabular form, and they are of use in plotting streamlines by means of arrows (see pp. 250–6).

Sunshine and Cloud

Duration of sunshine is measured by the Campbell–Stokes recorder, which produces a continuous record in the form of a line burnt upon a strip of cardboard. Tables of mean sunshine data are compiled upon the basis of absolute duration in hours per day, or month, or upon the basis of percentage of possible sunshine per day or month.[2]

Degree of Cloudiness. Cloudiness is related to sunshine, but discrepancies in records occur partly because sunshine recording instruments are not perfect, partly because the degree of cloudiness is observed usually only three times a day, while its measurement is extremely difficult because of the great variety of cloud. It is usually expressed in tenths, for example, one-tenth of the sky clouded, two-tenths and so on, or in eighths (*oktas*). Tables are compiled of mean cloudiness upon daily, monthly and yearly bases.

Humidity

Data concerning humidity are normally obtained from readings of dry- and wet-bulb thermometers at fixed hours, but thermohygrographs

---

[1] For details of computation, see V. Conrad and L. W. Pollak, *Methods in Climatology* (2nd edition, Cambridge, Mass., 1950).

[2] See *Averages of Bright Sunshine for the British Isles* (M.O. 408, 1936).

do provide continuous readings.[1] There is a variety of modes of expression of the data, but *relative humidity*, which represents the percentage degree of saturation of the air, is of major interest to the geographer because of its importance for human comfort and for plant growth.

Relative Humidity: Monthly Means and Diurnal Ranges

Aberdeen	Jan. %	Feb. %	Mar. %	Apr. %	May %	June %
Highest mean hourly value	81·8	81·6	83·0	84·4	86·1	86·4
Daily mean	80·7	79·6	78·7	78·0	78·5	78·1
Lowest mean hourly value	78·0	75·4	72·1	70·5	71·8	71·2
Diurnal range	3·8	6·2	10·9	13·9	14·3	15·2

Aberdeen	July %	Aug. %	Sept. %	Oct. %	Nov. %	Dec. %	Year %
Highest mean hourly value	86·4	87·1	86·3	86·0	83·8	83·4	84·6
Daily mean	78·5	79·5	80·3	82·4	82·1	82·3	79·9
Lowest mean hourly value	71·3	70·9	71·9	75·2	78·5	80·0	74·0
Diurnal range	15·1	16·2	14·4	10·8	5·3	3·4	10·6

Source: E. G. Bilham, *The Climate of the British Isles* (London, 1938).

Diurnal Variation of Relative Humidity is considerable, and mean daily figures therefore have not any great significance. Mean monthly statistics are compiled, but it is more usual to give monthly means for a specific time during the day, or to give additional information about mean diurnal ranges, as the above table illustrates.

Visibility

Visibility observations are usually taken several times a day according to specified tables. Mean data are usually compiled upon the basis of the old scale, but it should be noted that a new scale has recently been proposed by the World Meteorological Organization.[2] These observations are compiled on a frequency percentage basis, usually in the form of mean monthly visibility frequencies at 07, 13 and 18 hours

[1] See *Averages of Humidity for the British Isles* (M.O. 421, 1938, reprinted 1949).
[2] For the old scale, see *Instructions for the Preparation of Weather Maps, with Tables of the Specifications and Symbols* (London, 1946). For the new scale, see Meteorological Office, *International Meteorological Code* (London, 1950).

respectively.[1] Occasionally tables are compiled which give the frequency of different degrees of visibility, expressed as a percentage of
the total number of winds from each of several directions.

Synoptic Charts and Weather Summaries

Reference has already been made to the value of synoptic charts and
of weather summaries in providing an occasional salutary corrective
to the abstract concept of climate associated with the element-by-
element analysis of mean data. Weather charts are valuable in another
way, in that they provide a time record of total weather conditions;
this can be analysed in much the same way as mean data to provide a
more comprehensive concept of average weather in different places
at different times.[2]

Analysis of information about total weather conditions recorded in
weather charts since about 1900 for Great Britain, parts of Europe
and North America, undertaken in the first place by weather fore-
casters, has proved to be of great significance to the climatologist.
The relationship between pressure systems, direction of air-flow, and
types of weather can, it is true, be laboriously established by analyses
of mean data, but the problem is better dealt with by reference to
synoptic charts. The object of such analyses is to establish the 'ingredients' of climate in the form of characteristic types of weather, and to
find the seasonal and annual incidence of such types of weather.

The Calendar of Singularities. Singularities in the weather which are
sufficiently regular in their occurrence from year to year can be tabulated in the form of a calendar. Examples of such singularities in
British weather are as follows:

(a) Late Autumn Rains – last week in October and first fortnight in Novem-
    ber;
(b) Early Spring Anticyclones – third week in March;
(c) Northerly Weather – first fortnight in May.[3]

---

[1] Examples of such tables are given in Meteorological Office, *The Bay of Biscay and
the West Coast of Spain and Portugal* (London, 1944).

[2] See the following publications of the Meteorological Office: the *Daily Weather
Report;* the *Monthly Supplement* to the Daily Weather Report; the *Daily Aerological
Record* (contains information about upper air conditions); and *Monthly Weather Report*
(contains summaries from about 420 stations in the British Isles). A detailed analysis,
illustrated with 81 maps and diagrams, of the British Daily Weather Reports is provided by J. A. Taylor and R. A. Yates, *British Weather in Maps* (London, 1958).

[3] H. H. Lamb, 'Types and Spells of Weather in the British Isles', *Quarterly Journal
of the Royal Meteorological Society*, vol. 76, no. 330 (London, 1950).

No completely reliable calendar of singularities has yet been compiled, but work already done does indicate the episodal changes in our weather in contradistinction to broad seasonal changes.

Characteristic Types of Weather based on Pressure Systems. The frequency with which certain types of weather is associated with specific isobaric patterns suggests a profitable approach to the analysis of weather charts. E. G. Bilham (op. cit.) has indicated how E. Gold's classification of pressure systems for forecasting[1] may be simplified for climatological purposes into seven categories of weather, the average seasonal frequency of which can be determined by an examination of weather charts.

Spells of Weather. Analyses of daily weather charts have been made from time to time with the object of defining 'natural seasons' in terms of the frequency of long spells of weather. A 'natural season' in this sense is characterized by a high frequency of persistent weather, and 'average natural seasons' are demarcated by short spells of weather. H. H. Lamb, on the basis of such an analysis, suggests a seasonal division of the British Isles into five instead of four (see p. 222).

Data in daily and monthly weather charts may ultimately be utilized for the purpose of defining climatic regions. For example, J. R. Borchert has shown that such data might be applied to the definition of the North American Grassland.[2] In his case, the data extracted from the weather charts were used to determine the mean frequency of air-flow from different directions over the interior of the North American continent (see p. 253).

Measurement of Streamflow and other Hydrological Data

Data concerning streamflow and rates of run-off are usually returned in terms of cubic feet or metres of water per second for specific gauging stations. This information is published by various local water authorities. In Great Britain the data are compiled in an annual publication, *The Surface Water Yearbook*, published by the Ministry of Housing and Local Government and Scottish Office. Apart from basic information

[1] E. Gold, 'Aids to Forecasting', *Geophysical Memoir*, no. 16 (H.M.S.O., London 1920). See also two papers by R. B. M. Levick, 'Fifty Years of English Weather', and 'Fifty Years of British Weather', *Weather*, vol. 4, pp. 206–11, vol. 5, pp. 245–7 (London, 1949 and 1950).

[2] J. R. Borchert, 'The Climate of the North American Grassland', *Annals of the Association of American Geographers*, vol. 40, no. 1, pp. 1–39 (Lancaster, Pa., 1950).

concerning streamflow, hydrologists need a great deal of other data including all available climatological data within a specific catchment. Other vital measurements include (1) water-table level which depends to a large extent on accurate readings of well levels; (2) soil moisture content; there are various ways in which this measurement may be approached, either by direct measurement of the actual water content of the soil and its specific location, or water content may be derived indirectly through a measurement of soil moisture suction; (3) infiltration rates which involve the use of infiltrometers. These measure directly the rate at which water applied to a small clearly defined area is absorbed by the soil surface.

Hydrological studies are becoming increasingly dependent on the direct measurement in the field of a large number of variables. For this purpose experimental catchment areas are used. Some of the information from the Hull University experimental catchment in the Plain of Holderness is shown in Fig. 85, several variables being shown over a period of time.

ISOPLETH MAPS

Representation of climatic data by isopleths is the most important single cartographic method used by climatologists. It can be applied almost equally effectively to any aspect of climate. Isopleths may be interpolated (see p. 41) for places having the same mean values of temperature (isotherms), of rainfall (isohyets), of pressure (isobars), of sunshine (isohels), of frost (isorymes), of clouds (isonephs), of relative humidity and so on (Figs. 70–2).

Isobars

In the case of isobars, interpolation of observed daily values requires some appreciation of the structure of pressure systems, otherwise the delineation of fronts is likely to be obscured by over-simplification. The most practical manner of overcoming this particular difficulty is to attempt to draw the isobars from the tabular observations given in daily weather summaries, afterwards comparing the result with that given on the published weather map. Errors in interpolation may then be corrected and gradually the art of interpolation can be mastered.[1]

[1] For a full discussion of the problems of drawing isobars, see S. Petterssen, *Weather Analysis and Forecasting* (New York, 1940).

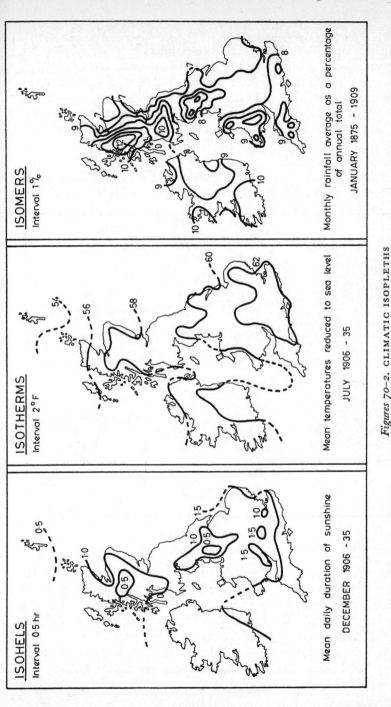

ISOHELS
Interval 0·5 hr

0·5

1·0

1·0

0·5

1·5

1·0

0·5

1·5

1·5

1·0

Mean daily duration of sunshine

DECEMBER 1906 - 35

ISOTHERMS
Interval 2°F

54

56

58

60

62

Mean temperatures reduced to sea level

JULY 1906 - 35

ISOMERS
Interval 1%

8

8

8

10

9

9

9

9

10

12

10

10

10

10

Monthly rainfall average as a percentage
of annual total

JANUARY 1875 - 1909

Figures 70–2. CLIMATIC ISOPLETHS

Based on (1) E. G. Bilham, *The Climate of the British Isles* (London, 1938); and (2) H. R. Mill and C. Salter, 'Isomeric Rainfall Maps
of the British Isles', *Quarterly Journal of the Royal Meteorological Society*, vol. 41 (London, 1915).

Isobars may also be drawn to depict mean pressure systems, but these are less valuable than those representing typical isobaric patterns associated with certain weather conditions. Attention may be drawn here to two maps showing the distribution of typical weather conditions over the surface of the earth at the time of the solstices.[1] The conditions are shown by isobaric patterns, with their fronts also delineated.

Isohyets

The interpolation of isohyets presents special problems. An example with reference to a specific region will serve to exemplify some of these

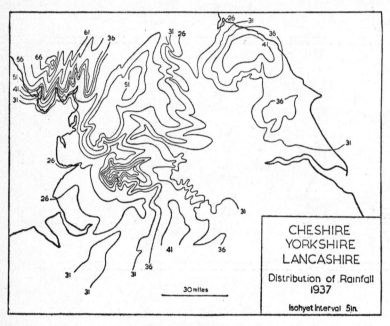

Figure 73. ISOHYETS
Source of data: Meteorological Office, *British Rainfall, 1937*, pt. 3 (London, n.d.).

problems. Fig. 73 shows the distribution of rainfall in 1937 in Lancashire, Yorkshire and Cheshire. The data were extracted from *British Rainfall, 1937*; there are many hundreds of stations given in the tables, but only those which received daily means were selected, and less reliable stations were ignored unless gaps had to be filled in the scatter of stations. The total rainfall of each station selected was plotted in

[1] M. A. Garbell, *Tropical and Equatorial Meteorology* (London, 1947).

pencil on a base-map on a scale of 10 miles to 1 inch. V. Conrad
advocates the use of base-maps on a smaller scale than this, but if
they are used it becomes difficult to make allowances for modifications
due to relief factors.[1]

Selecting the isopleth interval presents something of a problem. The
interval to be selected depends on the range of rainfall totals to be
plotted, the character of the region under consideration, that is, whether
coastal or interior, flat or mountainous, the scale of the base-map, and,
finally, the aim of the map, i.e. whether it is intended to show general
distribution or significant features of the distribution.

In Fig. 73 the general distribution of rainfall has been shown by
means of a progressive arithmetical interval of 5 inches. The isohyets
were carefully drawn with reference not only to the plotted values (see
p. 40) but also to the localized effects of relief. For this reason, the
draft map was superimposed on a good relief map and the following ten-
dencies, which apply generally in the British Isles, were allowed for:[2]

(a) total rainfall increases with height (average increase from one-and-a-half
    to two inches per hundred feet at the coast and rather less than one-half
    to three-quarters of an inch inland);
(b) maximum precipitation normally occurs just beyond the crest on the
    leeward side;
(c) increase in precipitation with height begins before the actual rise in
    elevation and the rate of increase is more than proportional to the
    increase of the gradients;
(d) valleys surrounded by mountains have high rainfalls;
(e) in the case of a valley running in the same direction as the prevailing
    rain-bearing winds, a tongue of lower rainfall will extend up the valley;
(f) in the case of a valley which lies transverse to the rain-bearing winds,
    rainfall will be heavier on the valley side facing the wind;
(g) in Britain the prevailing rain-bearing winds should be taken to be
    west-south-westerly.

Shading and Tinting. Finally, the distribution of the isohyets may be
clarified if intervening areas are shaded. Layer tinting is effective, but
if a variety of colour tints is employed the range and sequence of
colours need careful consideration. An effective scheme is one which
represents low rainfall in red and progressively heavier rainfall in
shades of brown, yellow, green, blue and white.[3] On the maps in

[1] V. Conrad, op. cit. (1944).
[2] See J. Glasspoole, 'The Rainfall of Norfolk', *British Rainfall, 1928* (London, 1929);
and 'The Rainfall of the Forth Valley and the Construction of a Rainfall Map',
British Rainfall, 1915 (London, 1916).
[3] See the rainfall maps in *Atlas de France*, Comité National de Géographie (Paris,
1933).

British Rainfall, 1943–45, four shades of red and four shades of blue are used, but unfortunately dark red is used for areas of lightest rainfall and light red for areas of heavier rainfall; the maps would have been more effective had the representation been reversed.

Isopleths of Duration

Isopleths can be used effectively to show the distribution of places experiencing a similar duration of particular mean weather conditions. For example, the mean duration of a growing season can be illustrated in this way, as revealed by the number of days when the average daily temperature exceeds 6° C. The growing season of any particular crop can be mapped in the same way, for example, cotton, where the critical temperature is 17° C. Similarly, the mean duration of the season of killing frosts, the mean duration of snow cover,[1] the mean duration of special rainfall intensities, and the mean daily duration of sunshine,[2] can also be illustrated by isopleths.[3]

Date Isopleths

Duration and seasonal changes of mean weather conditions may also be depicted by isopleths which join places experiencing similar changes in temperature, rainfall, etc., on the same date (Fig. 74). Date isopleths of this kind are effectively used in showing climatic distributions and seasonal changes over the great land masses.[4] Date isopleths are commonly used to show the progress of a monsoon in terms of mean dates of arrival of the first rains. They may also be used to delimit climatic zones.[5] Again, they can be used on the basis of median and

[1] See maps in *The Climatological Atlas of Japan* (Tokyo, 1949).

[2] See a series of maps in E. G. Bilham, *The Climate of the British Isles* (London, 1938).

[3] An interesting example of this type of map is by A. A. Miller, in *Transactions and Papers, Institute of British Geographers*, 1951, facing p. 17, which plots by means of isopleths, for the world, the numbers of months with mean temperatures of 43° F (6·1° C) or over, using colour for clarity. See also maps showing the length of the growing period in Australia and Argentina, by J. A. Prescott, J. A. Collins and G. Shirpurkar, 'The Comparative Climatology of Australia and Argentina', *Geographical Review*, vol. 42, p. 124 (New York, 1952).

[4] Good examples may be found in the United States Department of Agriculture, *Atlas of American Agriculture* (Washington, 1936). *The Great Soviet World Atlas* (Moscow, 1938) includes many climatic maps of this type; one series, for example, shows the mean dates of change of daily average temperature through −10° C, −5° C and 0° C in periods both of rising and falling temperatures.

[5] M. Jefferson, 'Standard Seasons', *Annals of the Association of American Geographers*, vol. 28, pp. 1–12 (Lancaster, Pa., 1938), makes use of them in this connection. See also H. M. Kendall, 'Notes on Climatic Boundaries in the Eastern United States', *Geographical Review*, vol. 25, pp. 117–24 (New York, 1935).

percentile values to indicate probability of certain weather conditions on particular dates, for example, a fifty per cent probability of frost, etc.[1]

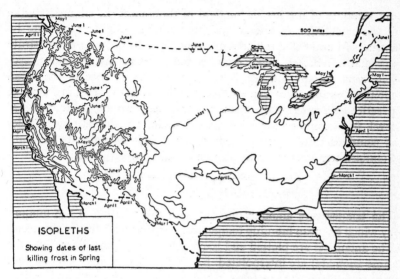

Figure 74. DATE ISOPLETHS FOR THE UNITED STATES

Based on United States Department of Agriculture, *Atlas of American Agriculture*, Section I, 'Frost and the Growing Season' (Washington, 1918).

The isopleths are interpolated on the basis of average dates of the last killing frost in spring compiled from records of about 4,000 stations, of which about 700 cover the full 20-year period (1895–1914).

Frequency Isopleths

These isopleths depict places which have the same mean frequency of climatic phenomena, expressed as the number of days or the number of occasions per year, or other period, on which they occur. Some of the more obvious examples include the mean number of wet days, rain-days and droughts, and of days of fog[2] and snow.[3] Less obvious examples include the mean number of days with ice-bound rivers, the

---

[1] See in this connection E. E. Lackey, 'A Variability Series of Isorymal (Equal Frost) Maps of Alaska', *Geographical Review*, vol. 26, pp. 135–8 (New York, 1936).

[2] See map in E. G. Bilham, op. cit. (1938), p. 275.

[3] See examples in Admiralty Geographical Handbooks, *France*, vol. 1, and *Germany*, vol. 1.

mean number of months per year with air-streams from specific directions,[1] and the mean frequency of characteristic types of weather.

Isanomals

An example of the use of 'method of differences' in climatic studies is the calculation of anomalous temperatures and their plotting by means of isanomals (isopleths of anomalies). In order to demonstrate seasonal differences, mean monthly temperatures are generally used. The *norm* for any latitude is taken to be the average of the mean monthly temperature experienced in that latitude. This is found in practice by averaging the mean monthly temperature, reduced to sea-level, for a number of well dispersed stations in the latitude. Tables of standard distribution of temperatures with latitude are helpful.[2] Standard temperatures may also be found by the formula:[3]

$$t = -17.8 + 44.9 \cos^2 (\theta - 6.5)^\circ C,$$
where θ *is the latitude.*

The temperature anomaly for any particular station is the amount by which its mean monthly temperature departs from that of the norm. The anomalies are plotted and isanomals (lines joining places with the same anomaly) are interpolated. Regions of high positive anomaly are known as *pleions* (*thermopleions* in the case of temperature) and regions of high negative anomaly as *meions* or *antipleions* (*thermomeions* in the case of temperature). Positive anomalies are sometimes tinted red for effect, and negative tinted blue.[4]

An application of this method is the plotting of isanomals to show anomalies of temperature conditions in a mountainous country, particularly between northward- and southward-facing slopes. In this case temperatures are not reduced to sea level, but are related to the mean temperatures of a number of selected stations dispersed over the countryside at different elevations.[5]

[1] P. R. Crowe makes use of frequency isopleths of 50, 70 and 90 per cent constancy to show the limits of the trade-winds in 'The Trade Wind Circulation of the World', *Institute of British Geographers. Transactions and Papers, 1949.* Publication no. 15, pp. 21–76 (London, 1951).

[2] Such a table is to be found in V. Conrad, op. cit. (1950).

[3] J. D. Forbes, 'Inquiries about Terrestrial Temperature', *Transactions of the Royal Society of Edinburgh*, no. 22 (1859).

[4] See B. C. Wallis, 'Geographical Aspects of Climatological Investigation', *Scottish Geographical Magazine*, vol. 30, pp. 356–69 (Edinburgh, 1914), in which the idea of anomalies is illustrated by twelve monthly world maps.

[5] See a map in V. Conrad and L. W. Pollak, op. cit. p. 278 (1950), entitled 'Isanomalies of Vegetative Period in Switzerland'.

Equipluves

Equipluves are lines joining places with the same pluviometric coefficients. The pluviometric coefficient for any month is arrived at by expressing the mean monthly rainfall total for a given station as a ratio of the hypothetical amount equivalent to each month's rainfall were the total rainfall for that station to be equally distributed throughout the year. For example, Ponta Delgada (37° 44′ N., 25° 40′ W.) has the following mean monthly rainfalls in inches:

J	F	M	A	M	J	J	A	S	O	N	D	Total
3·0	2·9	2·5	2·1	2·1	1·3	0·8	1·4	2·5	3·3	3·4	3·2	28·5

Using the above values:

$$The\ pluviometric\ coefficient\ for\ January = \frac{3\cdot0}{28\cdot5\left(\dfrac{31}{365}\right)} = 1\cdot4$$

The coefficients may be expressed as percentages in order to eliminate decimals.[1]

Equivariables

Isopleths may be used with effect in plotting the distribution of places with similar deviations from their average weather conditions.[2] For example, the mean annual rainfall figures for any given station do not reflect the reliability of rainfall at that station, for the figures may represent the means of a number of actual totals widely dispersed on either side of the mean. The exact degree of variability is difficult to compute, and various statistical formulae have been derived to represent it in the form of a *coefficient of variability* (see Appendix, p. 485). One such formula used in its calculation is as follows:

$$CV = \frac{\sigma}{\bar{x}}\ (100)$$

where CV = coefficient of variability, σ = standard deviation and x̄ = the mean value.

[1] B. C. Wallis, 'The Rainfall of Java', *Scottish Geographical Magazine*, vol. 33, pp. 108–19 (Edinburgh, 1917).

[2] For an example, see A. V. Williamson and K. G. T. Clark, 'The Variability of the Annual Rainfall of India', *Quarterly Journal of the Royal Meteorological Society*, vol. 57, pp. 43–56 (London, 1931). For cartographical portrayal of standard deviations in temperature, see A. R. Sumner, 'Standard Deviation of Mean Monthly Temperatures in Anglo-America' (with twelve monthly maps), *Geographical Review*, vol. 43, pp. 50–9 (New York, 1953).

Standard deviation (see Appendix, p. 486) is obtained by the formula,

$$\sigma = \sqrt{\frac{\Sigma(x - \bar{x})^2}{n}}$$

where σ = standard deviation, $(x - \bar{x})$ = deviations from the mean, and n = the total number of observations upon which the mean is calculated.

A simple example is considered below.

Year	Rainfall (inches)	x	x^2
1920	2·0	−2	4
1921	4·0	0	0
1922	8·0	+4	16
1923	2·0	−2	4
1924	4·0	0	0
Total	20·0	—	24·0

Arithmetic mean = 4·0 *inches*

$$\therefore \sigma = \sqrt{\frac{24}{5}} = \sqrt{4\cdot8} = \pm 2\cdot2 \; inches$$

$$\therefore CV = \left(\frac{2\cdot2}{4\cdot0}\right) 100 = 55 \; per \; cent$$

The coefficient of variability is usually expressed as a percentage as above. Generally speaking, it is found that places with low average rainfall have a high coefficient of variability, although not necessarily so.

Median values of rainfall (see p. 244) can also be used to measure variability. C. E. Hounam, in an analysis of Australian rainfall, uses the formula:[1]

$$CV = \left(\frac{Inter\text{-}quartile\ Range}{Median}\right) 100$$

Isopleths interpolated to join places with equal coefficients of variability might be termed *equivariables*.

In some cases it is of interest to show how the rainfall totals of one particular year differ from the means of previous years. Coefficients need not be worked out in this case, but the observed one-year totals

[1] Commonwealth of Australia, Meteorological Bureau, *Climate of the West Australian Wheat Belt with Special Reference to Rainfall over Marginal Areas* (Melbourne, 1945).

can be expressed simply as percentages of mean totals. Isopleths are then interpolated.[1]

Equivariables can be drawn equally well to show variability of the mean values of temperatures, variability of the mean date of killing frost, variability of the mean length of growing season and so on.[2] Such maps obviously have great value in their application to agricultural problems, and they are also of use in the delimitation of climatic regions.

Equicorrelatives

Isopleths indicating correlations of climatic data may be termed equicorrelatives. It is often difficult to correlate two or more sets of climatic data by inspection. For example, sets of annual totals of rainfall over a period of years for two stations might exhibit certain relationships, but such related characteristics can only be measured with a fair degree of accuracy by statistical methods. If the degree of relationship between two sets of rainfall statistics can be expressed in the form of a single index, then the distribution of places having similar relationships can be easily plotted on a map. Statisticians refer to such indices as a *correlation coefficient* (see Appendix, p. 504). They establish the degree to which two sets of statistics are related, the inference being in the case of rainfall statistics that some common cause is at work if the correlation is relatively close. Thus correlations can be used as a basis for climatic differentiation.

The Coefficient of Correlation. A short method of computing correlation coefficients for two or more sets of annual rainfall data is based on the formula:

$$r = \frac{\frac{1}{n}\Sigma(x - \bar{x})\,(y - \bar{y})}{\sigma_x\,\sigma_y}$$

where r = the coefficient of correlation, x = value of annual precipitation for one station, y = value of precipitation for the second station, $\bar{x}$ = mean annual value for the first station, $\bar{y}$ = mean annual value for the second station, σ_x and σ_y = standard deviations (see p. 486), and n = number of years of observations.

Isopleths may be interpolated to join places of equal correlation

---

[1] A number of maps of this type is to be found in the *British Rainfall* series, published by the Meteorological Office.

[2] Examples of such maps may be found in the *Atlas of American Agriculture*, op. cit. (1936).

coefficient of rainfall after such coefficients have been plotted on the map.[1] Coefficients with a value approaching $+1$ may be considered as having a close direct relationship; in the case of those approaching -1, an inverse relationship is to be assumed, while values near zero indicate chance relationships only.

Isomers

The term 'isomeric' is often used to describe a method of studying regional variations in the relative proportions of rainfall falling in each month of the year.[2] The mean monthly rainfalls for each station in the region under consideration are 'weighted' and expressed as percentages of the respective mean annual rainfalls. Twelve maps can then be drawn to show the distribution of comparative proportions of rainfall for each month by means of isopleths joining places with similar percentages (see Fig. 72). The method is clearly capable of wide applications in the case of other climatic data.

Isopleths of Temperature Range

The isopleth method can be used most effectively to show the distribution of the mean range of annual temperature,[3] the mean range of diurnal temperature,[4] and the range of extreme temperatures, diurnal and annual, together with the highest and lowest extremes recorded.

Isopleths of Accumulated Temperature

Not a great deal of work has been done by climatologists on accumulated temperatures,[5] but they may have significance in schemes for classification of climates, in the study of relationships between plant activity and temperature conditions, and in the analysis of climatic cycles. They are usually calculated from the mean daily temperatures experienced at any one station over a period of time. The values above

[1] For a good example of the application of this method, see E. E. Foster, 'A Climatic Discontinuity in the Areal Correlation of Annual Precipitation in the Middle West', *Bulletin of the American Meteorological Society*, vol. 25, pp. 299–306 (Milton, Mass., 1944).

[2] H. R. Mill and C. Salter, 'Isomeric Rainfall Maps of the British Isles', *Quarterly Journal of the Royal Meteorological Society*, vol. 41 (London, 1915).

[3] See maps in W. G. Kendrew, *The Climates of the Continents*, p. 70 (Oxford, 1927) and in E. G. Bilham, op. cit., p. 151, (1938).

[4] See an interesting map in E. G. Bilham, op. cit., p. 167 (1938).

[5] But see the full and well-illustrated article by S. Gregory, 'Accumulated Temperature Maps of the British Isles', *Transactions and Papers, 1954: Institute of British Geographers*, no. 20, pp. 59–73 (London, 1954).

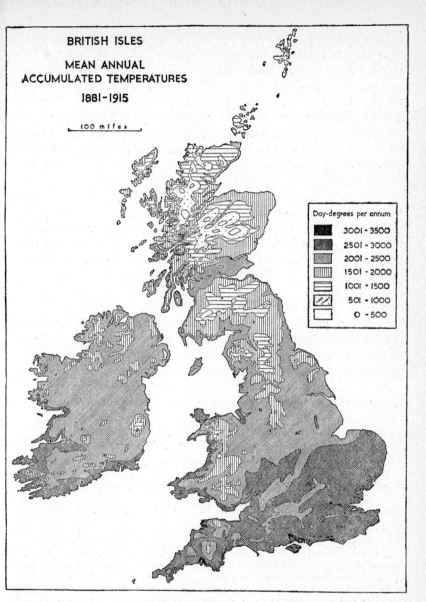

Figure 75. ISOPLETHS OF ACCUMULATED TEMPERATURE

Based on map compiled by S. Gregory, *Transactions and Papers, 1954: Institute of British Geographers*, no. 20, p. 64 (London, 1954). The data were extracted from Meteorological Office, *The Book of Normals of Meteorological Elements for the British Isles for periods ending 1915* (London, 1919). Values are given in day-degrees Fahrenheit above 42·8°. He also included eight maps of monthly values.

or below significant thresholds, such as 0° C, 6° C or 18° F, are added together for the period under consideration and expressed as a single accumulated value, usually in terms of day-degrees. Thus, if 6° F is taken as the threshold and the mean daily temperature for 1 January is 9° C, it would count as +3 day-degrees towards the final total. To cut down the laborious calculations necessitated by the use of daily means, estimates based on monthly means may be used. Thus if the January monthly mean is 8° F, it would count as

$$31 \times 2 = 62 \; day\text{-}degrees$$

towards the final total.[1]

Distribution of accumulated temperatures may be shown by isopleths interpolated in the usual manner (Fig. 75). They may be drawn to indicate accumulations for specific months or seasons, or for the whole year.[2]

Isopleths of Aridity and Moisture

Various ingenious formulae have been devised both by climatologists and botanists to give an indication of climatic aridity which might be of use both in the rational delimitation of climatic regions and in agricultural planning. Such indices have particular reference to the study of relationships between natural vegetation and climate. The simplest index is perhaps that suggested by R. Lang:[3]

$$Rain \; factor = \frac{Annual \; precipitation \; in \; millimetres}{Mean \; annual \; temperature \; in \; °C}$$

A. A. Miller also devised a simple, though effective, index of humidity in terms of $\frac{T}{R}$, where T is mean annual temperature in ° F, and R is mean annual rainfall in inches. He plotted values for 2,000

[1] For a more refined method of calculation of accumulated temperatures, see Meteorological Office, Form 3300, *Tables for the Evaluation of Daily Values of Accumulated Temperature above and below 42° Fahrenheit from Daily Values of Maximum and Minimum Temperature*. Graphical methods of calculating temperature accumulation by means of an *ogive* or cumulative temperature curve are discussed by C. E. P. Brooks, *Climate in Everyday Life*, p. 238 (London, 1950). Data required for exact measurement comprise hourly readings of temperature, but Brooks' method only requires certain selected means and extremes.

[2] See, for example, A. A. Miller's attractive folding map (in colour) of 'month degrees' above 43° F (6·1° C) in *Transactions and Papers, Institute of British Geographers*, 1951, facing p. 19.

[3] R. Lang, *Verwitterung und Bodenbildung als Einführung in die Bodenkunde* (Stuttgart, 1920).

stations, and drew isopleths of values 1 to 5 (<1, perhumid; 1–2, humid; 2–3, sub-humid; 3–5, semi-arid; >5, arid).[1]

Slightly more complex is de Martonne's index,[2] which has been used with great effect for the plotting of a series of pleasing maps in the French National Atlas.[3] De Martonne's index is as follows:

$$I = \frac{P}{T + 10}$$

where I is the index of aridity, T is the mean annual temperature in ° C, and P is the mean annual rainfall in millimetres.

Monthly indices are given by the formula:

$$i = \frac{p \times 12}{t + 10}$$

where t = mean monthly temperature and p = mean monthly rainfall.

The two formulae are made commensurate so that annual and monthly maps may be compared.

C. W. Thornthwaite has made convincing use of empirical relationships between rainfall, temperature and evaporation in the classification of the climates of North America, and his methods are being applied to the climates of other parts of the world. Thornthwaite's formula for finding the monthly precipitation-evaporation ratio, is:

$$P/E = 11 \cdot 5 \left(\frac{p}{t - 10} \right)^{\frac{10}{9}}$$

The summation of the twelve monthly values, gives the value for the year.[4] The index values range from 0 to about 150. They are divided into five categories – wet, humid, subhumid, semi-arid and arid. Thornthwaite has also produced corresponding temperature categories by devising a thermal efficiency factor, calculated from the formula,

$$i = \frac{t - 32}{4}$$

where i = thermal efficiency factor and t = mean monthly temperature (°F).

[1] A. A. Miller, in *Transactions and Papers, 1951: Institute of British Geographers*, no. 17, facing p. 17 (London, 1951); the isopleths are drawn in red, the infilling in five colours.
[2] E. de Martonne, 'Aréisme et Indice d'Aridité', *Comptes Rendus de l'Académie de Science de Paris*, vol. 182 (Paris, 1926).
[3] For another application, see J. Gottman, 'Une Carte de l'Aridité en Palestine', *Annales de Géographie*, vol. 45, pp. 430–3 (Paris, 1936).
[4] C. W. Thornthwaite, 'Climates of North America', *Geographical Review*, vol. 21, pp. 633–55 (New York, 1931).

Thornthwaite has further pointed out in one of his later papers that evaporation and transpiration from plants (evapo-transpiration) cannot be satisfactorily calculated from mean temperature figures, but that the potential rate of evapo-transpiration may be calculated with the aid of specially devised formulae, which can be solved with the aid of a nomograph. Moreover, there are certain adjustments to make to the result because of variations in the duration of daylight and in the lengths of months.[1] Once potential evapo-transpiration has been ascertained, it is comparatively easy to calculate the water surplus and deficiency for any station during the course of the year (see p. 227). Increased understanding of the physics of the evaporation process has enabled H. L. Penman to determine potential evapo-transpiration by calculating the heat balance at the surface of the earth[2] and his technique has found useful application in the calculation of irrigation requirements.[3]

The various indices discussed above lend themselves readily to cartographical expression by means of isopleths, and the completed maps effectively demonstrate fundamental climatic differences from one region to another.[4]

COLUMNAR DIAGRAMS

Columnar diagrams, because of their clarity and because of their ease of construction, are particularly effective in the depiction of certain aspects of climatic data. They can be used to show the rhythm of diurnal and seasonal changes, the distributional of regional variations by superimposition upon locational base-maps (Fig. 76), and the range and variability of various climatic elements.

---

[1] Formulae, nomographs and tables are to be found in C. W. Thornthwaite, 'An Approach towards a Rational Classification of Climate', *Geographical Review*, vol. 38, pp. 55–94 (New York, 1948).

[2] H. L. Penman, 'Evaporation over the British Isles', *Quarterly Journal of the Royal Meteorological Society*, vol. 76, pp. 372–83 (London, 1950).

[3] R. T. Pearl, 'The Calculation of Irrigation Need', *Ministry of Agriculture, Technical Bulletin*, no. 4 (H.M.S.O., 1954).

[4] J. A. Prescott, J. A. Collins, G. R. Shirpurkar, 'The Comparative Climatology of Australia and Argentina', *Geographical Review*, vol. 42, pp. 118–33 (New York, 1952), use the index $P/s.d.^{0.75}$, where P is the rainfall and $s.d.$ the saturation deficit in inches, to produce twelve monthly maps of Australia and Argentina to show the rainfall efficiency.

Simple Columnar Diagrams

In its simplest form each vertical column represents a number of units of rainfall, sunshine, etc., for a particular period of time, for

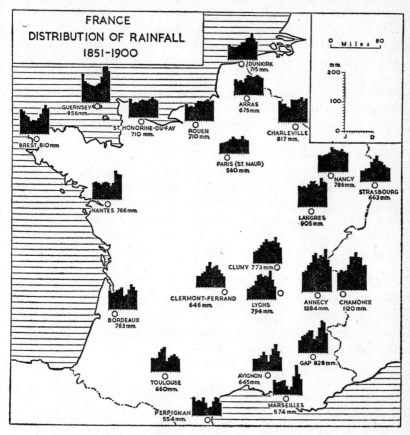

Figure 76. LOCATED COLUMNAR DIAGRAMS OF RAINFALL

Source of data: E. Alt, 'Klimakunde von Mittel und Südeuropa', vol. 3, pt 1, pp. 150–6, W. Köppen and R. Geiger, *Handbuch der Klimatologie* (Berlin, 1938).

example, by monthly periods to show seasonal variation, by 24-hour periods to show daily variation. In the latter case, columns representing hours of darkness are sometimes shaded black, and those representing hours of daylight are left white.[1] For seasonal comparisons, data relating

[1] See examples in Sir Napier Shaw, *Drama of the Weather* (2nd edition, Cambridge, 1939).

to calendar months may be used, but where some detailed analysis of seasonal distribution of data is being attempted, it becomes necessary to make allowances for differences in the number of days in each month (see p. 194).

Simple columnar diagrams may be constructed to show total amounts of rainfall (Fig. 76), of evaporation, of run-off, and of sunshine, or they may be adapted to show frequencies, for example, of the number of days with fog, snowfall or hail, the number of wet days, rain-days, and days with good visibility.[1] Accumulated temperatures are also often depicted in this way.[2] Significant heights, such as the height of freezing level, and the depth of frost penetration below ground, are also suitably depicted by means of the simple columnar diagram.

Percentage Columnar Diagrams

Where it is required to compare inter-regional mean monthly variations of specific weather elements, percentages are more effective than absolute totals. The vertical columns are made to represent, for example in the case of rainfall, the percentage of the yearly precipitation for each month, or, in the case of sunshine, the actual sunshine experienced in each month as a percentage of the possible sunshine for that month.

Superimposed Columnar Diagrams

Columnar diagrams of different stations may be superimposed for direct comparison (Figs. 77, 78). Superimposition of various elements can be made on the same diagram, for example, of mean rainfall, evaporation, percolation and run-off.[3]

Compound Columnar Diagrams

Columnar diagrams may be used with effect to show the 'make-up' of certain averages. Thus a column showing precipitation for December for a station in northerly latitudes may be subdivided to show the relative importance of snowfall compared with rainfall. Similarly, columns may be subdivided to show moderate and poor visibility, to

---

[1] Numerous examples of this type of columnar diagram are used to illustrate the regional monographs on weather published by the Meteorological Office, such as *Weather on the West Coast of Tropical Africa* (M.O., 492, 1949).

[2] See a good example in G. Manley, 'The Range of Variation in the British Climate', *Geographical Journal*, vol. 117, pp. 43–68 (London, 1951).

[3] Some good examples are in F. Shreve, 'Rainfall, Runoff and Soil Moisture under Desert Conditions', *Annals of the Association of American Geographers*, vol. 24, pp. 131–56 (Lancaster, Pa., 1934).

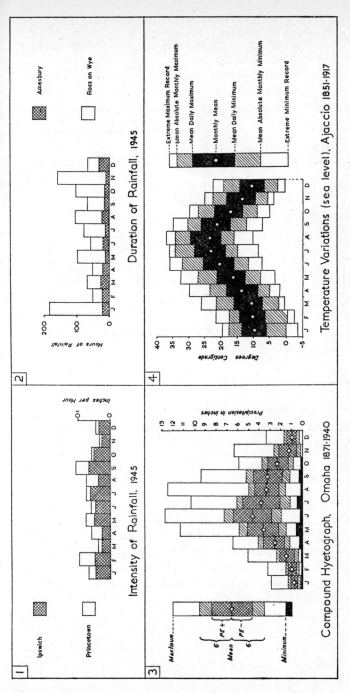

Figures 77–80. COLUMNAR DIAGRAMS OF RAINFALL AND TEMPERATURE

1 and **2** consist of columns superimposed for purposes of comparison. Source of data: Meteorological Office, *British Rainfall, 1943–1945* (London, 1949). It exemplifies a compound diagram to illustrate the maximum and minimum rainfall over a period of years for each month, together with the standard deviation (σ) and the probable error (*PE*). The source of data for **4** is *World Weather Records*, Smithsonian Miscellaneous Collections, vol. 79. It exemplifies a compound diagram to illustrate monthly variations in temperature over a period of years.

show wind direction, or to show the amount of rain falling at night compared with that in the daytime.

Deviations from mean conditions may also be effectively depicted by columnar devices. For example, columns may be constructed to show the maximum, the mean and the minimum rainfall in each

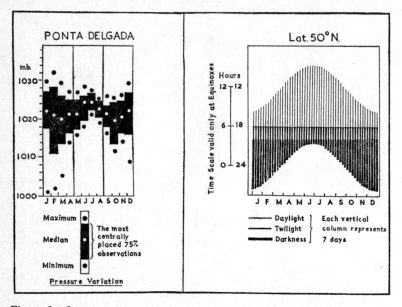

Figures 81, 82. COLUMNAR DIAGRAMS OF PRESSURE AND OF DAYLIGHT
Based on Meteorological Office, *Weather in Home Waters and the North-Eastern Atlantic*, vol. 2 (London, 1944).
    Figure 81 is a compound diagram to show monthly variations in pressure, and Fig. 82 is designed to give an impression of the weekly variations in daylight.

month, and still further elaboration may be introduced by indicating the standard deviation and the probable deviation of rainfall from the accepted mean (Fig. 79, and see p. 210). Such diagrams are known as *hyetographs*. Some information about dispersion of rainfall may be included by the indication of the position of median and quartile points (see p. 244) Fig. 81 shows a pressure variation diagram based on frequency of pressure conditions in which the maximum and minimum records are indicated, together with the median. A fair idea of the more normal range of variation is given by the area in solid black which covers the most centrally placed 75 per cent of the observations.

Yearly variations in monthly temperature conditions, although usually depicted by line-graphs (see p. 223), may occasionally be more effectively illustrated by columnar devices (Fig. 80). Columns depicting extreme temperatures recorded in each month, the absolute monthly maxima and minima, the daily maxima and minima, and the monthly means give a fairly comprehensive idea of both the range of temperature from month to month and of the range of diurnal temperature during each month.

Analyses of characteristic types of weather made from synoptic records are sometimes capable of graphic presentation in columnar form. It is possible, for example, to show average frequency of specific air-mass types in each month of the year by means of columns subdivided in accordance with the number of types selected for study.[1] Again, the make-up of seasons in terms of weather types may be shown by a compound columnar diagram (Fig. 83).

Special Columnar Diagrams

Columnar diagrams are useful in elucidating aspects of insolation and climate, for example, latitudinal variations in the length of the day, and in the angle of the sun's rays relative to the surface of the earth. A variety of diagrams of this type may be worked out with the aid of almanacs. In Fig. 82 variations in the duration of daylight, twilight and darkness for different weeks of the year have been shown by proportional columns.

Weather Integrals

An unusual use is made of the columnar diagram by Sir Napier Shaw in an illustration entitled 'Nature's Integrals'.[2] It is a type of cumulative polygraph in which the accumulated temperatures, the amounts of sunshine, amounts of rainfall, amounts of evaporation and amounts of daylight for each week in the year are all plotted, but arranged cumulatively so that each amount is represented by columns and successive columns are placed in steps (Fig. 84). The whole diagram summarizes admirably the average expectancy of the chief elements throughout the year.

[1] Thus E. M. Frisby and F. H. W. Green illuminated a quantitative analysis of air-mass types in south-east England with the aid of the columnar technique in 'Further Notes on Comparative Regional Climatology', *Institute of British Geographers, Transactions and Papers, 1949*, Publication no. 15, pp. 141–51 (London, 1951).

[2] *Drama of the Weather* (2nd edition, Cambridge, 1939).

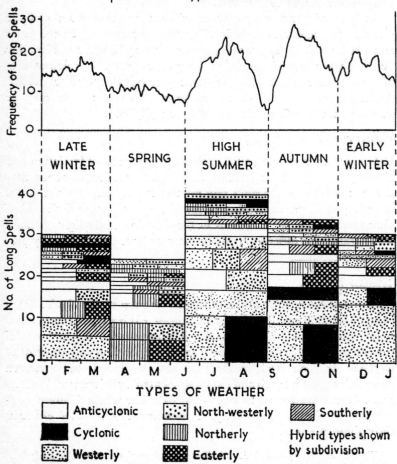

Figure 83. TYPES AND SPELLS OF WEATHER ANALYSED BY MEANS OF A
FREQUENCY CURVE AND DIVIDED RECTANGLES

Source of data: H. H. Lamb, 'Types and Spells of Weather in the British Isles',
Quarterly Journal of the Royal Meteorological Society, vol. 76, no. 330 (London, 1950).

LINE-GRAPHS

A wide variety of graphs is used in the illustration of different aspects of climate, and the methods selected for discussion below have been chosen largely with the idea of stimulating the ingenuity of the individual student rather than of covering the whole field. Once the principles of graphic illustration have been mastered, special graphs may be evolved for special purposes, and tried methods modified and adapted to serve new ends.

Continuous Tracings

Continuous records made by self-recording instruments constitute one type of graph. Such graphs may be of use to the climatologist in illustrating characteristic sequences of weather conditions, for example, those associated with the passage of a depression. Relationships between pressure, temperature, rainfall and wind are self-evident from such graphical records. But continuous tracings are by their very nature of more use in the illustration of weather than of climate.

Simple Line-graphs

In the conventional type of Cartesian graph, temperature, humidity, evaporation, and so on are plotted as ordinates, and months of the year, hours of the day, etc., as abscissae, to show seasonal or diurnal variations. Regional variations in climate can be demonstrated by comparing graphs which portray the conditions for selected type-stations.

Sometimes comparisons of seasonal variability are facilitated if temperature and other data are reduced to percentages (see below). Direct comparisons of varying conditions may also be shown on one graph – for example, by plotting mean January temperature as ordinates, and certain type stations, arranged either in order of magnitude or by geographical location, as abscissae.

Polygraphs

A series of multiple line-graphs, drawn on the same chart to show relationships between two sets of climatic statistics at one station, or between sets of similar statistics for two or more stations, comprise a polygraph.[1]

[1] A most interesting applied polygraph is one depicting, over a period of five months, the rain, sunshine, wind, temperature and dew-point for each day, with the daily

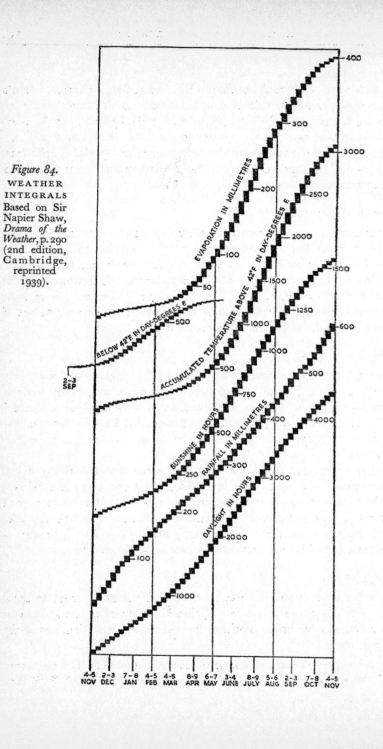

Figure 84.
WEATHER
INTEGRALS
Based on Sir
Napier Shaw,
*Drama of the
Weather*, p. 290
(2nd edition,
Cambridge,
reprinted
1939).

Climatic type summaries constitute the most familiar type of polygraph. These consist essentially of two line-graphs, to indicate temperature and rainfall regimes respectively. In practice, however, to enable a ready distinction to be made between the two, rainfall is shown by a columnar diagram.[1]

Composite rainfall and relief profiles comprise another type of polygraph; both rainfall and heights above sea level are plotted as ordinates, using a common horizontal scale.

Comparative Percentage Graphs. Where climatic elements for a number of contrasting stations have to be compared, the mean annual totals may be reduced to a percentage basis to facilitate comparison, and plotted on the same graph.[2]

Durational graphs show variability in the duration of snowfall, length of growing season, etc., at different stations on the same frame. This technique may be applied, for example, to the study of seasonal duration of snow in an Alpine valley. Mean height of snow-lines for the sunny (*adret, Sonnenseite*) and for the shady (*ubac, Schattenseite*) sides of the valley respectively, are plotted as ordinates, and months as abscissae.

Curve-parallels consist of superimposed graphs in series of related phenomena, for example, of rainfall with wind-run, relative humidity, vapour pressure and evaporation; curve parallels are also used to demonstrate similarities of climatic trends at different stations (Fig. 91). Diverse data including hydrological measurements for particular stations can also be shown in curve-parallels to facilitate comparative study (Fig. 85).

Adjusted Profiles. It may be profitable when studying regional variations in time-lag and in temperature variability to use temperature profiles, the shape of which is adjusted to make them directly comparable. One method of adjustment has been suggested by S. B. Jones in a study of the temperature regions of Hawaii.[3] This consists in reducing all temperature profiles to the same amplitude by adjusting

---

milk-yield at Llangeni and Valley, 1951, Anglesey. This was compiled by F. A. Barnes, and appears in 'Dairying in Anglesey', *Transactions and Papers, 1955: Institute of British Geographers*, no. 21, p. 151 (London, 1955).

[1] The *Great Soviet World Atlas* (Moscow, 1938) has a series of such graphs, distinctively coloured, and related to a world-map of Köppen's climatic regions.

[2] W. G. Kendrew, op. cit. (1927), makes frequent use of this device.

[3] S. B. Jones, 'Lag and Ranges of Temperature in Hawaii', *Annals of the Association of American Geographers*, vol. 32, pp. 68–97 (Lancaster, Pa., 1942).

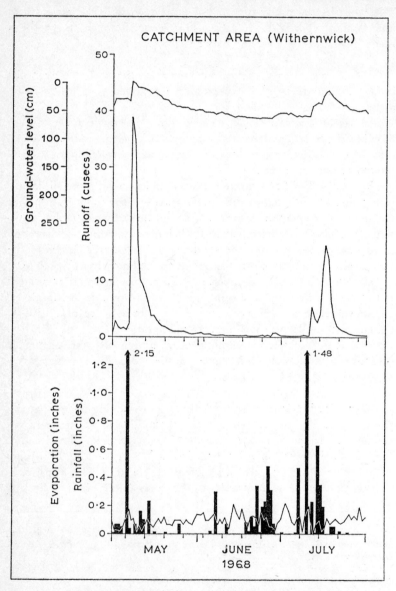

Figure 85. CURVE PARALLELS OF HYDROLOGICAL DATA

Hydrological data for Withernwick, East Yorkshire, are shown for the summer months of 1968. Rainfall is shown by columns.

the mean annual ranges of temperature to an equal size. Monthly and daily temperature profiles may be likewise reduced to the same amplitude. The adjusted profiles can then be placed one above another in a series or superimposed for comparative purposes (Fig. 86).

Diurnal and annual ranges of temperature may also be compared with the aid of adjusted profiles. For example, the mean monthly temperature curve of one station X can be made coincident with that for another Y, and the maxima and minima temperatures are then plotted; station X is plotted in the normal manner, but station Y in terms of deviations from the adjusted curve (Fig. 87).

A further method of facilitating comparison of temperature curves has been suggested by V. Conrad.[1] He made use of W. Köppen's concept of *relative temperatures* to produce a temperature curve in which the variations due to differences in the average temperatures and amplitudes at any two stations could be eliminated. The amounts by which the consecutive monthly averages at any station exceed the temperature of the coldest month are expressed as percentages of the differences in average temperatures between the coldest and the warmest months. These values are then plotted as ordinates and the months as abscissae. The horizontal scale is so graduated as to enable the relative temperatures for the first six months of the year to be compared directly with those for the last six (Fig. 88).

Water surplus graphs are line-graphs showing the superimposition of the mean monthly evapo-transpiration and the mean monthly rainfall. Thus they constitute a valuable means of analysing local climate in terms of seasonal moisture surplus and rainfall efficiency (Fig. 89, and see also pp. 214–16).

Sociographs. A number of ingenious graphical devices in the form of polygraphs have been evolved for the purpose of relating the seasonal rhythm of the climate to the amount and nature of the work done in particular regions. L. Garrard has called these graphs *sociographs.*[2] Graphs showing seasonal rhythm and cropping are of a similar nature.[3]

---

[1] V. Conrad, *Fundamentals of Physical Climatology* (Harvard, 1942).

[2] L. Garrard, 'Sociographs of the Kulymans, Andamanese, etc.', *Studies in Regional Consciousness and Environment*, edited by I. C. Peate (Oxford, 1930). The paper includes a sociograph in which temperature curves, incidences of snow, floods, polar night, etc., are shown with reference to seasonal activities such as fishing, boat-making, marriage festivals, forest activities and so on.

[3] A. Geddes, 'India: (1) The Chota Nagpur Plateau and its Bordering Plains', *Comptes Rendus du Congrès International de Géographie*, vol. 2, 3c, pp. 365–80 (Amsterdam, 1938).

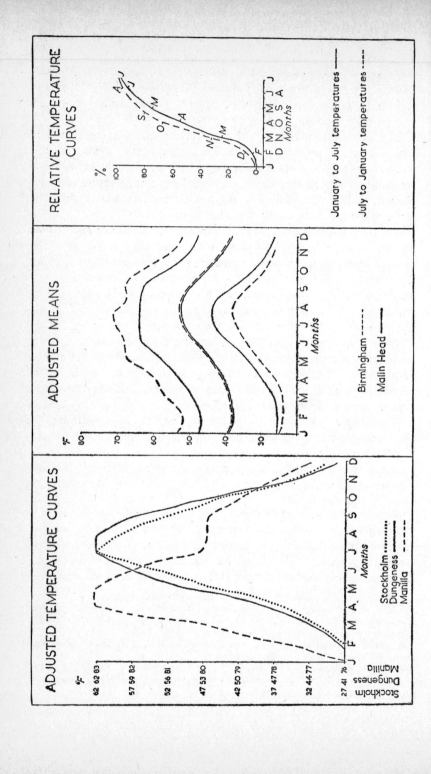

Circular graphs are sometimes better fitted to carry this type of information (see p. 298).

Trend-graphs

The accurate determination of trends of climate is dependent upon the availability of reasonably complete and reliable records over a long period of time. The data have to be scrutinized, and corrections made for possible instrument error, for possible anomalies arising out of the location of the recording stations, and for possible error due to the faulty siting of instruments. Unless data are duly corrected for such errors the finished graphs will record variability in the observation of data rather than genuine trends.[1]

Running Means (see Appendix, p. 490). Trends in temperature and rainfall naturally attract the attention of the climatologist. Graphs, in

Calculation of Three-year Running Means for Temperature (° F)

Date	Annual Means	Running Means		
		1	2	3
1860	62·1	62·1		
1861	69·3	69·3 }65·2	69·3	
1862	64·2	64·2	64·2 }68·4	64·2
1863	71·6		71·6	71·6 }65·1
1864	59·4			59·4

[1] See S. Gregory, 'Regional Variations in the Trend of Annual Rainfall over the British Isles', *Geographical Journal*, vol. 122, pp. 246–53 (1956); and E. C. Barrett, 'Regional Variations of Rainfall Trends in Northern England, 1900–59', *Transactions of the Institute of British Geographers*, pp. 41–58 (London, 1966).

Figures 86–8. ADJUSTED TEMPERATURE PROFILES

Figure 86 (*left*) shows three temperature curves adjusted to the same amplitude to facilitate comparison of the temperature regime.

Figure 87 (*centre*) shows temperature conditions at two stations, Birmingham and Malin Head. The mean curve of one, Birmingham, has been made coincident with that of Malin Head so that variability may better be compared.

Figure 88 (*right*) shows relative temperature curves for Bismarck, N.D., and is based on V. Conrad, *Fundamentals of Physical Climatology* (Harvard, 1942). The amounts by which the consecutive monthly average temperatures exceed the temperature of the coldest month are each expressed as a percentage of the difference between the average temperature of the warmest and coldest months.

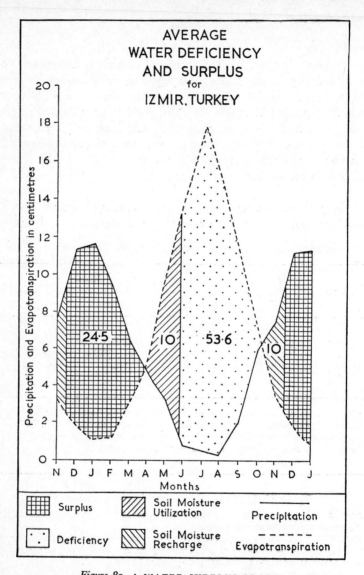

Figure 89. A WATER-SURPLUS GRAPH

Based on S. Erinç, 'Climatic Types and Variation of Moisture Regions in Turkey,' *Geographical Review,* vol. 40, p. 228 (New York, 1950).

UNITED STATES, WEST OF ROCKIES

TRENDS IN AVERAGE PRECIPITATION
1895 - 1940

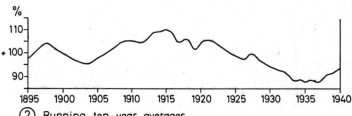

① Yearly percentage departures from mean rainfall

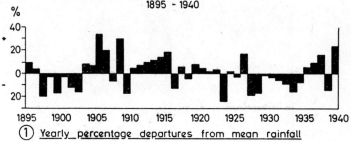

② Running ten-year averages

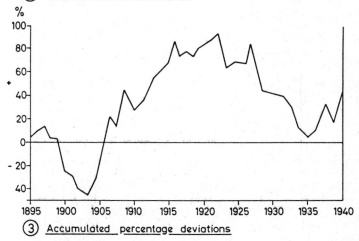

③ Accumulated percentage deviations

Figure 90. GRAPHS OF RAINFALL TRENDS

1 and **2** are based on diagrams in United States Department of Agriculture, *Climate and Man* (Washington, 1941). The data used were average 'weighted' precipitations, the mean value of which was 17·9 inches. **3** has been drawn on the basis of the same data.

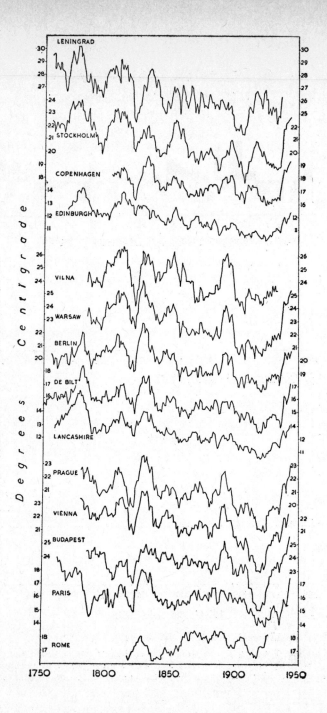

Degrees Centigrade

which individual yearly means are plotted chronologically, are sufficient to show well-marked trends. But in order to smooth out fluctuations in the curves, running (or moving) means may be plotted, on either a three, a five, a ten or a twenty-year basis (Figs. 90, 91). A method of calculating running means is given in the Table on p. 229. The running means are plotted to represent the midpoint of the three annual means, for example, 65·2 is plotted against 1861, 68·4 against 1862, 65·1 against 1863 and so on.

Curves to represent fluctuations in amplitudes of temperature over a period of years can also be smoothed out in this manner (Fig. 91).

Deviational Graphs. Cyclic fluctuations and trends in temperature and rainfall may be represented graphically by methods other than the chronological plotting of mean values. Cumulative graphs are effective, for example, in showing cyclic fluctuations in rainfall.[1] For this type of graph, annual deviations from the mean annual rainfall or from base are obtained by subtraction, and the accumulated deviations by addition as shown in the Table on p. 234.

The accumulated deviations are then plotted, +2·9 against 1860, +10·3 against 1861 and so on. Cyclic fluctuations in rainfall for adjacent stations are easier to compare if the accumulated deviations are expressed as percentages (Fig. 90).

[1] See, for example, (*a*) C. B. Saville, 'Some Rainfall Variations, England and New England (U.S.A.)', *Quarterly Journal of the Royal Meteorological Society*, vol. 60, pp. 313–31 (London, 1933), for graphs of rainfall trends, 183-year series, by accumulated sums of departure from base; (*b*) Commonwealth of Australia Meteorological Bureau, *Bradfield Scheme for Watering the Inland – Meteorological Aspects* (Melbourne, 1945), for rainfall trends by accumulated deviation from mean; (*c*) R. W. Longley, 'The Variability of Mean Daily Temperature', *Quarterly Journal of the Royal Meteorological Society*, vol. 73, pp. 418–25 (London, 1947), for graphs of temperature by standard deviations.

Figure 91. CURVE PARALLELS OF TEMPERATURE

Based on I. A. Labrijn, *Onderzoek naar Klimataatschommelingen in het Stroomgebied van de Rijn* (Amsterdam, 1948).

The graph shows variations in the amplitudes of temperature (i.e. differences between the mean July temperatures and those of the preceding January). Hence the curve for Stockholm indicates a tendency towards a decline in amplitude during the nineteenth century but a marked increase in recent years. Large amplitudes are usually associated with higher summer temperatures and colder winter conditions so that an increase in amplitude denotes a tendency towards greater extremes of climate. The strikingly close relationship between long-term temperature trends at different stations in Europe is clearly revealed by placing the curves in parallel.

D. Brunt used a form of deviational graph which he called a *periodo-gram*, in connection with the determination of climatic cycles.[1] In this graph amplitudes of temperature in the form of deviations are plotted as ordinates and various periods of time as abscissae. Thus a periodo-gram for Stockholm derived from monthly temperature values for 100 years indicated a thirteen-month period as one with the largest amplitude.

Accumulated Deviations (Rainfall in Inches)

Year	Rainfall	Mean Rainfall for the Series 1860–1940 = 23·0 inches	
		Deviation	Accumulated Deviation
1860	25·9	+2·9	+ 2·9
1861	30·4	+7·4	+10·3
1862	17·9	−5·1	+ 5·2
1863	24·6	+1·6	+ 6·8
1864	25·2	+2·2	+ 9·0

Frequency Graphs

Graphs in which the frequency of extreme temperatures, droughts, spells of heavy rainfall, thunderstorms, characteristic types of weather and similar climatic phenomena is plotted are of great value to the climatologist. They demonstrate aspects of climate which graphs of mean values may conceal, and they are of particular use in the delimita-tion of seasons.

A *histogram* is a diagram in which the frequency percentage (see pp. 34, 478) of amounts of rain for example, is plotted as ordinate and the amounts of rainfall as abscissa, and the peak of the curve represents the highest frequency. In practice, partly owing to limited data and partly in order to reduce the labour involved in their compilation, these graphs are reduced to step-diagrams by grouping values together (Fig. 92).[2]

Seasonal Frequencies. Seasonal variations in frequency may be shown on a monthly as well as on an annual basis (Fig. 92). Superimposition of January and June frequencies suffices to show the contrast between

---

[1] D. Brunt, 'Climatic Cycles', *Geographical Journal*, vol. 89 (London, 1937).

[2] D. Christodoulou, *The Evolution of the Land Use Pattern in Cyprus*, p. 21 (London, 1959), gives some revealing histograms of rainfall incidence and amount for Nicosia.

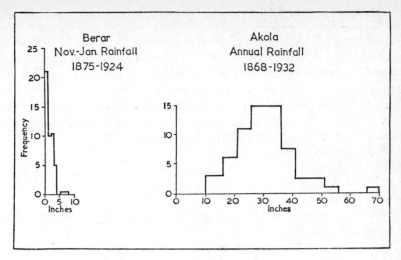

Figure 92. HISTOGRAMS OF RAINFALL

Based on N. Carruthers, 'An Analysis of the Variations in Rainfall at Akola, Berar, Central India', *Geography*, vol. 30, p. 70 (London, 1945).

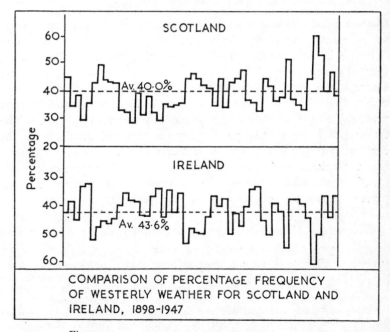

Figure 93. A COMPARABLE FREQUENCY GRAPH

Based on R. B. M. Levick, 'Fifty Years of British Weather', *Weather*, vol. 5, no. 7, p. 247 (London, 1950).

summer and winter rainfall frequencies. Seasonal frequencies may also be shown by plotting the number of frequencies or percentage frequencies as ordinates and months as abscissae.[1] Frequency of long spells of weather may be grouped in this way (Fig. 83). It is sometimes useful when seasonal frequencies of types of weather are being computed to invert the frequency curve for one station, so that comparison of irregularities in the frequency curves of both stations is facilitated (Fig. 93).

Use of Logarithmic and Probability Graph-paper. Either because of a very great range in frequency, or because of great ranges in the climatic data being plotted, it is advisable to use logarithmic or semi-logarithmic graph-paper for certain types of frequency graph. Particularly is this the case where pluviographic records of rainfall are being analysed upon a frequency basis (Fig. 94). Probability graph-paper also has its uses for this kind of work.[2]

Circular Graphs

Circular graphs in which values of temperature, rainfall, wind or of other elements are plotted from a central origin along axes radiating outwards in different directions, the plotted points then being joined to form closed curves, have certain advantages over the conventional Cartesian graph.[3] In the depiction of seasonal changes they show continuity from the end of one year through to the beginning of the next year which cannot be shown conventionally without duplication. The division of a circle through 360 degrees nearly coincides with the division of a year into 365 days, and monthly data for most purposes may be depicted by intervals of about 30 degrees. For more exact work monthly data have to be adjusted to represent thirty days. Circular percentage graph-paper may be used for specific types of data.

Climatographs. The climatograph devised by E. N. Munns[4] and

[1] Note A. T. Doodson and H. J. Bigelstone, 'The Frequency Distribution of Rainfall at Liverpool Observatory, Bidston', *Quarterly Journal of the Royal Meteorological Society*, vol. 60, pp. 403–41 (London, 1934), for analysis of rainfall based on frequency.

[2] See H. Landsberg, *Physical Climatology*, pp. 76–80 (latest edition, Dubois, Penn., 1960), for a summary, with examples, of the use of probability graph-paper.

[3] Their possibilities were first discussed by S. Friedman, 'Graphische Darstellung der Jährlichen Temperatur eines Ortes durch geschlossene Curven', *Mitteilungen der k.-k. geographischen Gesellschaft in Wien*, vol. 6 (Vienna, 1962).

[4] E. N. Munns, 'The Climatograph, a New Form of Chart for Climatic Phenomena', *Monthly Weather Review*, vol. 50 (Washington, 1922).

elaborated by R. Hartshorne[1] provides a clear illustration of the use of circular graphs (Fig. 95). Mean monthly temperatures are plotted from the centre with the aid of a graduated table. The distance from

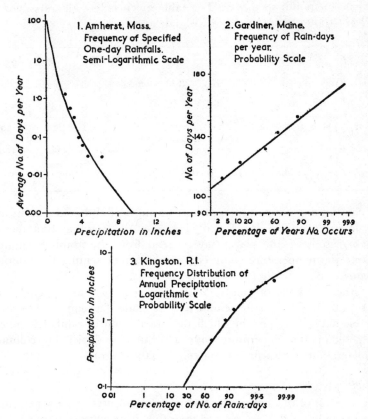

Figure 94. VARIETIES OF RAINFALL FREQUENCY GRAPHS
Based on graphs in E. E. Foster, *Rainfall and Run-off* (New York, 1949).
Note the use made of logarithmic and probability graph-paper.

the centre of the circle to 100° F is taken as ten times the distance from the centre to 0° F. If the latter difference is X, then the difference Y for any temperature $t°$ F is given by the formula:

$$Y = X \frac{(Colog.\ t)}{100}$$

[1] R. Hartshorne, 'Six Standard Seasons of the Year', *Annals of the Association of American Geographers*, vol. 28, pp. 165–78 (Lancaster, Pa., 1938).

This formula is used in order that temperatures below zero can be plotted, and it has the advantage of producing a temperature curve the slope of which correctly represents the degree of change in temperature from month to month. The table given below allows values of *Colog. t/*100 to be interpolated with fair accuracy.

°F	110	105	100	95	90	85	80	75
*Colog. t/*100	12·59	11·72	10·0	8·91	7·94	7·08	6·31	5·62
°F	70	65	60	55	50	40	30	20
*Colog. t/*100	5·01	4·47	3·98	3·55	3·16	2·51	2·0	1·59
°F	10	0	−10	−20	−40	−60	−100	
*Colog. t/*100	1·26	1·00	0·79	0·63	0·40	0·25	0·10	

If the limiting temperatures of hot, warm, cool and cold seasons are assumed to be 68° F, 50° F and 32° F, then the length and number of such seasons at any place may be read from the graph by noting where the temperature curve cuts the lines representing the limiting temperatures (Fig. 95).

Other methods of constructing circular graphs are discussed at length by J. B. Leighly.[1] Mention may be made of a compound circular graph devised by him in which the distribution of rainfall is superimposed upon a temperature curve. The rainfall is depicted by columns which radiate outwards from the centre of the diagram.

Isopleth Graphs

In a graph of this nature the hourly values of pressure, temperature, etc., are plotted as abscissae and their time of occurrence in the month by ordinates. Similar values are then joined by isopleths. Sir Napier Shaw refers to these groups as 'chrono-isopleth diagrams'.[2] The procedure is made clear in Fig. 96. Similar diagrams rather ponderously

---

[1] J. B. Leighly, 'Graphic Studies in Climatology, II. The Polar Form of Diagram in the Plotting of the Annual Climatic Cycle', *University of California Publications in Geography*, vol. 2, no. 13 (Berkeley, 1929).

[2] *Manual of Meteorology*, vol. 1 (Cambridge, 1926). Chapter 6 deals with the development of arithmetical and graphical manipulation of climatic statistics. The method is applied in W. T. Gehrke, 'The Wind Flow Diagram', *Annals of the Association of American Geographers*, vol. 34, pp. 63–6 (Lancaster, Pa., 1944). Extensive use is made of this method in R. Geiger, *The Climate Near the Ground* (Harvard, 1950).

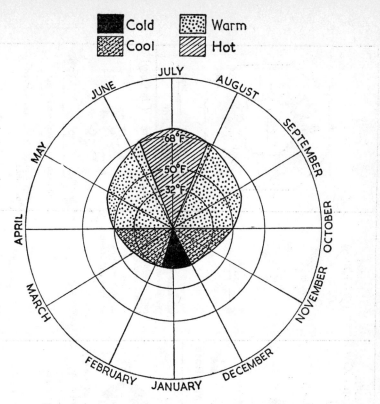

Cold   Warm
Cool   Hot

JULY
JUNE   AUGUST
MAY   SEPTEMBER
APRIL   OCTOBER
MARCH   NOVEMBER
FEBRUARY   DECEMBER
JANUARY

68°F
50°F
32°F

Figure 95. A CLIMATOGRAPH FOR BUDAPEST

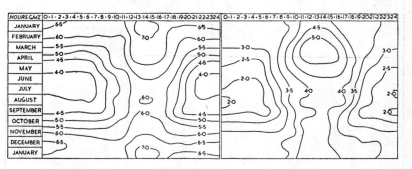

Figure 96. ISOPLETH GRAPHS OF WIND SPEED

Based on Sir Napier Shaw, *Manual of Meteorolcgy*, vol. I, p. 267 (Cambridge, 1926).
    The diagram shows mean wind speeds in metres per second at two stations, Valencia, Ireland (*left-hand*) and Kew, London (*right-hand*). The wind speeds are plotted for hourly periods, month by month. The interpolation of isopleths facilitates direct comparison of wind-speed conditions.

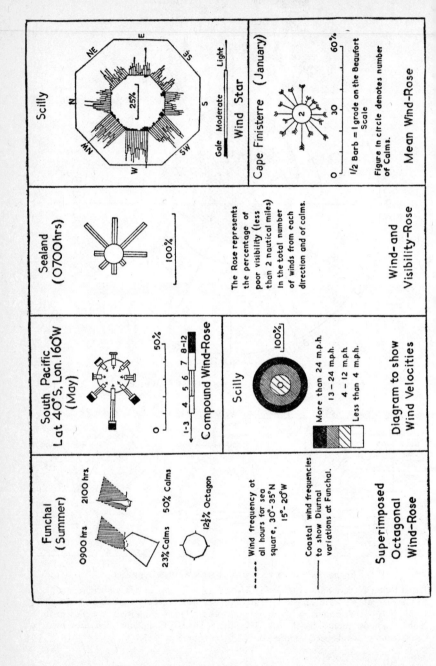

Scilly

NE
E
ES
25%
N
S
NW
W
SW

Gale Moderate Light

Wind Star

Cape Finisterre (January)

2

0   30   60%

1/2 Barb = 1 grade on the Beaufort Scale

Figure in circle denotes number of Calms.

Mean Wind-Rose

Sealand (0700 hrs)

100%

The Rose represents the percentage of poor visibility (less than 2 nautical miles) in the total number of winds from each direction and of calms.

Wind- and Visibility-Rose

South Pacific Lat 40°S, Lon. 160°W (May)

0   50%

1-3   4   5   6   7   8-12

Compound Wind-Rose

Scilly

100%

More than 24 m.p.h.
13 – 24 m.p.h.
4 – 12 m.p.h.
Less than 4 m.p.h.

Diagram to show Wind Velocities

Funchal (Summer)

0900 hrs   2100 hrs.

23% Calms   50% Calms

12½% Octagon

------ Wind frequency at all hours for sea square, 30°-35°N 15°-20°W

——— Coastal wind frequencies to show Diurnal variations at Funchal.

Superimposed Octagonal Wind-Rose

termed hypso-chrono-isopleths can be used to show variations of elements of the weather with both height and time.

WIND-ROSE DIAGRAMS

The wind-rose is a form of star-diagram which is peculiarly suited to show the average frequency and direction of wind at one place. In Fig. 97 a variety of types of wind-rose is depicted. Usually wind direction is defined in terms of the cardinal points, although ordinal points may be used where greater detail is required. All wind is assumed to come from the directions of the points selected. Calms are considered separately.

Octagonal Wind-roses

This diagram is designed to reflect the total mean monthly conditions both of wind frequency and of direction at one station. Two concentric octagons are constructed, so that the distance between the corresponding sides of each octagon represents a frequency of $12\frac{1}{2}$ per cent. Each side represents one of the eight cardinal wind directions, and the twelve mean monthly frequencies of wind from each of these directions are plotted as columns, so that if winds were equally frequent from each direction, eight sets of twelve equal columns would result, each column having its base on the inner octagon and its apex on the outer octogon. Thus frequencies greater than average will be shown by columns extending over the outer polygon and vice versa. Calms are represented diagrammatically within the centre of the inner octagon.

Simple Wind-roses

Octagonal wind-roses are difficult to interpret and the conditions they represent may be shown more simply by twelve diagrams, each representing the conditions for one month. Two concentric circles, the circumferences of which are set at a distance representing $12\frac{1}{2}$ per cent,

---

Fig. 97. TYPES OF WIND-ROSE AND WIND-SPEED DIAGRAM

Particular examples have been drawn from the following sources: (**1**) Meteorological Office, *Weather in Home Waters and the North-eastern Atlantic*, vol. 2, pt 1 (London, 1944) and pt 4 (London, 1940); (**2**) E. G. Bilham, *Climate of the British Isles* (London, 1938); (**3**) Sir Napier Shaw and L. G. Garbett, 'A New Sort of Wind Rose', *Quarterly Journal of the Royal Meteorological Society*, vol. 59, pp. 38–44 (London, 1933).

are drawn, and columns representing the percentage frequency of wind from each of eight directions are plotted upon this basis; the percentage frequency of calms is indicated by a number placed within the smaller circle. Average annual frequencies of wind from each direction may also be depicted by this method when seasonal variations are not required.

Compound Wind-roses

Wind-rose columns indicating frequency may be subdivided to show the frequency of wind strengths associated with the particular direction represented (Fig. 97). Four divisions indicating the following velocities respectively are usually sufficient for most purposes:

> 1. More than 24 m.p.h.
> 2. 13 to 24 m.p.h.
> 3. 4 to 12 m.p.h.
> 4. Less than 4 m.p.h.

Wind-roses of this type are frequently employed in the analysis of upper winds, the velocities of which are much stronger than those of surface winds.

Superimposed Wind-roses

The idea of wind-roses may be adopted for the purpose of showing unusual diurnal variations of wind. For example, to demonstrate changing directions of wind between day and night at any place, wind-roses, each representing mean conditions at different times of observation during the day, may be superimposed (Fig. 97). In this case, for the sake of clarity, frequencies are plotted from a central point and joined to form irregular octagons.

Wind- and Visibility-roses

Relationships between visibility and wind direction may be expressed in the form of wind-roses. Observations are often made at the same time, so that frequencies of bad and good visibilities may be correlated with wind direction. The percentage frequency of bad visibility, for example, may then be plotted for different wind directions to form a rose (Fig. 97).

Wind-stars

Wind-roses constructed to show monthly frequencies of wind from various directions, together with mean velocities, have been termed

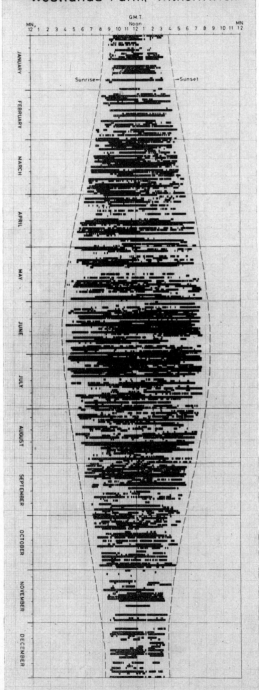

SUNSHINE HOURS 1967
Westlands Farm, Withernwick

Figure 98. ANNUAL SUN-
SHINE RECORD AT WITH-
ERNWICK IN THE EAST
RIDING OF YORKSHIRE
(After D. P. Brachi)

The sunshine recording for
each day during the whole
course of the year can be summarized in a graphic manner
by simply blocking in the extent of sunshine period by
period throughout the day for
each day of the year. This
method not only shows the
daily incidence of sunshine in
relation to length of day but
also allows visual comparison
of sunshine records from year
to year.

wind-stars (Fig. 97).[1] These stars are ingeniously devised, but they are rather laborious to construct. Moreover, they do not reduce very well and they cannot be located on a base-map to show regional variations.

SUNSHINE RECORD AND OTHER DURATION DIAGRAMS

A variety of diagrams can be produced to show daily and seasonal incidence of sunshine. Perhaps one of the most graphic methods is to show the daily incidence of sunshine period by period throughout the year for any given year (Fig. 98). This diagram shows the variation in sunshine throughout the year and also enables visual comparisons to be made from year to year or place to place. The diagram also shows sunshine in relation to the duration of daylight.

Another type of diagram which shows duration has been used by A. J. W. Catchpole to show the duration of snow-cover at Winnipeg. In this case seasonal occurrence is plotted against the period of the year in months from September to May. There are three categories of snow cover distinguished, each shaded differently: (1) continuous cover over 5 inches; (2) continuous trace; and (3) discontinuous trace.[2]

RAINFALL DISPERSION DIAGRAMS

In recent years rainfall dispersion diagrams have become an important tool in the analysis of rainfall distribution. The median and other percentile values derived from the diagrams have greater significance in some cases than mean values, or than coefficients derived from mean values, in arriving at a more rational estimation of rainfall variability, and of classifying rainfall regimes (see Appendix, p. 481).[3]

---

[1] Sir Napier Shaw and L. G. Garbett, 'A New Sort of Wind Rose', *Quarterly Journal of the Royal Meteorological Society*, vol. 59, pp. 39–44 (London, 1933).

[2] A. J. W. Catchpole, 'Solar Control of Diurnal Temperature at Winnipeg', *The Canadian Geographer*, vol. 13, p. 264 (Toronto, 1969).

[3] For a discussion of the value of the median, see American Geographical Union, 1940 Committee, 'Report of Committee on Median *v.* Arithmetical Average', *Transactions of the American Geographical Union*, pt 1 (Washington, 1941). See also an interesting graphic comparison of the distribution of rainfall as computed by means of the coefficient of variability, and as actually revealed by a dispersion diagram in E. E. Lackey, 'Annual Variability Rainfall Maps of the Great Plains', *Geographical Review*, vol. 27, pp. 665–70 (New York, 1937).

Construction of the Diagram

Diagrams may be made of the dispersion of annual amounts of rainfall for any one station simply by plotting dots of suitable size, each of which represents one year's rainfall, against a vertical scale uniformly graduated so that the base is zero and the top is equivalent to the maximum rainfall total (Fig. 99).[1] It is not advisable to attempt analysis unless at least a thirty-five-year record of rainfall is available.

Annual dispersion graphs have only a limited value; of greater significance are monthly diagrams. For these, the vertical scale is graduated from zero to the highest monthly rainfall in the series. Monthly values are then plotted to give the individual rainfall dispersion for each month in the year (Fig. 99).

Median and Percentile Values

The median or middle value of rainfall for any dispersion diagram is that which lies midway between the two extremes of rainfall at either end of the diagram. Thus for a thirty-five-year series the eighteenth value reckoned from the minimum is the median value. The lower quartile value is that which lies midway between the median and the minimum, i.e. in the above case it will lie halfway between the ninth and tenth figures reckoned from the minimum on the diagram. Similarly, the upper quartile value will lie between the ninth and tenth value reckoned from the maximum on the diagram. These values may be marked on the diagram by short horizontal lines (Fig. 99). Half the recorded values in the series will thus lie between the upper and lower quartile values.

Major, Minor and Graded Breaks

Discontinuities in rainfall at any one station may be estimated from an inspection of the relative positions of quartile and mean values for adjacent months. If the interquartile band of one month is clear of that for an adjacent month, a *major break* in rainfall is generally indicated. The month with the upper band will on an average be wetter than that with the lower band. The chances that the latter will be

---

[1] The idea of applying dispersion diagrams to rainfall analysis was conceived by P. R. Crowe, 'The Analysis of Rainfall Probability', *Scottish Geographical Magazine*, vol. 49, pp. 73–91 (Edinburgh, 1933); and 'The Rainfall of the Western Plains', *Geographical Review*, vol. 26, pp. 463–84 (New York, 1936). See also H. A. Matthews, 'A New View of Some Familiar Indian Rainfalls', *Scottish Geographical Magazine*, vol. 52, pp. 84–97 (Edinburgh, 1936).

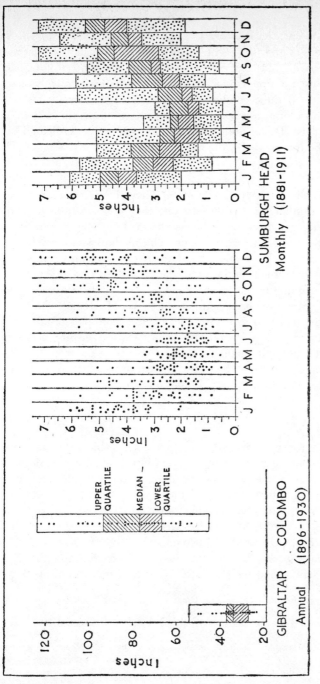

Figure 99. RAINFALL DISPERSION DIAGRAMS

The data for the left-hand diagram were derived from *World Weather Records*, Smithsonian Miscellaneous Collections, vols 79 and 90. The inter-quartile ranges have been distinctively shaded.
The middle diagram is based on *Great Britain, Rainfall, Annual Average, 1881–1915* (Ordnance Survey, 1949).
The right-hand diagram is drawn from the same data but only the percentile range, the median and the quartile values are indicated by means of columns shaded for the purpose of comparison.

wetter are only one in eight. *Minor breaks* which mark semi-discontinuities are said to exist if the median and lower quartile of one month lie above the upper quartile and the median respectively of an adjacent month. If these conditions are satisfied for alternate instead of for adjacent months, a *graded break*, major or minor as the case may be, is said to exist. Discontinuities in rainfall regimes, as indicated by breaks, may be mapped with the aid of isopleths.

Merits and Demerits

Dispersion graphs are easy to construct. They show at a glance the dispersion of rainfall for any one station; they give an indication of seasonal distribution and actual variability without the necessity of making lengthy statistical calculations. The median value, as far as rainfall is concerned, has many advantages over the mean value, since the latter may give a false impression. Particularly is this so for places with a dry season. On the other hand, some disadvantages of the dispersion diagram have been pointed out by W. H. Hogg,[1] who drew attention to the fact that discontinuities for the same pair of adjacent months is not always a satisfactory criterion for distinguishing between rainfall zones, and that median values of rainfall cannot be validly used on a percentage basis because the twelve monthly medians of any station do not add up to the annual median.

CLIMOGRAPHS

A climograph (or climogram) is a diagram in which the data for elements of climate at any one station are plotted against one another, and the shape and position of the resultant graph provides an index to the general climatic character of the place.[2] Usually a number of such diagrams is plotted on one chart for comparative purposes. The diagrams may be used to summarize variations in world climatic conditions, and they were used in this connection by W. Köppen.[3] He devised a chart in which temperatures of the coldest month were plotted as abscissae and temperatures of the warmest month as ordi-

---

[1] W. H. Hogg, 'Rainfall Dispersion Diagrams', *Geography*, vol. 33, pp. 31–7 (London, 1948).

[2] This type of diagram was first conceived by J. Ball, 'Climatological Diagrams', *Cairo Scientific Journal*, vol. 4 (Cairo, 1910).

[3] W. Köppen, 'Klassifikation der Klimate nach Temperatur, Niederschlag und Jahrslauf', *Petermanns Geographische Mitteilungen*, vol. 64, pp. 193–203, 243–8 and Tafel 11 (Gotha, 1918).

nates. The framework of the chart was subdivided to indicate the relative positions of tropical, mesothermal, boreal and tundra climates, etc.

J. B. Leighly's Climographs

J. B. Lieghly expanded on Köppen's ideas to produce a number of climographs which could be used for comparison of the climates of different parts of the world.[1] These included one based on Köppen for analysing temperature. Another was specifically designed to record critical values of temperature and rainfall with regard to soil moisture conditions. The framework of this chart was graduated horizontally in ° F and vertically in inches of rainfall, so that mean annual temperature could be plotted as abscissae and mean annual rainfall as ordinates. As a further refinement critical axes, derived from formulae which took into account various regimes of the distribution of rainfall in the definition of dry climates, were added to the chart.

Further graphs were constructed to enable subdivision of rain or forest climates to be made according to seasonal distribution of rainfall. In one of these graphs the rainfall of the driest month was plotted against the annual rainfall to indicate the presence of a dry season, and in a second, to illustrate seasonal contrasts in rainfall regime, maximum rainfall was plotted against minimum rainfall on an appropriate axis according to season of occurrence.

G. Taylor's Climographs

Climographs may be adapted to show the influence of climatic conditions on human activity. G. Taylor, for example, has used climographs for this purpose. In one of his climographs he indicates the physiological effects of climate on man by plotting relative-humidity values in relation to wet-bulb temperatures. The framework of the chart consists of a graduated vertical side-scale showing wet-bulb temperatures from −10° F to 90° F, and a horizontal bottom-scale showing percentage relative humidity from 20 to 100 per cent. The north-west corner of the chart he marked as *Scorching* (high wet-bulb, low relative humidity), the north-east corner as *Muggy* (high wet-bulb, high relative humidity), the south-west as *Keen* (low wet-bulb, low relative humidity), and the south-east as *Raw* (low wet-bulb, high

[1] J. B. Leighly, 'Graphic Studies in Climatology. I. Graphic Representations of a Classification of Climates', *University of California, Publications in Geography*, vol. 2, no. 3 (Berkeley, 1926).

relative humidity). The mean monthly data are plotted on this chart
for a particular station, each month being marked by a letter, and the
plotting points are then joined.[1] For a variation on this type of climograph, see Fig. 100.

The Hythergraph is another type of climograph used by G. Taylor,
in which mean monthly temperature values are plotted as ordinates
and mean monthly rainfall values as abscissae.[2] They are principally
used in summarizing broad climatic differences in relation to human
activity, more particularly with reference to settlement.

E. E. Foster's Climograph

E. E. Foster has devised a climograph (Fig. 102) with the aid of
Thornthwaite's scheme of climatic classification (see p. 215).[3] This
consists of a chart formed by a grid system of rectangular co-ordinates.
The vertical side-scales are graduated from $-20°$ F to $100°$ F, the hori-
zontal bottom-scale from 0 to 18 inches of rainfall. The chart consists
of six temperature zones, *Frigid* ($-20°-0°$ F), *Cold* ($0°-32°$ F), *Cool*
($32°-50°$ F), *Mild* ($50°-65°$ F), *Warm* ($65°-80°$ F) and *Hot* (over $80°$ F).
The Cool, Mild, Warm and Hot divisions are further subdivided into
Arid (limiting grid points, $32·4°$ F, $0·32$ inch of rain; $83·2°$ F, $1·03$
inch of rain); *Semi-arid* (limiting grid points, $32·4°$ F, $0·59$ inch;
$83·2°$ F, $1·93$ inch); *Semi-humid* (limiting grid points, $32·4°$ F, $1·10$ inch;
$83·2°$ F, $3·6$ inches); *Humid* (limiting grid points, $32·4°$ F, $2·05$
inches; $83·2°$ F, $6·73$ inches).

Special Climographs

Climographs may be used to demonstrate special aspects of climatic
differentiation – for example, economic aspects by plotting length of
growing season in hours per month against effective rainfall. They
may even be adapted for showing the relationships between climate
and soil-type (Fig. 101). An interesting example is by P. N. Hore, who
constructed a series of climographs in a study of rice yields and irrigation needs in West Bengal.[4] He constructed his diagrams with the vertical axis giving the annual rainfall, the horizontal axis the number of

[1] For an example, see A. A. Miller, *Climatology* (London, 1942).
[2] See examples in G. Taylor, *Urban Geography* (London, 1949).
[3] E. E. Foster, 'A Descriptive Graph of Climate', *Transactions of the American
Geographical Union*, pt II (Washington, 1944).
[4] P. N. Hore, 'Rainfall, Rice Yields and Irrigation Needs in West Bengal', *Geography*, vol. 49, pp. 114–21 (London, 1964).

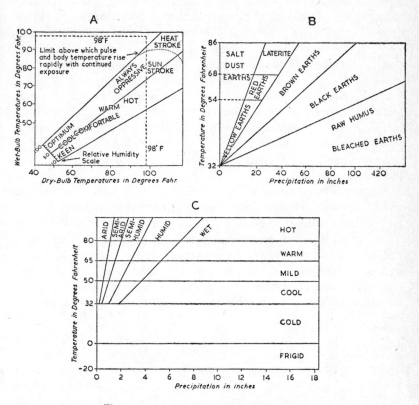

Figures 100–2. TYPES OF CLIMOGRAPH

A (Fig. 100) is based on United States Department of Agriculture, *Climate and Man*, p. 249 (Washington, 1941). This is a type of climograph favoured by G. Taylor to describe climatic conditions in terms of human comfort.

B (Fig. 101) is derived from R. Lang, *Verwitterung und Bodenbilding als Einfuhrung in die Bodenkunde* (Stuttgart, 1920) and is of help in the elucidation of soil-types in terms of climatic conditions.

C (Fig. 102) is based on E. E. Foster, *Rainfall and Runoff* (New York, 1949). It makes use of Thornthwaite's formula and is designed to illustrate his system of climatic classification.

In the case of diagrams A and C monthly values are plotted and joined by a line, but only annual values are required in the case of Lang's climograph.

rainy days. For each of the thirty-five years under discussion, a dot was placed according to these scales; beside each dot was given the rice-yield per acre for the respective year, and isopleths were drawn on the climographs to indicate the concentration of points of high yield. A further set of climographs was constructed to show the relationship between the rainfall in the sowing-transplanting season (vertical axis) and that in the late growing season, again with the addition of the rice-yield.

Climographs can also be utilized to show physiological climates. W. H. Terjung has devised a comfort index in terms of dry-bulb temperatures and relative humidity, combined with a wind effects index designed to measure skin cooling, and has used these indices to produce maps of the United States to show the distribution of physiological climates.[1]

Another interesting application is that used by J. R. Mather and G. A. Yoshioka to show the relationship between certain climatic features and natural vegetation.[2] In this instance potential evapo-transpiration is plotted against moisture index to provide a framework. The natural vegetation at representative stations is then plotted by means of a symbol. Peripheral symbols are joined by a line to show the location within the framework of specific areas of natural vegetation such as Douglas fir, spruce, birch, oak, chestnut, tropical rain-forest, etc.

ARROWS

Arrows are used conventionally to show the horizontal movement of air over the surface of the earth. They may be used either to show the trajectories[3] of air or, more frequently, wind direction and direction of streams of air.[4]

[1] Werner H. Terjung, 'Physiological Climates', *Annals of the Association of American Geographers*, vol. 56, pp. 141–79 (Lancaster, Pa., 1966).

[2] John R. Mather and Gary A. Yoshioka, 'The Role of Climate in the Distribution of Vegetation', *Annals of the Association of American Geographers*, vol. 58, pp. 29–41 (Lancaster, Pa., 1968).

[3] See a map after G. I. Taylor, in H. B. Byers, *General Meteorology* (New York, 1944), showing the computed trajectory of air in the vicinity of the Grand Banks, July 16–25, 1913. The calculation of trajectories is discussed fully in W. J. Saucier, *Principles of Meteorological Analysis*, p. 312 (Chicago, 1955).

[4] A large number of maps with arrows to indicate air-mass streamlines is used by R. A. Bryson, 'Air Masses, Streamlines and the Boreal Forest', *Geographical Bulletin*, vol. 8, no. 3, pp. 228–69 (Ottawa, 1966); the arrows were obtained by the detailed computation of trajectories.

Wind systems, either planetary or local, can be depicted quite simply by short arrows, 'flying with the wind', and indicating prevailing wind directions at different times of the year. They may be drawn by inference from mean isobars and are not necessarily based on tabulated observations.[1]

Where observations are available, the frequency and force of such winds may roughly be indicated by thickening the shaft of the arrows according to average wind-force and by pecking the shaft of the arrow to show frequency. Alternatively, a system of tail-feathers and of differential shafts may be employed (Fig. 103).[2]

On synoptic charts, where wind conditions at a particular time have to be plotted, frequency is not required. Wind direction, therefore, is shown by point-tipped arrows, or arrows flying into the station symbols, the tails of which indicate wind-force by means of the number of feathers they carry.[3]

Climate is associated more with the typical movement of various types of air-mass from specific regions than with surface winds pure and simple. The plotting of average surface winds by means of arrows has its uses in the local differentiation of climate, but has little value in the analysis of the fundamental causes of climatic phenomena. A close study of the chronological sequence and average frequency of isobaric patterns and their associated weather conditions affords one method of determining the source and type of air-mass at any one place. Such analyses are properly undertaken by the professional weather forecaster, and findings are plotted by marking the hypothetical source regions of characteristic air-masses and by using arrows to indicate the generalized movement of air from such source regions.[4]

The mean movement of air, representative of different seasons, can be shown by *streamlines*, which take the form of elongated arrows curving to denote changes in direction (Fig. 104). Streamlines are usually drawn to resultant winds (see p. 198) at *gradient level*, which is at about 500 metres above the surface. Streamlines can be drawn to surface

---

[1] Given the pressure conditions, the computation of wind directions and wind speeds involves elaborate calculations demanding some knowledge of thermodynamics. Rule-of-thumb formulae may, however, be employed, and these are discussed at length in H. B. Byers, *General Meteorology* (New York, 1944), and S. Petterssen, op. cit. (1940).

[2] For an application, see Meteorological Office, *A Barometer Manual*, plates 7 and 8 (11th edition, London, 1932).

[3] For scale of velocities, see Meteorological Office, *The Weather Map* (3rd edition, London, reprinted 1950).

[4] For such maps, see S. Petterssen, op. cit. (1940).

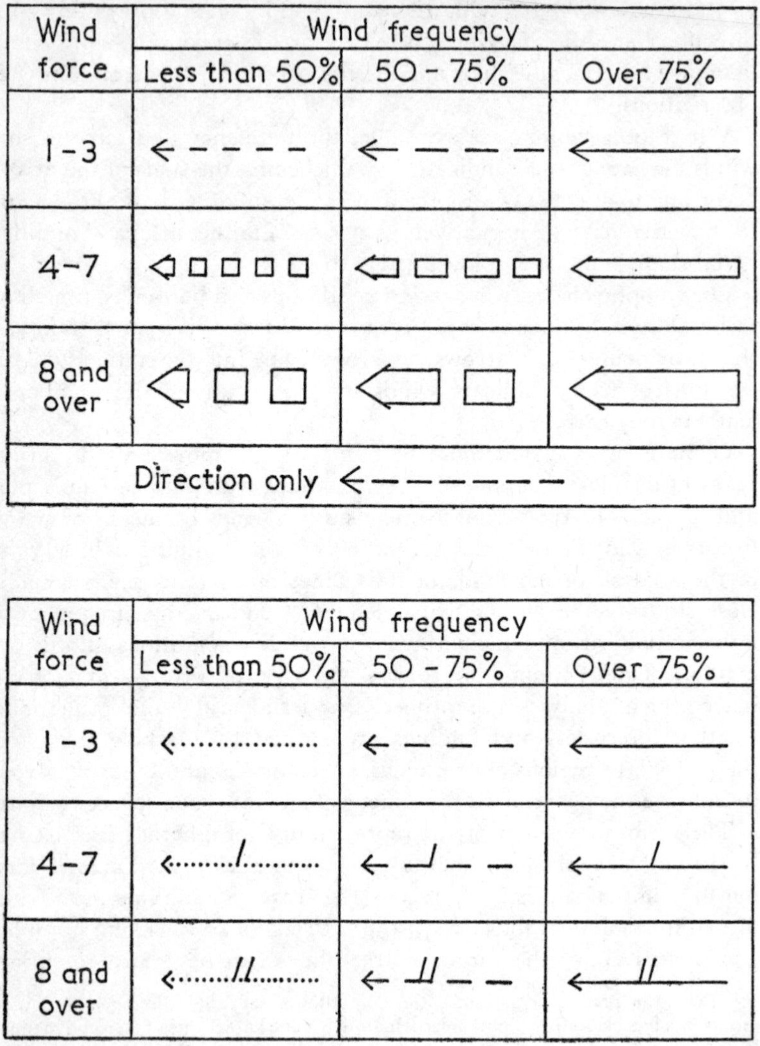

Wind force	Wind frequency		
	Less than 50%	50 - 75%	Over 75%
1 - 3	← - - - -	← — — —	← _____
4 - 7	◁□□□□	◁□□□□	◁_____
8 and over	◁□□	◁□	◁_____
Direction only ← — — — — — —			

Wind force	Wind frequency		
	Less than 50%	50 - 75%	Over 75%
1 - 3	←··················	← — — —	← _____
4 - 7	←·······/········	← ⌐— —	← ___/____
8 and over	←·······//·······	← ⌐⌐ — —	← ___//___

Figure 103. ARROW SYSTEMS

Various systems for showing the distribution of wind direction and speed can be devised and the two given above, exaggerated for the sake of clarity, are only meant to serve as examples.

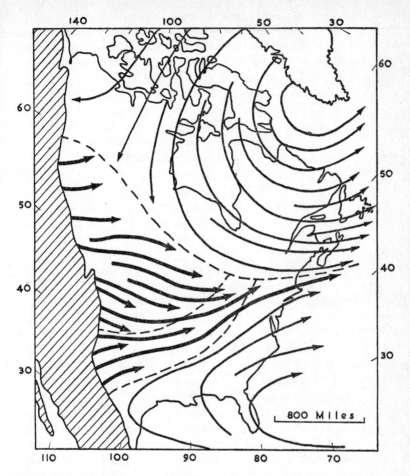

Figure 104. STREAMLINES FOR NORTH AMERICA EAST OF THE ROCKY
MOUNTAINS

Based on J. R. Borchert, 'The Climates of the Central North American Grassland',
Annals of the Association of American Geographers, vol. 40, p. 19 (Lancaster, Pa., 1950).

The streamlines show January resultant mean flow of air at gradient level derived
from data for the years 1928–40. Air streams from different directions have been
differentiated by means of thick and thin arrow-shafts. The pecked lines denote the
axial belts of maximum occurrence of low-pressure centres recorded on the January
(*Principal*) *Surface Weather Map* (Weather Bureau, Washington, D.C.). The data are
conveniently extracted from manuscript maps which incorporate the number of low-
pressure centres according to the daily weather map (0730 hours, Eastern Standard
Time). The records are based on some 420 areas of 3,000 square miles each, covering
North America and adjacent waters, 1930–9.

winds, but since turbulence at the surface due to friction is so great, and also because direction of wind varies according to local obstructions, the findings are not likely to be significant. Streamlines for very high altitudes can also be constructed, but they have not the same significance, from the point of view of life and activity on the surface as the 500-metre streamlines.[1] Also, pilot-balloon observations are fairly frequent and reasonably accurate at about 500 metres, whereas above that height both inaccuracy increases and observations are lacking.

In Fig. 104 streamlines based on resultant winds for January at gradient level reveal a strong westerly circulation over the greater part of the continent, but three distinctive air streams enter into its composition – a dry continental air stream from the base of the Rockies, a warm moist airstream from the sub-tropical Atlantic, and a cold one from the Arctic. The different origins of these air streams account for many peculiarities in the climate of parts of the North American continent.[2]

As J. R. Borchert emphasizes in his paper, the drawing of streamlines to resultant winds is a most valuable method of climatic analysis, but has been so far comparatively neglected. It must be pointed out, however, that some of the pioneers in the fields of dynamic meteorology advocated analyses along these lines many years ago.[3]

To illustrate relationships between streams of air at different altitudes and the weather conditions on the surface, three-dimensional sketches may serve a useful purpose, although, of course, it is necessary to resort to simplification and generalization (Fig. 105).

A further use of arrows is to depict the tracks of depressions and storm-paths. These are marked usually by elongated arrows which follow the average direction of movement of the centres of the depressions under consideration. The data are extracted from a series of synoptic charts covering the region concerned.[4] The speed of movement may also be indicated by dots placed along the arrows to represent intervals of time, and it is often the practice to plot a whole series of tracks covering a season, or even a number of years (Fig. 106).

[1] See a series of maps showing resultant wind directions at different altitudes in H. R. Byers, op. cit. (1944).
[2] J. R. Borchert, 'The Climate of the Central North American Grassland', *Annals of the Association of American Geographers*, vol. 40, pp. 1–39 (Lancaster, Pa., 1950).
[3] J. W. Sandström, *Dynamic Meteorology and Hydrography*, vol. 2 (Washington, 1910), for example, discusses graphical methods of constructing streamlines.
[4] M. A. Garbell (op. cit., 1947) has an excellent arrow-map showing the general geographical distribution of tropical cyclones.

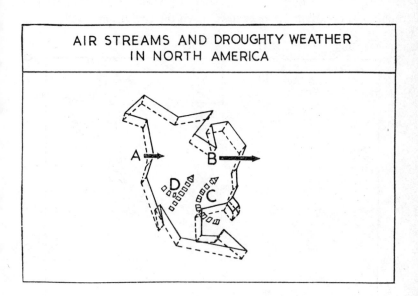

AIR STREAMS AND DROUGHTY WEATHER IN NORTH AMERICA

Movement of air at higher altitudes
Movement of surface air
Surface outline of North America
Upper level outline of North America

Figure 105. A THREE-DIMENSIONAL SKETCH OF AIR STREAMS

Based on I. R. Tannehill, *Drought, its Causes and Effects*, p. 94 (London, 1947).

The diagram shows the main components of air currents over North America during times of drought. On the left the short arrow, **A**, indicates slow movement of air into the continent due to the relative warmth of the continent. The long arrow, **B**, indicates rapid movement of air out of the continent towards the relatively cool Atlantic. Arrow **C** shows surface rain-bearing air flowing into the continent to make up for the inequality of components **A** and **B**. This surface air may normally bring rainfall into the central plains. If, however, the inequality between **A** and **B** is too great, dry surface air will be drawn into the continent from the direction of the Mexican plateau – arrow **D**. Air current **C** will then be deflected north-eastwards and the interior plains will suffer drought as a result. There is little or no surface air coming in from Canada during this period. It is of interest to compare this diagram with Fig. 104.

From such diagrams it is possible to produce generalized tracks representing mean conditions of movement. Some idea of the frequency with which cyclones follow particular paths may be given by elongated arrows, the shaft-thicknesses of which are made proportional to the number of occasions cyclones have taken the routes indicated.[1]

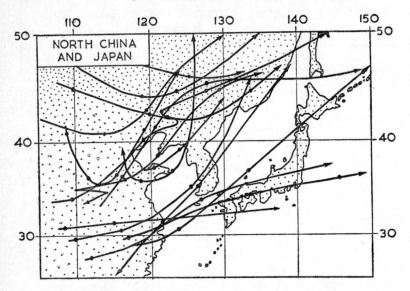

Figure 106. TRACKS OF CYCLONIC DEPRESSIONS, JULY

Based on Meteorological Office, *Weather in the China Seas and in the Western Part of the North Pacific Ocean*, vol. 1, pt. 1, p. 48 (London, 1938).

The tracks are taken from the annual charts published by the observatory at Zikawei, and have been amplified in northerly regions by reference to the synoptic charts published by the Meteorological Services of Nanking and Tokyo. The black dots on the tracks mark the successive positions of centres of depressions as identified from consecutive synoptic charts. The distance apart between two dots in the same track is thus equal to the distance travelled by any particular depression in 24 hours. The data used refer to the five years 1929–31 and 1933–4.

SYMBOLS

Symbols are used chiefly in the plotting of weather data on synoptic charts.[2] They may also be used to show the distribution of various types of weather-station, according to reliability, and according to the nature of function.

[1] See an example in W. G. Kendrew, *Climatology*, p. 342 (Oxford, 1949).

[2] For details consult *The Weather Map* (op. cit.), and *The International Meteorological Code* (op. cit.).

Proportional symbols may be specially adapted to show distributions of weather and climate. Proportional squares, for example, may be effectively employed to show the distribution of rainfall associated with the passage of a depression. The squares are made proportional to the amount of rainfall at the particular places, and may be shaded according to the time, or duration of the fall.[1]

Proportional circles are occasionally used to show regional variations in mean wind velocities. Uniform circles are drawn divided into concentric circles, so that circular bands of varying thickness are shaded to represent percentage frequencies of winds of more than 24 m.p.h., 12 to 24, 4 to 12, and under 4 m.p.h. (Fig. 97).

Symbols are also used to show distribution of freak weather over a number of years – heavy thunderstorms, record rainfall intensities and record droughts.[2]

SCHEMATIC DIAGRAMS

Considerable use is made of schematic diagrams in climatology to illustrate generalized schemes of climatic distribution, behaviour of air streams, varying physical properties of the atmosphere with altitude and variations of insolation with latitude and season. Very good diagrams of this type are to be found in such textbooks as those of A. N. Strahler[3] and R. G. Barry and R. J. Chorley.[4] As an example, one can refer to a diagram by A. N. Strahler to show physical properties of the atmosphere in which heights are shown in miles and kilometres, related to the atmospheric pressure, temperature zones, wind speeds, water vapour, the occurrence of clouds, performance of satellites, rockets and balloons, the location of the ionosphere and the chemosphere, the reflection of radio waves, the ozone layer, the penetration of soft X-rays, ultra-violet rays and light rays, the light at which radio waves are reflected and the occurrence of the tropopause, together with a variety of other features. In another diagram A. N. Strahler sketches the location of climate groups in relation to frontal zones, air

[1] See a map by J. H. G. Lebon, 'The Interpretation of the Daily Weather Report', *Geography*, vol. 32, pp. 53–66 (London, 1947).

[2] For a map utilizing such symbols, see S. S. Visher, 'Regionalization of the United States on a Precipitation Basis', *Annals of the Association of American Geographers*, vol. 32, pp. 355–65 (Lancaster, Pa., 1942).

[3] A. N. Strahler, *Physical Geography* (New York, 1960); ibid., *The Earth Sciences* (New York, 1963).

[4] R. G. Barry and R. J. Chorley, *Atmosphere, Weather and Climate* (London, 1968).

masses and source regions; the base map in this case is a hemisphere showing the location of an idealized continental land mass.

These diagrams can also be used effectively to explain and illustrate interactions within the atmosphere; as an example, R. G. Barry and R. J. Chorley use a 'tephigram' to illustrate the different cases of stable and unstable air. This diagram shows an idealized vertical section through the atmosphere in relation to temperature distribution shown by isotherms and pressure distribution. Dry and saturated adiabats are also shown. An arrowed line shows the path curve followed by a rising parcel of air with known surface temperature of 10° C and a dewpoint of 6° C at 1,000 mb.

Reference should also be made to simple but effective diagrammatic cross-sections of the atmosphere by P. R. Crowe to show important features in relation to marked horizontal contrasts in temperature in winter in the northern hemisphere.[1]

[1] P. R. Crowe, 'The Geographer and the Atmosphere', *Transactions of the Institute of British Geographers*, no. 36, pp. 1–19 (London, 1965).

4

Economic Maps and Diagrams

The economic geographer is concerned with the spatial distributions and inter-relations of various forms of economic activity, which involve primarily a study of the production, distribution and consumption of commodities in their regional settings. The geographical relationships of the distributions, forms and patterns thus involved lend themselves to a wide range of cartographical representation. As V. C. Finch wrote: 'No other of the social studies than economic geography insists upon so rich a symbolization of its facts and concepts in cartographic form.'[1] Many of the methods used must of course conform to certain carto-graphical conventions, but the economic geographer can frequently devise modifications of method or even wholly original presentations of his data. In view of the manifold possibilities, it must be realized that the methods discussed in this chapter and the maps and diagrams used in illustration are, perhaps even more than usual, merely examples and suggestions.

DATA

With the exception of maps of simple areal distributions, most economic maps and diagrams involve some precise depiction of amounts, values, areas, ratios, distances and rates, and their compilation necessitates the handling of much statistical material. In fact, the economic geo-grapher finds that such data form the major part of his sources; from them he selects his factual information, presents his analysis, and draws his conclusions. So he assembles and tabulates his data, considers the cartographical method most suited to the problem involved – whether isopleth map, choropleth map, dot map, graph or diagram – and then converts his tables into the chosen medium.

[1] V. C. Finch, 'Training for Research in Economic Geography', *Annals of the Association of American Geographers*, vol. 34, p. 213 (Lancaster, Pa., 1944).

Available Sources[1]

The available statistical sources may be described under four heads. These comprise international publications, produced by such agencies as the former League of Nations or the United Nations Organization, official government publications, other published information, and a vast amount of unpublished material.

International Publications. International agencies are able to collect, compile and publish statistics in a very convenient form for the economic geographer. Between the wars, the Economic Intelligence Service of the League of Nations produced some very valuable statistical summaries. Its base-volume was the *Statistical Year-Book*, which gave figures for the most recent year available (and retrospectively for a decade) of areas, population, employment and wages, production and consumption, transport and trade, for almost every country in the world. Countries were grouped alphabetically under their respective continents, with text in both English and French, and with very useful notes on sources, which could be followed up if greater detail were required.[2]

Since the beginning of 1947, the responsibility for the publication of international economic statistics has been assumed by the Statistical Office of the United Nations, which is under the Department of Economic Affairs. This agency collects information from the member nations and from the U.N. Special Agencies. The basic publication is the *Monthly Bulletin of Statistics*, which provides averages for each year from 1937, and monthly figures for two years previous to the date of issue. It is not confined to the member states of UNO, but there are some obvious gaps. Users of the *Monthly Bulletin* should have the *Supplement: Definitions and Explanatory Terms*, which provides a vast amount of information concerning the statistics and their derivations, and gives useful leads to more detailed sources. This is especially

[1] A detailed summary of material before 1959 is given by F. J. Monkhouse, 'Current Sources of Reference Material and Statistics', being Chapter 3 (pp. 46–62) of *A Geographer's Reference Book* (Sheffield, 1955; revised edition 1959).
[2] Other League of Nations publications included *Raw Materials and Foodstuffs: Production by Countries, 1938* (Geneva, 1940); *International Trade in Certain Raw Materials and Foodstuffs* (Geneva, 1939); *World Economic Survey, 1938–9* (Geneva, 1939); *Review of World Trade* (annually); and *International Trade Statistics* (annually). A complete biographical survey of the work of the League, 1920–47, is given by H. Aufricht, *Guide to League of Nations Publications* (London, 1951).

essential when comparing figures for a number of countries, because the bases and formulae from which the published statistics are derived often differ appreciably from one country to another. In addition to the *Monthly Bulletin*, there appeared in 1949 the first *Statistical Year-Book of the United Nations Organization*, referring to 1948, and successive volumes have appeared annually. They contain retrospective information for a number of years, as well as that for the current year.

The U.N. Statistical Office publishes also a wide range of other economic summaries; these are too numerous to list here, but details can be obtained from the various U.N. agents. Many other international agencies, for the most part directly or indirectly associated with UNO, publish statistical abstracts and year-books; these agencies include the Economic Commission for Europe, the Food and Agricultural Organization, and the International Labour Office.[1]

An immense amount of statistical information is put out by the Organization for Economic Co-operation and Development in Paris (O.E.C.D.), the successor to the Organization for European Economic Co-operation (O.E.E.C.), by the European Economic Community in Brussels, and by the European Coal and Steel Community in Luxembourg. In fact, the spate of material available is almost embarrassing.

Official Government Publications. Most countries publish statistics of their national resources and economic activities, some in more detail than others. The *Supplement* produced by the U.N. Statistical Office has a most useful Bibliography which gives the main official statistical publications of sixty-six countries. Most of these countries have a single annual publication, which takes the form of some kind of Abstract or Year-book.[2]

These national statistical year-books are extremely useful, as they provide for the economic geographer the detailed information of production and distribution of commodities on a regional administrative basis: by counties, *départements* and provinces. Most countries, too, produce statistical summaries at more frequent intervals than the

---

[1] H.M.S.O. publish annually a most useful *International Organizations Publications*. A catalogue of all U.N. publications is *Ten Years of United Nations Publications, 1945–55.*

[2] Three examples are the *Annual Abstract of Statistics*, published by H.M.S.O. for the United Kingdom Central Statistical Office; the *Annuaire Statistique de la France*, published by the Institut Nationale de la Statistique et des Etudes Economiques for the Ministère de l'Economie Nationale; and the *Jaarcijfers voor Nederland*, published by the Centraal Bureau voor Statistiek at The Hague.

year-books. Finally, the statistical offices of many countries issue a range of more specialized publications.[1]

Other Published Statistical Information. A large amount of statistical material is issued by industrial concerns, both by individual firms and by cartels, such as the Esso Petroleum Company, the Burmah Group and Unilever. Banks and commercial houses are also prolific sources.[2] Finally, mention must be made of the various periodicals devoted to economic surveys, which often provide most convenient and authoritative data.[3]

The statistical sources already mentioned may appear overwhelming, especially as they represent only a few examples. The economic geographer, however, selects only the material he requires for his particular work. If it be a general survey of the world output of some commodity, he requires only summary figures, and these can be obtained from some general international source. If it be a more detailed study of some aspect of the economic life of a particular country, he can begin with the relevant year-book, and as his work develops he will have to consult more detailed and necessarily more specialized sources.

A particularly useful and convenient work is *The Geographical Digest,*[4] published annually, which summarizes statistics for the most recent year of population, production, trade, etc.

Unpublished Statistical Information. For most research work it is often essential to obtain access to unpublished statistical material. Thus to construct the agricultural maps of north-eastern Belgium (Figs. 116–19) on a commune basis, it was necessary to consult many hundreds of manuscript returns in the archives of the Institut National de Statistique in Brussels, and to draw the diagram of freight-movement along

[1] The output of statistical information in the United Kingdom is so vast that the student can only be referred to the *Catalogue of Government Publications* and the *Sectional Lists,* such as those produced for the Ministry of Agriculture, Fisheries and Food, and the Ministry of Technology; these lists are issued at intervals by H.M. Stationery Office. Similarly, the Institut National de la Statistique of France has issued no less than 160 *Etudes Speciales.*

[2] For example, the *Rotterdamsche Bankvereeining Quarterly Review* contains authoritative data concerning many aspects of the economic life of the Netherlands. Industrial directories, such as the *Indicateur des Produits Belges,* published by the *Fédération Nationale de Chambres de Commerce et d'Industrie de Belgique,* are indispensable for a study of the industry of any country.

[3] See, for example, the numerous Supplements of *The Times* and the *Guardian,* and the *Bulletin de l'Institut de Recherches Economiques,* published at Leuven.

[4] Edited by H. Fullard, and published by G. Philip & Son Ltd., London.

the Juliana Canal (Fig. 131); the unpublished figures were supplied by the Centraal Bureau voor de Statistiek at The Hague. A study of the Kempen coalfield and the construction of a series of maps to illustrate the production and export of coal from that field entailed visits to each of the seven collieries and to the coal-port at Genk on the Albert Canal. The gradual tracking down of this information as the work proceeds is indeed one of the fascinations of an enquiry into some aspect of economic geography.

Agricultural Statistics

Statistics of agricultural areas, yields and values are compiled, tabulated and published by the Ministry of Agriculture, Fisheries and Food,[1] or the equivalent government department, for most countries.[2] The raw material of these summaries consists of the collated census forms, completed by each farmer in every parish, commune or similar administrative unit.

For England and Wales and for Scotland, summaries are published on a county basis of the area of arable land, grassland, land under each of the major crops, the yield per acre of each of these, the numbers and categories of animals, and details about farm-labour. The total areas of each county are also tabulated, so that densities can be calculated. If more detailed figures than on a county basis are required, it is necessary to extract the information for each parish from the files of the Ministry of Agriculture, but details for each individual farm are confidential.

Belgium may be quoted as a second example. Agricultural censuses have been held there at intervals since 1846; the results of each of these censuses were published in several large volumes in immense detail on a commune basis, and are invaluable source-books. To replace these large-scale publications, which have become increasingly expensive to produce, an annual summary has been brought out since 1939; up to 1945 information was presented on a basis of *cantons*, but since

[1] Reference should be made to *Sectional List No. 1*, and to two annual publications, *Agricultural Statistics: England and Wales*, and *United Kingdom Agricultural Censuses and Production*.

[2] One of the earliest official agricultural returns is the *Acreage Returns for 1801* (P.R.O., H.O. 87); the systematic collection of agricultural returns, apart from these, began much later in 1866. See D. Thomas, 'The Statistical and Cartographical Treatment of the Acreage Returns to 1801', *Geographical Studies*, vol. 5, no. 2, pp. 15–25 (London, 1959). This discusses the nature of the material, and the methods which may be used to handle them. A full bibliographical note summarizes much of the work done on them.

then the increasing need of economy has limited the publication to a basis of *arrondissements* only. The unpublished details for the individual communes for each year are available in the files of the Institut National de Statistique in Brussels.

Agricultural data are on a formidable scale, and increasingly use is made of calculating machines and computers in their collation.[1] J. T. Coppock, for example, utilized the 350 National Agricultural Advisory districts, with the 1958 agricultural census figures, to prepare material for an agricultural atlas. In order to prepare 120 items, such as proportion of tillage acreage, densities of livestock, livestock units, ratios between crops and livestock, etc., between 450 and 1,100 calculations were required for each district. This represented one year's work by manual calculation on the basis of a continuous eight hours day; a desk calculator would have reduced this, but a computer enabled him to complete the work, despite programming difficulties and various hitches, in eight weeks. Moreover, this vast body of data remains available on tape for later work.

Industrial Statistics[2]

The range of industrial statistics includes the production and consumption of fuel and power and of raw materials, and the output of semi-finished and finished manufactures. Returns vary enormously, and if comparisons are being made between different countries, care must be taken to examine the bases of the figures used. This problem may be illustrated by the consideration of one useful comparative figure, the index of industrial production. An overall index is calculated for each country, which can be used to measure changes in the physical volume of industrial output, and is converted to relate to the base of 100 for a given year. The index numbers are computed according to the weighted aggregate method; the basic data used are the actual quantities produced by a select series of industries, while the weighting

[1] See J. T. Coppock, 'Electronic Data Processing in Geographical Research', *Professional Geographer*, vol. 14, no. 4, pp. 1–4 (New York, 1962). He also made a valuable summary, with examples of the range of techniques, in 'The Cartographic Representation of British Agricultural Statistics', *Geography*, vol. 50, pp. 101–13 (Sheffield, 1965). See also R. H. Best and J. T. Coppock, *The Changing Use of Land in Britain* (London, 1962); and J. T. Coppock, 'The Statistical Assessment of British Agriculture', *Agricultural Historical Review*, vol. 4, pp. 4–21, 66–79 (London, 1956).

[2] D. R. Macgregor, 'The Mapping of Industry', *International Yearbook of Cartography*, vol. 7, pp. 168–85 (London, 1967), who discusses a wide range of problems (range and choice of data, map and symbol scales, the use of colour, etc.) in mapping industrial features.

may consist of the gross value of production, or of the number of workers, or of man-hours.[1]

In some countries, notably in Great Britain, useful indications of industrial activity are provided by the numbers of people employed in a particular industry.[2] Thus the numbers of operatives in various Lancashire textile mills, or the number of men employed in the collieries of a particular coal-field, can be used as a basis for a detailed distribution map of the textile industry or of the coal-field (Fig. 122). The problems involved in handling these occupational statistics are discussed on pp. 318–20.

Transport and Communication Statistics

Trade statistics, i.e. figures of imports and exports, which reflect so much of the economic life of a country, are available for most countries on a commodity basis in terms of both weight and value, and of the origin and destination of the commodities concerned. Various countries use different classifications and groupings of items, and values are subject to very intricate definitions because of Customs requirements.

An analysis of the activity of a port, which is a most interesting aspect of economic geography, must necessarily include a close study of its external trade.[3] In addition, the entrances and clearances of national and foreign sea-going shipping in terms of net registered tonnage, and of passenger traffic for the 'ferry ports', are required. It is essential to examine the criteria upon which the statistics of the ports of the particular country are based; thus French returns include deep-sea fishing vessels; some countries include vessels in ballast, while others exclude them; some omit vessels below a certain minimum size; and so on.

The statistics available for a study of rail, road and water transport vary enormously. Rail figures, including track- and route-length, numbers of locomotives and rolling-stock, and freight and passengers conveyed, are as a rule available for operating regions or districts, as in France and Germany; in France, the returns are published for the four regions of the *Société Nationale des Chemins de Fer français*; in Germany before 1939 returns were available in somewhat more detail

---

[1] See, for example, the *Monthly Bulletin of Statistics, Supplement: Definitions and Explanatory Notes* (Statistical Office of the United Nations, New York, 1948); on pp. 22–48 a comparison is made between the industrial returns of various countries, and the differences are emphasized.

[2] See, for example, *List of Mines* (H.M.S.O.).

[3] See, for example, *Annual Statement of the Trade of the United Kingdom* (London, annually).

for the forty-one traffic districts. But for a country like Belgium only total figures for the whole country can be obtained, which makes the analysis of freight-movement a difficult or even impossible task.

Few statistics are published relating to road transport. Occasionally traffic censuses are taken; thus in Belgium three such censuses have been held, in 1908, 1926 and 1933, when continuous observation was maintained at a thousand points for twenty-four hours. Other sources of information, such as the analysis of motor-bus time-tables, are sometimes helpful (see pp. 400, 444). For the analysis of rail traffic in the United States, 'One per cent sample Carload Waybill Statistics' are published by the Interstate Commerce Commission. Edward L. Ullman has made graphic use of these in his copiously illustrated publications.[1]

The statistics referring to individual inland waterways are usually published in considerable detail, especially for west European countries, since the numbers of vessels and the volume of freight can be easily recorded at the locks; thus it is possible to find the total freight conveyed and the major freight categories, either in terms of tons absolute or of ton-kilometres. The latter figure is derived from the product of each load in tons and the actual distance it travels. It is therefore applicable to the whole length of the waterway, unlike figures of absolute tonnage, which are usually measured at a particular lock, and so may include both short and long-distance loads. The figure of ton-kilometres makes adjustments between the different distances travelled, and so gives a reasonable impression of 'work done'. To compare activity between one waterway and another, an index of ton-kilometres per kilometre of length can be calculated, which represents a value adjusted in proportion to the actual length of the route. Thus a long waterway would have a large return of ton-kilometres because of its length, but might be no more busy than a short waterway. Ton-kilometres per kilometre, therefore, produces a strictly comparable impression of the relative importance of each waterway.[2] Figures can

[1] Special reference may be made to his *American Commodity Flow* (Seattle, 1957), which affords a description and interpretation of traffic-flows in American foreign and domestic trade. See also *Waybill Statistics, Their History and Uses* (Washington, D.C., 1954), compiled and published by the Interstate Commerce Commission, Bureau of Transport Economics and Statistics.

[2] For a full discussion of this problem, with particular examples of the maps which can be drawn from the various statistics available, see F. J. Monkhouse, 'Coal Movement in Belgium, with Special Reference to the Kempen Field', *Publication No. 17*, pp. 99–109, *Transactions and Papers, 1950: The Institute of British Geographers* (London, 1952).

often be obtained of actual loadings and unloadings at ports along each waterway in tons absolute.

NON-QUANTITATIVE MAPS

The Chorochromatic Technique

The most obvious type of economic map is that which shows areal distributions without any quantitative indications. These areal distributions may be either *simple*, merely showing the extent of a single element for definition purposes, or *compound*, differentiating between a series of associated elements on the same map. Basically, this method implies the drawing of bounding lines to delimit specific areas, within which is applied some distinctive shading or colouring. These areal distribution maps can be drawn on any scale, showing a single farm, or a parish,[1] a county, a country or even a continent; obviously the smaller the scale, the more generalized and the less accurate must the map become, as small-scale maps in economic atlases frequently show.

Land-utilization and land-classification maps are probably the best-known and most commonly used of this type of agricultural map.[2] The detail is surveyed on the ground and plotted on a large-scale outline, field by field, to distinguish between arable land, temporary or permanent pasture, orchards and so on, or even in more detail to show the arable land under various crops. Such a map can be drawn on a 6-inch scale and then reduced photographically to a 1-inch or smaller scale. Moreover, material can be extracted from a general land-use

[1] See, for example, J. T. Coppock, 'The Changing Arable in the Chilterns', *Geography*, vol. 42, pp. 217–29 (London, 1957); he uses a series of chorochromatic maps for the Chilterns as a whole to show permanent grass, arable and tillage for 1875, 1931 and 1951, and some most interesting parish maps to show the extent of arable for the same three dates.

[2] The Land Utilisation Survey of Great Britain, directed by L. D. Stamp, issued sheets on 1-inch, ¼-inch and smaller scales. A new Survey, directed by A. F. Coleman, is publishing sheets on the scale of 1 : 25,000. Miss Coleman discusses the problems cf producing these sheets in 'Some Technical and Economic Limitations of Cartographic Colour Representation on Land Use Maps', *Cartographic Journal*, vol. 2, no. 2, pp. 90–4 (London, 1965). The World Land Use Survey was established as the result of the recommendations of the Commission on a World Land Use Survey set up in 1949 by the International Geographical Union. Among its publications are *Occasional Papers* and *Regional Monographs*, edited by L. D. Stamp. The standard land-use classification recognizes nine main categories, with a number of sub-divisions, each denoted by a specific category. See 'A World Land Use Survey', *Geographical Journal*, vol. 115, pp. 223–6 (London, 1950).

survey to portray particular elements, such as the extent of woodland (Fig. 107), or of heathland, or of irrigated lands.[1] At the other end of the scale of generalization is the broad distribution map, such as one

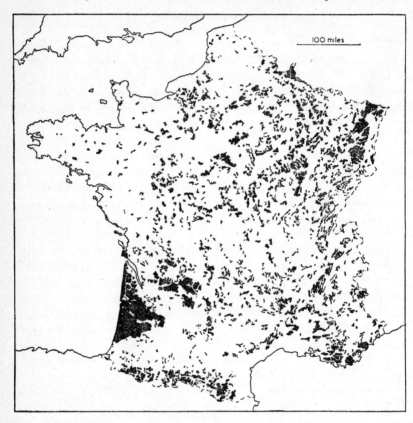

Figure 107. A CHOROCHROMATIC MAP OF WOODLANDS

Based on the *Atlas de France*, plate 38 (Paris, 1935).

This map refers to 1931. The significance of the Landes and the upland areas (note especially the indication of the escarpments bounding the eastern Paris Basin) is brought out clearly.

showing the extent of the American wheat, corn and cotton 'belts', which is of value only in an elementary textbook.

In this connection, a topic of special interest is the mapping of vegetation, on both small and large scales. Maps of 'natural' and 'wild'

[1] L. D. Stamp, *The Land of Britain, its Use and Misuse* (London, 1948), presents many examples of extraction and reduction from large-scale field surveys.

vegetation are extremely difficult to draw, involving particularly the problem of transitional boundaries on large-scale maps. Some of these problems are discussed in detail by A. W. Küchler.[1] Undoubtedly, the most attractive chorochromatic vegetation maps yet produced are French, notably the four regional sheets on a scale of 1 : 1 million entitled *Tapis Végétal*, the single sheet maps on a scale of 1 : 4 million of *Régions Florales* and *Etages et Zones de Végétation*, and a 1 : 2,500,000 map of *Elements floristiques et Limites d'Especes végétales*. All these maps use a brilliantly successful technique of a wide range of colour tints, shading and superimposed symbols. The magnificent *Carte de la Végétation de la France*, on a scale of 1 : 200,000, is in course of publication.[2] The first sheets, of Perpignan and Toulouse, appeared in 1946.

Industrial maps can also be constructed on the same principle. These will include maps of coal-fields, ore-fields and oil-fields, the limits of which can be extracted either from a geological map or from a plan of the concessions (Fig. 108), and maps of land devoted to industrial uses, which can be plotted in the field. Point symbols (see below) can be inserted over the chorochromatic shading to locate the actual collieries, oil-wells and so on, as is done on Fig. 108, thus producing what may be called a *choroschematic* map.

The outline of the distribution area is drawn in, and then shading or colour-tinting is applied. Solid black is preferable for widely scattered small areas of a single element (Fig. 107), particularly if any

---

[1] A. W. Küchler, 'Some Uses of Vegetation Maps', *Ecology*, vol. 34, pp. 629–36 (Brooklyn, N.Y., 1953); 'A Comprehensive Method of Vegetation Mapping', *Annals of the Association of American Geographers*, vol. 44, pp. 404–15 (Lancaster, Pa., 1955); and 'Classification and Purpose in Vegetation Maps', *Geographical Review*, vol. 46, no. 2, pp. 155–67 (New York, 1956). In 'Analysing the Physiognomy and Structure of Vegetation', *Annals of the Association of American Geographers*, vol. 56, pp. 112–25 (Lawrence, Kansas, 1966), the same writer gives a workable classification for describing and analysing vegetation, in terms of rapid and accurate field-work. Valuable instruction in the construction of the maps, classifications of types, choice of tints and symbols, etc., is given by P. L. Wagner, 'A Contribution to Structural Vegetation Mapping', *Annals of the Association of American Geographers*, vol. 47, pp. 363–9 (Lancaster, Pa., 1957). See G. McGrath, 'The Representation of Vegetation on Topographic Maps', *Cartographic Journal*, vol. 2, no. 2, pp. 87–9 (London, 1965); and also 'Further Thoughts on the Representation of Vegetation on Topographic and Planimetric Maps', *Cartographic Journal*, vol. 3, no. 2, pp. 74–8 (London, 1966), with a bibliography of nineteen items.

[2] These maps are a triumph of field-work, compilation and printing, a tribute to the work of H. Gaussen. The printing involves fourteen colours, and a vast number of symbols and letters. Six smaller maps of the same area (1 : 1,250,000) afford summaries of soil, rainfall, '*adversités agricoles*', etc., and there is a large amount of printed marginal information.

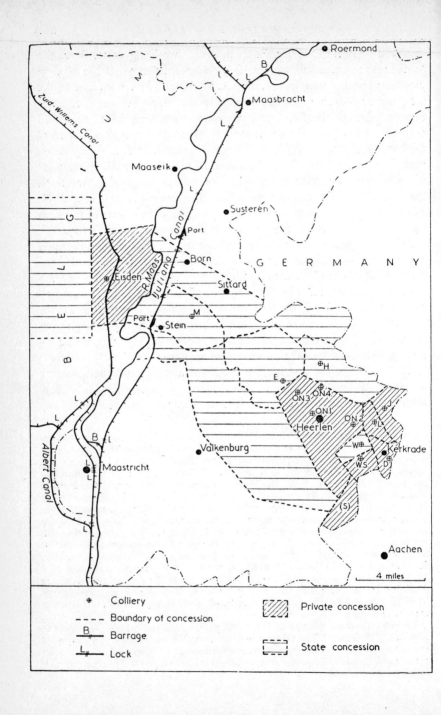

reduction is intended, but if several forms of land-use appear on the same map distinctive shading or tinting must be used (Fig. 109).

Delimitation of Hinterlands

A category of areal map of special interest to the economic geographer is that which seeks to delimit an area of influence such as a port hinterland, and this obviously involves a problem of some complexity. A pattern-map can show railways and waterways converging upon a port, a columnar diagram can depict the shipping and freight handled by the port, while a map can be drawn on which appear concentric circles to represent specific distances from port centres; the last is useful when transport charges on a distance basis form a high proportion of the selling prices of goods.[1]

F. W. Morgan produced a series of maps to depict the pre-1939 hinterlands of the major German ports.[2] He was able to draw these maps because of the detailed nature of the German statistics. Thus railway dispatches and loadings were returned for forty-one traffic-districts within Germany and for twenty-six outside, and returns of traffic were available for each waterway. By studying the freight-flow

[1] A. C. O'Dell, 'Port Facilities and the Dispersal of Industry: the Problem in Scotland', *Geographical Journal*, vol. 97 (London, 1941), shows on p. 115 the relation of ports to the industrial areas of the Midland Valley of Scotland by drawing distinctive circles at distances representing both 25 miles and 10 miles from a major port, and 10 miles only from a minor port.

See also an interesting map of the hinterlands of Brazilian ports, by Preston E. James and S. Faissol, 'The Problem of Brazil's Capital City', *Geographical Review*, vol. 46, p. 305 (New York, 1956).

A whole range of maps, illustrating the use of blocks, divided circles and flow-lines in the study of ports and their hinterlands, is used by N. R. Elliott, 'Hinterland and Foreland as illustrated by the Port of the Tyne', *Transactions of the Institute of British Geographers*, no. 47, pp. 153-70 (London, 1969).

[2] F. W. Morgan, 'The Pre-War Hinterlands of the German North Sea Ports', *Transactions and Papers, 1948: The Institute of British Geographers*, no. 14, pp. 45-55 (London, 1949); and 'The Pre-War Hinterlands of the German Baltic Ports', *Geography*, vol. 34, pp. 201-11 (London, 1949).

Figure 108. A CHOROSCHEMATIC MAP OF THE SOUTH LIMBURG COALFIELD

Based on P. R. Bos and J. P. Miermeyer, *Schoolatlas der Geheel Aarde*, plate 11B (Groningen, 1936).

The following letters are used to indicate the names of individual collieries: D. Domaniale; E. Emma; H. Hendrik; J. Julia; L. Laura; M. Maurits; O.N. 1, 2, 3, 4. Oranje Nassau 1, 2, 3, 4 collieries; W. Wilhelmina; W.S. Willem and Sophia.

Both shading (to indicate private and State concessions) and point symbols (to indicate collieries, barrages and locks) are utilized.

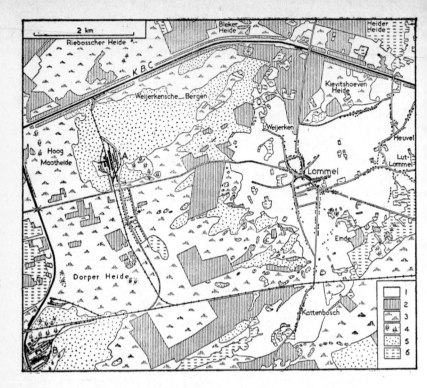

Figure 109. A CHOROCHROMATIC MAP OF LAND UTILIZATION IN THE
LOMMEL DISTRICT IN NORTH-EASTERN BELGIUM

Based on manuscript commune maps and fieldwork.

The numbers in the key are as follows: **1.** arable land; **2.** coniferous plantations;
3. heathland; **4.** marsh; **5.** bare sand and dunes; **6.** permanent pasture. Waterways
are shown by a thick black line, roads by a double line, and railways by a single
barbed line.

The abbreviations are as follows: **A.** factory of the *Cie des Métaux d'Overpelt-Lommel
et Corphalie*; **B.** factory of the *S.A. des Mines et Fonderies de Zinc de la Vieille-Montagne*;
B.B.C. *Beverloo Branch Canal*; **K.B.C.** *Kwaadmechelen-Bocholt Canal.*

Lommel is situated near the Netherlands frontier in north-eastern Belgium. The
main part of the village has grown up at the point of convergence of five roads and
along their inter-connections, and has spread along these roads towards a ring of
small hamlets lying a mile or two from the centre of the village. Surrounding the
group of houses is a series of almost concentric zones of different land utilization. First
there is an area of predominantly arable land, only about a mile in width to the west
but nearly five times as wide to the east. There are small scattered patches of permanent pasture, and more continuous tracts in the valleys of the numerous small
streams. On the edge of the cultivated land lies an incomplete belt of woodland, one
to two miles from the commune centre. Beyond this is the heathland, which in parts
is exceptionally bare and desolate. To complete this picture of a typical Kempen
commune, in the extreme west there are large zinc and chemical works, each with its
housing estate.

by rail and water to and from the major ports, Morgan was able to delineate 'commodity hinterlands'. On his maps he indicated the hinterland of each port by a line bounding traffic districts which dispatched to and received from the respective port at least 50,000 tons of freight; this figure was of course an arbitrary one, but it was chosen as being of a significant and representative order of magnitude. As a rule, statistics are not available for such a detailed depiction of hinterlands.

An interesting modification of port hinterland maps is used by J. H. Bird in a study of British seaports.[1] In this case arcs have been inserted at distances of 25 miles representing 1 per cent of the total export tonnage of that port originating within 25 miles, and at 100 miles representing those ports with 20 per cent of their export tonnage from origins beyond 100 miles. This method allows first-, second- and third-order ports to be identified.

Linear Patterns

The networks of road, railway and waterway systems can be extracted by tracing from detailed topographical maps. The only problem is the classification of specific categories, which may be indicated by different types of line (Fig. 3) or thicknesses of line. Thus on a railway map, it may be necessary to distinguish single, double, multiple and electrified lines, or to differentiate between main-line and light railways, or between broad, standard and narrow-gauge tracks. Waterways are usually classified on the basis of the size of vessel accommodated, while road maps should show the various widths, grades of surface, and official classifications. Compare, for example, the road patterns in three parts of Jugoslavia (Figs. 110–12). The visual impression of main and local road density derived from these patterns is a striking reflection of the diverse geographical features of these three parts of the country. In addition to these 'system maps', detailed plans can also be drawn of sidings and marshalling yards, city 'lay-outs', seaports and canal ports.

The patterns of such route-ways as ocean highways and airways can be drawn by linking scheduled points en route. Care must be taken on the small-scale maps with the projection. The actual route on many projections is not the shortest line between two points, as frequently appears on many small-scale maps in elementary textbooks.

[1] J. H. Bird, 'Traffic Flows to and from British Seaports', *Geography*, vol. 54, pp. 284–302 (London, 1969), contains maps showing the import and export hinterlands of twenty-five British ports for 1964.

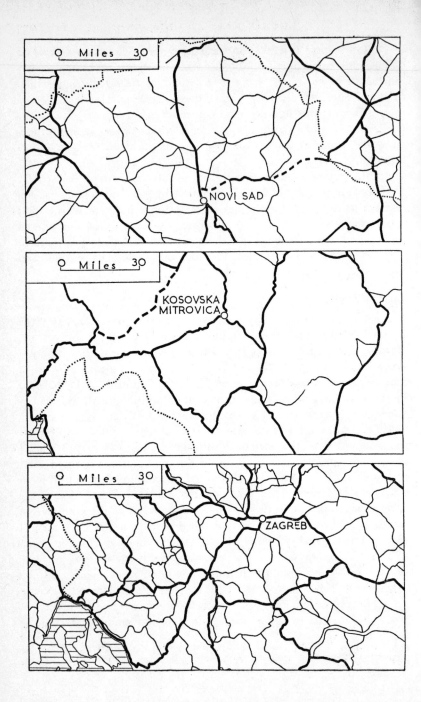

Frequently, however, such route-maps are made in a generalized and diagrammatic form.

Symbols

Non-quantitative distributions centred at a point may be shown quite clearly by symbols, whether geometrical, pictorial or literal (i.e. shown by an initial letter). Thus Fig. 113 gives some indication of the distribution of the various branches of the Belgian textile industry; a black disc was placed over each wool-manufacturing centre, a black triangle over a cotton centre and so on. Its advantage is limited in that there is no indication of relative importance; thus the square at Gent, the centre of the industry, is the same size as the one at Tielt or Ronse. But it does show the concentration in the Lys and Scheldt valleys, with the outstanding exception of the group of woollen towns near Verviers. If the symbols are drawn proportionally to scale, a quantitative element is introduced;[1] these media are discussed in the text.

A recently published article[2] discusses the use of symbols for agricultural distributions on large-scale maps. Q. D. Innis produced two land-use maps, one on a scale of 1 : 288 (1 inch = 8 yards), another of a ½-acre field on a scale of 1 : 144 (1 inch = 4 yards). On these maps, depicting a most complex and intensive land-use, every tree (banana, citrus, coffee), shrub and patch of plants (yam, ginger, sugar) was depicted by one of a large range of symbols.

[1] A wealth of symbols, though placed in a diagrammatic table rather than on a map, is used by G. W. S. Robinson, 'The Resorts of the Italian Riveria', *Geographical Studies*, vol. 5, no. 1, facing p. 32 (London, 1958). For 30 resorts he presents in symbolic form information about their size, numbers of visitors, seasons, classes of accommodation, character, and natural conditions influencing them (e.g. nature of beach, direction of on-shore winds).

[2] Q. D. Innis, 'The Efficiency of Jamaican Peasant Land Use', *Canadian Geographer*, vol. 5, no. 2, pp. 19–23 (Toronto, 1961).

Figure 110–12. ROAD-PATTERN MAPS OF JUGOSLAVIA

Based on *Automobiliska Karta*, 1 : 1,000,000, published by the *Jugoslovenska Standard-Vacuum-Oil Company* (Beograd, n.d.). The map refers to the pre-war situation.

The main roads are shown by heavy lines, main roads under construction by heavy pecked lines and minor roads by fine lines. The frontier is indicated by a dotted line.

The maps are of (*top*) areas in the Vojvodina, (*middle*) the Region of Kosmet, and (*bottom*) Croatia. The main roads are evenly distributed over the whole country, the minor roads are an indication of local density of population and of prosperity.

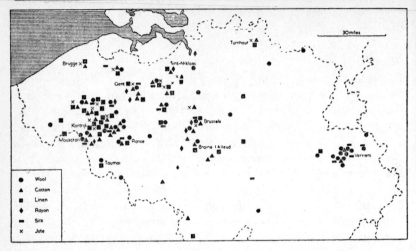

Figure 113. A NON-QUANTITATIVE SYMBOL MAP OF THE BELGIAN TEXTILE
INDUSTRY

Based on various industrial directories, including *Indicateur des Produits Belges*
(Bruxelles); *Comité central industriel de Belgique: Liste des Etablissements industriels affiliés*
(Bruxelles); and *Official Directory of Belgian Exporting Manufacturers* (Brussels).

The map shows the general distribution of the various branches of the industry,
particularly its concentration in Flanders, with the outlying woollen centre of Verviers.

ISOPLETH MAPS

Agricultural Isopleths

If sufficiently detailed data are available, isopleths can be used with
effect to express ratios between associated agricultural phenomena.
Frequently, in fact, these quantitative ratios are much more revealing
and significant than the absolute values of the related elements as

Figures 114–15. ISOPLETH AND CHOROPLETH MAPS OF BELGIAN HEATH-
LANDS, 1866

The data of heathland in each commune were obtained from *Agriculture: Recensement
Général* (Bruxelles, 1866). The boundaries of the communes were taken from the *Carte de
Belgique – 1 : 320,000, Comportant la Subdivision Administrative du Territoire* (Bruxelles,
1938). The area of each commune (from which the densities were calculated) was
taken from the *Recensement Général de la Population, 1866* (Bruxelles, 1870).

The figures in the key indicate the percentage of the total area of each commune
under heath as follows: **1.** over 40; **2.** 30 to 40; **3.** 20 to 30; **4.** 10 to 20; **5.** 1 to 10;
6. under 1.

The towns are indicated by abbreviations, as follows: **A.** Antwerp; **H.** Hasselt;
L. Leuven: **Ma.** Maaseik: **Me.** Mechelen; **T.** Turnhout.

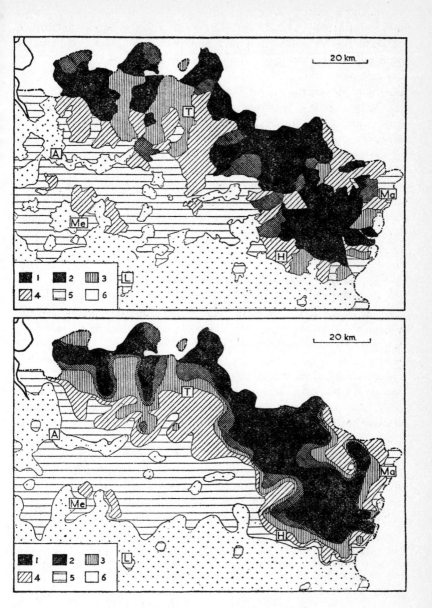

portrayed separately. The simplest application is when the ratio is
between the total area of a unit and that area under some specific
land-use, such as crop-acreage ratios, or the proportion under permanent pasture. Fig. 114 shows an isopleth map of the distribution of
heathland in north-eastern Belgium in 1866; it is useful to compare
this with Fig. 115, which is a choropleth map constructed from the
same data. Other ratios include the number of animals per square
mile, or livestock units per square mile, or such interesting correlations
as the yield per unit and the rainfall total during the growing season,
the amount of milk produced compared with the area under all crops
expressed in gallons per annum per acre,[1] and many more. In addition,
climatic isopleths, such as isotherms, isohyets, lines showing the length
of the growing season or the date of the last killing frost (Fig. 74), may
be superimposed with very great effect upon a choropleth or dot map
of agricultural distributions.

The Drawing of Agricultural Isopleth Maps. The method of drawing
these isopleths depends upon the nature of the statistics available and
the size of the units involved. Where the simple interpolation of isopleths is desired, figures must be available for sufficiently small units
to allow a number of control points (see p. 43) to be plotted at the
centre of each unit, as in the case of Fig. 114, for which proportional
values were calculated for 666 communes. Similarly, W. D. Jones[2]
placed his plotted points in the geometrical centre of each township.
R. R. Rawson, on the other hand, dealing with mainly square administrative units in the Dakotas, shaded those squares representing counties
with more than 15 per cent of the total farmland under wheat, and
then somewhat arbitrarily smoothed off the square boundaries to form
isopleths.[3] A third technique was used by E. R. Payne, who first constructed a series of dot-maps, to show beef-cattle, dairy-cattle, and
actual areas of arable land, and superimposed over each map a grid,
each square of which represented 4 square miles on the ground.[4] She

---

[1] W. D. Jones, 'An Isopleth Map of Land under Crops in India', *Geographical Review*,
vol. 19 (New York, 1929); and 'Ratios and Isopleth Maps in Regional Investigation
of Agricultural Land Occupance', *Annals of the Association of American Geographers*,
vol. 20 (Lancaster, Pa., 1930). The very interesting map of milk-crop ratio appears
in the second reference, on p. 193.

[2] W. D. Jones, op. cit. (1929).

[3] R. R. Rawson, 'The Agricultural Geography of the Dakotas', *Geography*, vol. 25,
pp. 6–17 (London, 1940).

[4] E. R. Payne, 'The Agricultural Regions of the Market Harborough–Rugby Area',
Geography, vol. 31, pp. 98–105 (London, 1946).

counted the number of dots falling in each square, next delimited those particular squares containing certain critical quantities, and finally smoothed out the square boundaries and replaced them with generalized isopleths.

The features of the agricultural geography of the United States may be presented in outline on the basis of states, or in more detail on the basis of counties. Over much of the country, notably in the Mid-West, the rectangular land-survey system affords a pattern of townships, six miles square, for which details (though unpublished) are available in manuscript form from the Bureau of the Census, U.S. Department of Commerce. Immensely detailed work has been carried out by J. C. Weaver,[1] who has produced a large number of isopleth maps in black and white to show the distribution of harvested cropland, individual crops, their increase and decrease, crop combinations, etc. On a map of the last, each township square is coloured in one of forty-four different colours, shadings or stipple, together with combinations of fourteen initial letters. The results are most impressive, attractive and informative.

A further variant was used by P. W. Porter,[2] in producing a map of Minnesota showing harvested cropland as a percentage of total area. He began by inspecting air photo mosaics, on which cropland can be readily distinguished from lake, forest, bog and waste, and from them plotted unit areas of very dense, dense, moderately scattered, very scattered and no cropland. Using the statistical returns of the Census of Agriculture, he placed the dots (ratio, 1 dot = 1,000 acres of cropland) within the outline county maps, using the areas of varied density as a guide. Over this dot distribution, Porter then superimposed a hexagonal grid (see p. 43), each cell of which represented to scale 100,000 acres. The number of dots in each cell were counted, thus giving the percentage of that cell covered with cropland. Each point used to plot the isopleths was selected according to the position of the majority of the dots – in the centre of a dominant cluster, or at the geometrical centre of the hexagon if the dots were evenly spread. The isopleths were then drawn at 10 per cent intervals for the state of Minnesota.

[1] J. C. Weaver, 'Changing Patterns of Cropland Use in the Middle West', *Economic Geography*, vol. 30, pp. 1–47 (Worcester, Mass., 1954); 'Crop-Combination Regions in the Middle West', *Geographical Review*, vol. 44, pp. 175–200 (New York, 1954); and 'The County as a Spatial Average in Agricultural Geography', *Geographical Review*, vol. 46, pp. 536–65 (New York, 1956).

[2] P. W. Porter, 'Putting the Isopleth in its Place', *Proceedings of the Minnesota Academy of Science*, vol. 35, pp. 372–84 (St Paul, 1958).

'Accessibility' Isopleths

Isopleths can be used to supplement linear-pattern maps of transport systems. One type of 'distance map' can be constructed where isopleths represent values calculated in terms of miles of road per hundred square miles of area. From an inspection of a road-pattern map, the country is divided into areas within each of which the density of the road network is broadly uniform, and that density is calculated in terms of miles per 100 square miles, by measuring the length of road within each area with an opisometer and finding the area with a planimeter. The values at the centre of each area are plotted, and isopleths interpolated at various intervals.

As an alternative method, the route-map may be gridded, the length of route measured in each square, and an average figure obtained, which is then placed in the centre of the square. Each of these methods bring out the areas of high, average and low density, and so helps to correct the impression of even density which a pattern-map might otherwise convey.

Another type of 'distance-isopleth' map was constructed by L. D. Stamp, who drew lines to indicate areas more than five miles from a railway and the same distance from an 'A' road respectively.[1] The areas thus defined were filled in black for emphasis.

A group of 'accessibility isopleth' maps involve both time and distance; these are often known as 'travel-speed' maps. E. G. R. Taylor used generalized isopleths (to which is given the name of *isochrone*) (see p. 438) to divide England and Wales into four zones in respect of accessibility by rail to and from London, Leeds, Liverpool, Newcastle, Manchester and Birmingham.[2] A different application of this idea was used to show the travel time in days from Boston in the years 1790–8,[3] while yet another depicted New York as a centre of travel in 1800, 1830, 1857 and 1930, stressing the progressive acceleration in travel.[4] S. W. Boggs elaborated this principle by drawing a series of *isotachic maps*, to illustrate the distances that may be feasibly travelled by surface means in 1940, in all directions.[5] He delimited distances of 50–100,

---

[1] L. D. Stamp, *The Land of Britain: Its Use and Misuse*, p. 209 (London, 1948).

[2] E. G. R. Taylor *et al.*, 'Discussion on the Geographical Distribution of Industry', *Geographical Journal*, vol. 92, pp. 22–39 (London, 1938).

[3] E. Staley, *World Economy in Transition* (New York, 1939).

[4] Edited J. K. Wright, *Atlas of the Historical Geography of the United States* (New York, 1932).

[5] S. W. Boggs, 'Mapping the Changing World: Suggested Developments in Maps', *Annals of the Association of American Geographers*, vol. 31, pp. 119–28 (Lancaster, Pa., 1941).

100–250 and 250–500 miles per day in terms of available roads, and also added a category of over 1,000 miles by rail. He assumed that travelling was carried out by the fastest possible means, whether by horse, automobile or railway, but excluded aircraft, and he used the positions of the roads and railways as 'ribbons of land dedicated to movement' to help to define his areas. Four grades of shading were added between the isotachs.

Edward L. Ullman constructed a map of the United States showing isoline maps of equal delivered costs and equal freight-rates per 1,000 board-feet of lumber shipped by rail from the north-western Douglas Fir region and from Hattiesburg, Mississippi.[1] Akin to this are maps of *isophers*, on which isopleths connect points of equal freight-rate from some centre.[2]

Isopleths and Economic Regions

The linear quality of isopleths lends itself readily to the delimitation of a region, whether it be climatic, agricultural or industrial. Isopleths can be superimposed to give a quantitative impression of all the factors bearing upon the definition of a particular region.

Agricultural Regions. As Hartshorne and Dicken wrote in 1935, 'the great advance which Köppen made in the study of climatic regions by the use of the statistical method can also be expected in the similar study of agricultural regions based on statistical criteria, thereby putting on a scientific basis a significant aspect of the study of the cultural landscape'.[3] These geographers drew a series of maps of Europe and the United States on which appeared a variety of isopleths based on crop-acreage ratios, and then they chose the most significant of these isopleths to delimit their regional types of agriculture.

Two further examples of this method deserve mention. Rawson[4] constructed isopleth maps based on quantities established for each county in the two states of North and South Dakota, and a super-imposition of various isopleths enabled the agricultural divisions to

[1] Edward L. Ullman, *American Commodity Flow*, p. 58 (Seattle, 1957).

[2] See also J. W. Alexander, S. E. Brown and R. E. Dahlberg, 'Freight Rates: Selected Aspects of Uniform and Nodal Regions', *Economic Geography*, vol. 34, pp. 7–18 (Worcester, Mass., 1958).

[3] R. Hartshorne and S. N. Dicken, 'A Classification of the Agricultural Regions of Europe and North America on a Uniform Statistical Basis', *Annals of the Association of American Geographers*, vol. 25, pp. 99–120 (Lancaster, Pa., 1935).

[4] R. R. Rawson, op. cit. (1940).

be delimited. E. R. Payne[1] superimposed isopleths upon maps of the Market Harborough–Rugby area, and so delimited agricultural regions by assessing the overlap. Thus one of her regions was defined as having over 850 beef-cattle, less than 400 acres of arable land, and less than 375 dairy-cattle per unit of 4 square miles.

Industrial Regions. E. G. R. Taylor produced a map upon which all areas in Great Britain unsuitable for industrial location were indicated in solid black.[2] These areas were determined by superimposing isopleths representing certain specific factors, such as contours, density of population isopleths, distance isopleths from nodal cities and so on. This process was termed 'sieving-out' and the resultant maps are sometimes referred to as '*sieve-maps*'.

CHOROPLETH MAPS

Agricultural Choropleths

Choropleth maps of economic distributions are of very wide usage. Many statistics, particularly of agricultural returns, are published on a basis of administrative divisions. Figures of areas, total yields, average yield per unit of area, total values and the like are presented in tabular form in agricultural censuses. From these, various ratios and proportions for each administrative division can be calculated, a scale of values chosen, and a graduated system of shading applied. Such maps include the percentage of the total area of each division under arable, the average size of holding in each, the average yield per unit of area for each division, the average number of animals per unit of area for each division, the value of farmland per unit of area, which gives a comparable impression of agricultural prosperity, and many more.[3] An interesting variation on a map showing the number of animals per unit of area is the *stock-index* map, on which the average number of

---

[1] E. R. Payne, op. cit. (1946).

[2] E. G. R. Taylor, op. cit. (1938).

[3] G. T. Trewartha, 'Ratio Maps of China's Farms and Crops', *Geographical Review*, vol. 28, pp. 102–11 (New York, 1938), uses fifteen ratio maps, including the relation of cultivated to the total area, crop acreage per farm household, total crop acreage in relation to the cultivated area, the irrigated to the total area, and the proportion of the total area under various crops. See also an interesting series of agricultural choropleth maps for the state of Nebraska, on a county basis, dealing with the critical problem of wheat failure, by L. Hewes and Arthur C. Schmieding, 'Risk in the Central Great Plains: Geographical Patterns of Wheat Failure, 1931–51', *Geographical Review*, vol. 46, pp. 375–87 (New York, 1956).

'stock units' per unit of area for each administrative division are indicated.[1] Similar ratio maps may be used to depict changes in land-use areas or yields over a period of time, as in Figs. 116–19, which show on a comparable basis the striking alteration in the proportions of arable land, permanent pasture, heathland and woodland in north-eastern Belgium between 1866 and 1946. The total area of each of 666 communes, together with the areas of the various land-use categories, were tabulated for the two dates, and the change over the period was expressed as a percentage of the total area for each commune. Particular care had to be taken in cases where the area of a commune had changed, or where communes had been created or had been merged with others; separate values had to be calculated for those pieces of land not comparable for the two dates as whole communes.

An interesting series of agricultural choropleth maps has been constructed by J. C. Weaver,[2] using the township values as a basis for his distributions. He has produced a large number of maps, in black and white and in colour, to show the distribution of first- and second-ranking crops, their increase and decrease, combinations, etc. On one coloured map of the last (op. cit., 1956, folded map in pocket), he used forty-four different colours, shadings and stipples, together with combinations of fourteen initial letters.

R. N. E. Blake has developed a valuable method of constructing choropleth land-use maps based on a grid-network. Few regional or countrywide surveys contain land-use maps based on quantitative data, largely because of the lack of this data and of a simple acceptable method of resolving complex patterns into manageable and significant units. The new 1 : 25,000 Land-use Survey (see p. 267) provides the necessary source material, and the kilometric National Grid is both a suitable unit for resolution and is in harmony with trends towards other data processing by grid-computer methods (see p. 82). He made a test-analysis of East Suffolk by drawing a transparent square kilometre grid over the Six-inch Series sheet-line, and put four dots in

---

[1] W. Smith, *An Economic Geography of Great Britain*, p. 249 (London, 1948), has a map of northern Northumberland, compiled by J. C. Dunn, in which stock units of cattle and sheep per 100 acres of the total area of each parish are computed, and a choropleth grading devised. Stock-units are here defined in a special index in relation to the specific conditions of the area as follows: cows and heifers in milk, cows in calf, 1; heifers in calf, $\frac{3}{4}$; other cattle, over 2 years, $\frac{7}{8}$; other cattle, 1–2 years, $\frac{2}{3}$; calves, $\frac{1}{5}$; lowland ewes with lambs, $\frac{1}{4}$; lowland yearling sheep, $\frac{1}{5}$; hill ewes with lambs, $\frac{1}{7}$; hill hogs and others, $\frac{1}{8}$. The U.S. Department of Agriculture uses a different basis: 1 unit equals 1 horse, 1 mule, 1 cow, 7 sheep, 7 goats and 5 pigs.

[2] See p. 279, with references cited.

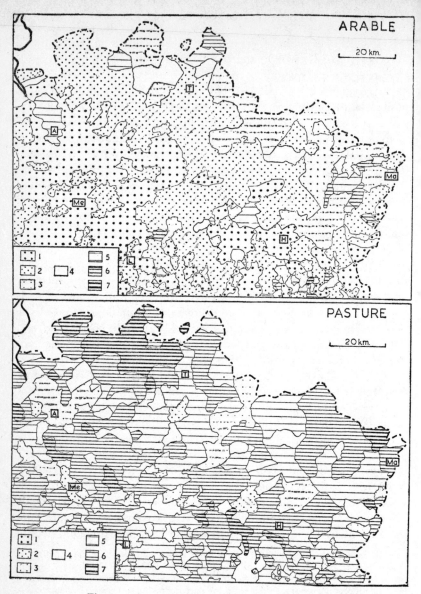

Figures 116–19. CHOROPLETH MAPS OF LAND-UTILIZATION

The change in each commune, for each of the four categories, was calculated from data obtained from *Agriculture : Recensement Général* (Bruxelles, 1866), and from results of the unpublished cadastral survey of 1942, made available by the *Institut National de Statistique* in Brussels.

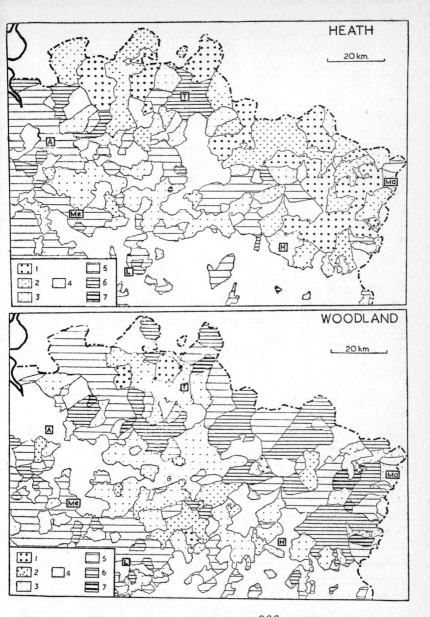

CHANGES IN THE BELGIAN KEMPENLAND, 1866-1942

The figures in the keys indicate the change in the areas of the four categories, between the two surveys, expressed as a percentage of the total area of each commune, as follows: **1.** decrease exceeding 25; **2.** decrease of 10-25; **3.** decrease of 2-10; **4.** little or no change, i.e., less than plus or minus two; **5.** increase of 2-10; **6.** increase of 10-25; **7.** increase exceeding 25.

each grid-square, i.e. 100 dots per sheet; these were spaced at regular intervals conforming to the corners of squares 0·25 sq. km in area located in the centre of the kilometre grid squares, thus avoiding the visual discomfort of taking readings from line intersections or sheet edges. The transparent grid-sheet was placed over each Six-inch L.U.S. sheet, and the land-use falling under each dot was recorded, square by square. The classification adopted was the first-order subdivision in the land-use hierarchy, i.e. settlement, farmland, woodland and 'wildland'. For statistical accuracy the sampling was scrupulously followed, even if insignificant uses did not seem to typify the square in which the dot occurred. The results were recorded on a pro-forma for each 6-inch sheet, with the four land-use categories along the horizontal axis and the twenty-five squares along the vertical. Thus the sampling grid gave a total value of 4 to each kilometre square, which were mapped as percentages for each of the four categories on a square kilometre grid choropleth map. The method is claimed to be very accurate at county levels, and reasonably so at district level. The summary of this survey gave the following land-use for East Suffolk in the mid-60s (Fig. 120).

	Acres	Percentage of total
Settlement	56,620	10·1
Farmland	445,605	80·0
Woodland	34,485	6·2
Wildland	20,650	3·7
Total	557,360	100·0

Industrial Choropleths

Choropleth maps are infrequently used for industrial maps, since 'quantity in area' is rarely involved. However, one example of such a map was compiled and drawn by R. E. Murphy and H. E. Spittal,[1] which was intended to show 'coal-mining intensity' in terms of tons per square mile on a county basis. The average density was calculated for each county in the Appalachian bituminous coal region, regardless of the fact that in many cases a portion of the county was known to be non-producing. Obviously such an industrial map is of limited value.

[1] R. E. Murphy and H. E. Spittal, 'A New Production Map of the Appalachian Coal Region', *Annals of the Association of American Geographers*, vol. 34, pp. 164–72 (Lancaster, Pa., 1944).

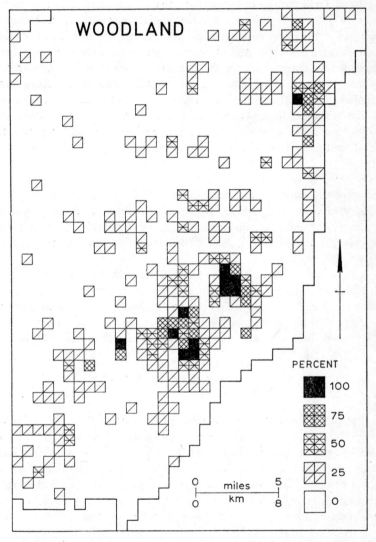

Figure 120. DENSITY OF WOODLAND OVER PART OF EAST SUFFOLK
After a map by R. N. E. Blake (formerly of the East Suffolk Planning Department).

B. L. C. Johnston,[1] seeking to depict industrial distributions in the
West Midlands of England, used the 429 kilometre squares of the
National Grid system covering the area under consideration, partly
because the confidential nature of the official returns on which he based
his work made exact plotting impossible, partly because '. . . the kilometre square is sharper and more manageable than the variously sized
administrative areas used by the population census or by employment
exchanges for their published enumerations'. Thus he shaded squares
containing more than 400 workers per square kilometre, inserted three
grades of symbol for factories within their correct squares, worked out
the percentage of total workers round in each square, and so produced
significant patterns of the Black Country.

Transport Choropleths

Occasionally choropleth maps can be used for the maps showing the
density of transport systems, where official returns of the length of
routeway for each administrative unit are available. Such a map,
however, is less revealing than an isopleth map, constructed as described above.

A remarkable example of transport choropleth maps is provided by
D. Neft,[2] who studied the frequency of the passenger services and made
maps (with eight shaded 'frequency zones', according to the number of
trains per operating day) for New York, London and Paris. A frequency
zone was defined as a maximum radius of $2\frac{1}{2}$ miles from each station.

QUANTITATIVE SYMBOLS

Dots

Perhaps the most convenient method of illustrating the distribution of
some absolute agricultural figures is by means of dots, each with an
identical value, placed within the boundaries of the administrative
areas to which they refer. The chief objection is that yields and numbers per unit of area are not accurately shown, although a reasonable
visual impression of density of distribution is given for stock; thus if
two counties have the same number of cattle, but one county is twice

[1] B. L. C. Johnston, 'The Distribution of Factory Population in the West Midlands
Conurbation', *Transactions and Papers, 1958: Institute of British Geographers*, no. 25,
pp. 209–23 (London, 1958).

[2] D. Neft, 'Some Aspects of Rail Commuting: New York, London and Paris',
Geographical Review, vol. 49, pp. 151–63 (New York, 1959).

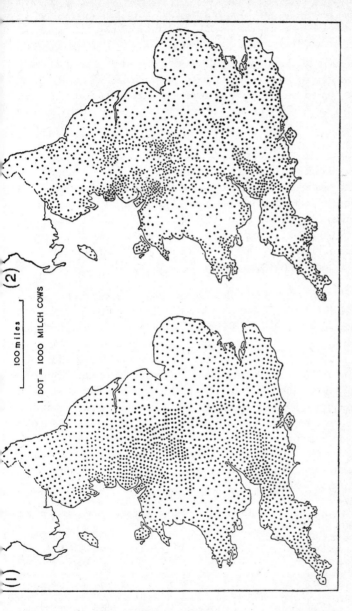

100 miles

1 DOT = 1000 MILCH COWS

Figure 121. TWO METHODS OF PLACING THE DOTS ON A DISTRIBUTION MAP

The data were obtained from Ministry of Agriculture and Fisheries, *Agricultural Statistics, 1930* (London, 1931).

On the left-hand map, the dots are placed objectively, so that they are evenly spaced within each county (or county division).

On the right-hand map, the dots are placed subjectively. An attempt has been made to show geographical distribution by taking into consideration factors of relief and land utilization as far as the scale allows.

as large, then the density of the dots will be half as great (Fig. 121).
Geographical factors can be allowed for when placing each dot to give
a subjective impression (see p. 27).[1]

Dots can be used very effectively when plotting values in detail on
a large scale. E. R. Payne, for example, drew a most detailed dot map
of the Market Harborough–Rugby area, on which each dot represented
10 animals, based on the 1935 returns for each parish, and another
map to show the acreage of arable land, on which one dot represented
5 acres.[2]

'Mille' Maps. L. D. Stamp used a variation of the dot principle to
emphasize changes in spatial distributions between two dates.[3] He
called these 'mille' maps, because each map bears 1,000 dots, each
dot therefore representing 0·1 per cent of the total. Thus one pair of
maps compared the distribution of arable land in Britain in 1874 with
that in 1938; on the first map each of the thousand dots represented
0·1 per cent of 18,089,000 acres, on the second each represented the
same proportion of 11,861,000 acres.

Colour-dots. An extremely effective method of indicating various
agricultural distributions, though one requiring immense patience and
care, is by means of dots of different colours. This has been developed
particularly by G. F. Jenks,[4] who sought to make a map of the Mid-
West on a scale of 1 : 2,500,000, using dots of 0·01 inches radius each
representing 10,000 acres under a specific crop, using data from the
Census of Agriculture, plotted county by county. The dots, with
adhesive reverse sides, were patiently stuck in their correct positions.
This work was extended, and culminated in the publication in 1959 of
a map, 'Crop Patterns in the United States', on a scale of 1 : 5,068,800.
This forms a sheet of the National Atlas of the United States.[5] He chose
his eleven colours carefully, with some psychological relationship to the

---

[1] There are many striking examples in the various county volumes of the Land
Utilisation Survey, and in L. D. Stamp, op. cit. (1948), while the *Agricultural Atlas of
England and Wales*, published by the Ordnance Survey (Southampton, 1932), makes
extensive use of the method.

[2] E. R. Payne, op. cit. (1946).

[3] L. D. Stamp, op. cit. (1948). The pairs of maps represented arable land (p. 102),
permanent grass (p. 103), wheat (p. 104), oats (p. 105), cattle (p. 106) and sheep
(p. 107).

[4] G. F. Jenks, '"Pointillism" as a Cartographic Technique', *Professional Geographer*,
vol. 5, no. 5, pp. 4–6 (New York, 1953).

[5] For sale by the Superintendent of Documents, U.S. Government Printing Office,
Washington 25, D.C.

crops they represent, such as yellow (small grains), orange (corn), light green (hay), brown (peanuts), etc. High value, low acreage crops were shown in intense hues for emphasis, such as purple (fruit) and truck-crops (black). The result is most striking; the concentrations of dots emphasize the well-known 'belts', their blending results in transitional zones which avoid the problem of boundaries which a solid wash would involve, and justice is done to minor crops.

Commodity Origin and Destination Maps. A remarkable series of maps using located dots denoting specific values has been constructed by Edward L. Ullman,[1] in making a geographical interpretation of rail and water traffic based on principles of inter-state exchange. In all, he published in the volume mentioned below (in addition to many other articles) maps for twenty states. Using one dot to represent specific quantities of freight (varying according to the nature of the commodity), grouped in tens for easy appreciation, he plotted pairs of maps showing origin and destination respectively for each state examined. In the maps of origin, the number of dots is proportional to the tonnage movement from that state to the state indicated in the title. In the maps of destination, the number is proportional to the amount received from the title state. The maps show both total movements, and major commodity groupings (forest products, petroleum, other minerals, agricultural products, animals and manufactures).

Proportional Symbols

Proportional symbols are used very effectively to illustrate quantities which can be located specifically at a point. The exact location of the mine or factory can usually be found from a large-scale map or by an actual visit. Then the symbol must be chosen, whether square, rectangle, circle,[2] other geometrical figure, block-pile,[3] literal[4] or

---

[1] Edward L. Ullman, *American Commodity Flow* (Seattle, 1957).

[2] J. C. Weaver, 'United States' Malting Barley Production', *Annals of the Association of American Geographers*, vol. 34, pp. 97–131 (Lancaster, Pa., 1944), which employs dots and circles of various sizes.

[3] See E. Raisz, 'Geographical Distribution of the Mineral Industry of the United States', *Mining and Metallurgy* (New York, 1941), in which he devotes six plates to block-pile illustrations. See also J. H. Bird, 'Seaports and the European Economic Community', *Geographical Journal*, vol. 133, p. 313 (London, 1967), who uses located block-piles (1 square = 100,000 tons of cargo) to show cargo handled by French ports, 1965.

[4] For a striking example of this method, see G. B. Cressey, *China's Geographical Foundations*, p. 108 (New York, 1934), where minerals are denoted by initials which are drawn approximately proportional to the square root of the value they represent.

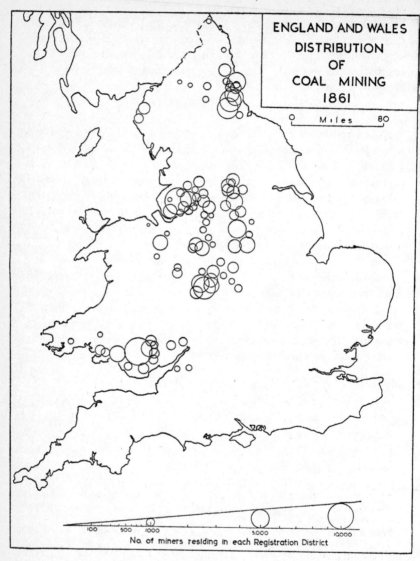

Figure 122. A DISTRIBUTION MAP OF COAL MINING IN ENGLAND AND
WALES, 1861

The data were obtained from *Census of England and Wales 1861: Appendix to the General
Report* (London, 1863).

pictorial symbol, and the size or area of the symbol calculated in proportion to the quantity which it represents. A scale of symbols can be constructed, as shown in Fig. 10. The relative importance of the various mining areas in England and Wales in 1891, for example, is

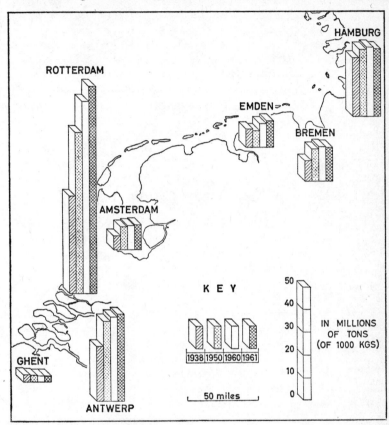

Figure 123. LOCATED BLOCK–PILE MAP TO SHOW FREIGHT HANDLED BY PORTS IN WESTERN EUROPE FOR FOUR SELECTED YEARS

Based on *Rotterdam Europoort*, Quarterly, 1962/no. 2, published by the Port of Rotterdam, p. 13 (Rotterdam, 1962).

shown on Fig. 122 by means of open circles; it will be noticed that the relative importance of the mining districts is indicated in terms of the numbers of workers. Figure 123 depicts the relative importance of the major ports of Western Europe in terms of freight handled for a number of years; the increasing pre-eminence of Rotterdam is manifest.

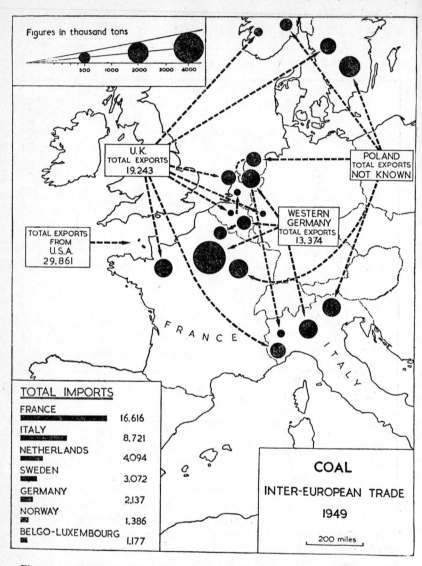

Figures in thousand tons

500 1000 2000 3000 4000

U.K.
TOTAL EXPORTS
19.243

POLAND
TOTAL EXPORTS
NOT KNOWN

WESTERN
GERMANY
TOTAL EXPORTS
13.374

TOTAL EXPORTS
FROM
U.S.A.
29.861

F R A N C E

I T A L Y

TOTAL IMPORTS	
FRANCE	16.616
ITALY	8.721
NETHERLANDS	4.094
SWEDEN	3.072
GERMANY	2.137
NORWAY	1.386
BELGO-LUXEMBOURG	1.177

COAL

INTER-EUROPEAN TRADE

1949

200 miles

Figure 124. A SYMBOLIC MAP TO ILLUSTRATE THE SOURCES AND DESTINA-
TIONS OF COAL IN INTER-EUROPEAN TRADE, 1949
Based on data supplied by the British Iron and Steel Federation.

Quite apart from point locations, proportional symbols can be placed on a map within the administrative region to which they refer. Fig. 124 uses proportional discs, together with arrows to show not the actual routes (which would be virtually impossible) but the countries of origin and destination; together they strikingly demonstrate the complicated interdependence of European heavy industry.

Symbols are less valuable for agricultural distributions, since point locations are rarely involved, except for such features as co-operative dairies or beet-sugar refineries, which are really industrial distributions. Proportional symbols placed within an administrative unit can be used for totals of animals, but should not as a rule be employed for total crop yields, since yields per unit of area, which are more significant than absolute totals, are not indicated. Axel Sömme drew an interesting map of the present importance and recent development of the transhumance of milch-cows in southern Norway.[1] Two contrasting symbols were used, one to represent an increase of 100 cattle on each *seter* between 1907 and 1938, a second to represent a similar decrease.

GRAPHS

Line-graphs

Line-graphs, either simple, multiple or compound, may be used to show absolute or percentage values of agricultural or industrial output, freight transport, trade statistics and so on, over a period of time. Absolute and percentage figures of the quantities involved may be placed on the same graph, using different weights of line and an absolute scale on the left side of the graph, with a percentage scale on the right; thus a graph of coal output of the southern and northern basins of Belgium from 1917 to 1946 is rendered all the more useful if the rapidly increasing proportion of the total contributed by the northern field is emphasized by a superimposed percentage graph. Figure 125 is a compound graph, used to analyse the total and individual outputs of the seven collieries in northern Belgium from 1919; as each colliery came into production, its output was added in the form of another graph-line, and the section was shaded distinctively.

Semi-logarithmic graph-paper can be used to show rates of change of output, or where the range of absolute values to be represented is considerable. Figure 126 graphs on three-cycle paper the output of

---

[1] A. Sömme, 'Norwegian Agriculture and Food Supply', *Geography*, vol. 35, pp. 215–27 (London, 1950).

French coal and iron ore for a century, from 1849 until 1949. The range of absolute figures for coal production is from 4 million tons in 1849, to a maximum of 55,057,000 tons in 1930. Not only are the widely fluctuating trends of production accurately presented, but the result is of manageable size; a graph of the same quantities on an arithmetic scale, allowing one inch of vertical scale to 100,000 tons (i.e. the length

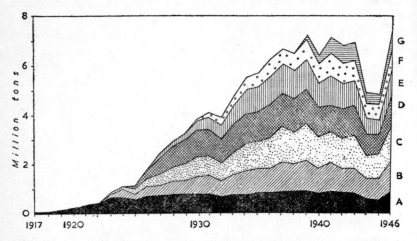

Figure 125. A COMPOUND GRAPH OF COAL OUTPUT FROM THE KEMPEN-
LAND COLLIERIES, 1917–46

Based on data obtained from successive volumes of the *Annales des Mines* (Bruxelles).
The collieries are shown in chronological order of production, as follows: **A.** Winter-slag; **B.** Beringen-Koersel; **C.** Eisden; **D.** Waterschei; **E.** Zwartberg; **F.** Helchteren-Zolder; **G.** Houthalen.

of the lowest division of the log-scale used) would be over four feet in length.

Ergographs

Ergographs, or curves showing the amount of work done at various times of the year, is a term coined by A. Geddes and used by A. G. Ogilvie in a paper on regional techniques.[1] The information can be plotted as a curve showing the amount and nature of work done each month, either in the conventional Cartesian manner,[2] or more properly

[1] A. G. Ogilvie, 'The Technique of Regional Geography with special reference to India', *Journal of the Madras Geographical Society*, vol. 13, p. 121 (Madras, 1938).
[2] See examples in A. Demangeon, 'France: Economique et Humaine', pp. 2–3, *Géographie Universelle*, vol. 6, pt 2 (Paris, 1946).

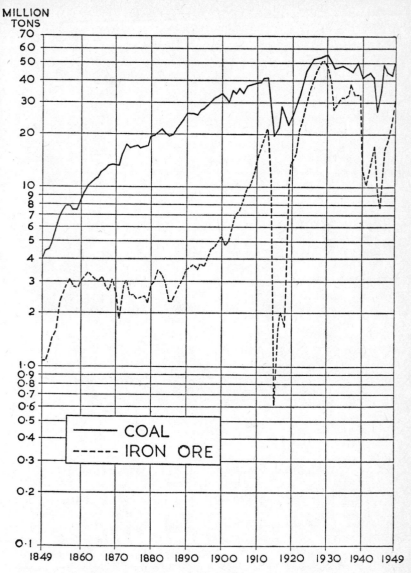

Figure 126. A SEMI-LOGARITHMIC GRAPH OF COAL AND IRON ORE
PRODUCTION IN FRANCE, 1849–1949

Based on *Annuaire Statistique, Résumé Rétrospectif, 1946,* and subsequent volumes (1947–9)
of the *Annuaire Statistique* (Paris, annually).

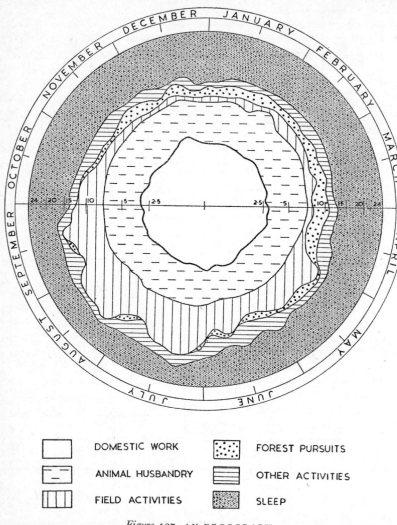

DOMESTIC WORK   FOREST PURSUITS

ANIMAL HUSBANDRY   OTHER ACTIVITIES

FIELD ACTIVITIES   SLEEP

Figure 127. AN ERGOGRAPH

Compiled by W. R. Mead, from statistical data derived from M. Sipilä, *Maatalouden työajankäyttoo ja työntutkimus* (Helsinki, 1946).

This diagram has been drawn to show employment of working hours on a weekly basis for a group of sample farms in Finland.

in the form of a circular compound graph (Fig. 127). This circular form is well adapted to show the continuous rhythm of seasonal activities.

Akin to the ergograph is the 'circular-time diagram', which can be used as a kind of 'clock-face' on which events are placed according to a time-scale. An interesting example is by P. W. Lewis,[1] showing the dates of opening and closing of paper-mills in the Maidstone, Kent, area from 1670.

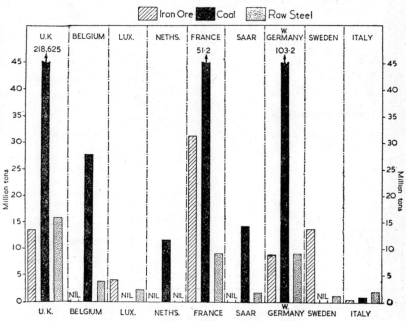

Figure 128. THE OUTPUT OF COAL, IRON ORE AND STEEL BY COLUMNAR DIAGRAM FOR NINE EUROPEAN COUNTRIES, 1949

The statistics were obtained from the *Statistical Year-Book of the United Nations Organization, 1949* (New York, 1950).

COLUMNAR DIAGRAMS

Columnar diagrams, or bar-graphs, also have a wide use in representing economic statistics, and in fact are perhaps the most frequently

[1] P. W. Lewis, 'Changing Factors of Location in the Papermaking Industry as Illustrated by the Maidstone Area', *Geography*, vol. 52, p. 289 (Sheffield, 1967).

used economic diagram. Figure 128 is a three-fold bar-graph, drawn to summarize the output of iron-ore, coal and raw steel in 1949 for nine European countries, which reveals the critical shortages in some of these countries. A simple form of horizontal bar-graph, representing

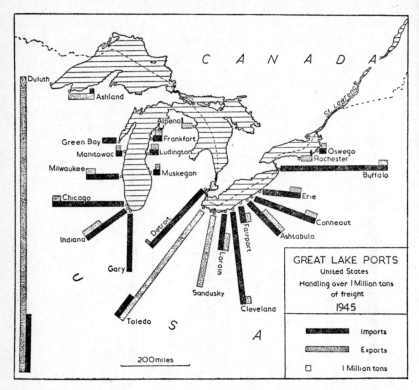

Figure 129. LOCATED BAR-GRAPHS OF FREIGHT HANDLED BY THE GREAT LAKES PORTS, 1945

The data were obtained from *Statistical Abstract of the United States* (Washington, 1946).

merely a total figure, is used in the inset on Fig. 124 to supplement the information on the map itself. Bar-graphs may also be placed conveniently on a location map; thus Fig. 129 summarizes the freight handled by the Great Lakes ports in 1945. The head of each double column was placed as near as possible to the port it represents, and a distinction was made between imports, the column representing which was filled in black, and exports, represented by a stippled column. One can see immediately the vast exports of Duluth from the ore-fields

of the Superior region, and the corresponding imports of the ore by the
Erie ports to serve the Lake-side and Pennsylvanian industrial areas.[1]

DIVIDED RECTANGLES AND CIRCLES

Divided Rectangles

Divided rectangles may be drawn, like bar-graphs, proportional in
length to the values they represent, or, where comparisons are involved,

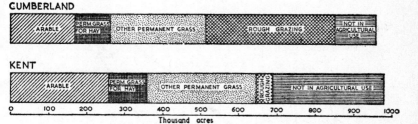

Figure 130. DIVIDED RECTANGLES TO SHOW LAND UTILIZATION IN
CUMBERLAND AND KENT

The data were obtained from Ministry of Agriculture and Fisheries, *Agricultural
Statistics, 1939* (H.M.S.O., 1940).

two or more identical rectangles may be divided on a percentage basis.
Figure 130 compares the land utilization in two counties, Cumberland
and Kent, of nearly similar size; the diagrams bring out the striking
contrasts one would expect, but in addition there are some rather
unexpected similarities. Figure 131 uses divided rectangles, analysing
north- and south-bound freight on the Juliana Canal in 1948, in conjunction with a location map which depicts the exact lock at which
the freight returns under consideration were recorded. There are many
published examples of divided rectangles. Hartshorne and Dicken
delimited[2] six type-regions based on crop and livestock associations.
For each region, they constructed divided rectangles to illustrate the
crop and livestock proportions. In the European examples, the crop
rectangles were made the same length and divided on a percentage

---

[1] Compare A. G. Ballert, 'The Coal Trade of the Great Lakes and the Port of
Toledo', *Geographical Review*, vol. 38, pp. 194–205 (New York, 1948), in which he
analyses the coal trade of the Great Lakes ports, using proportional circles to show
shipments and receipts.

[2] R. Hartshorne and S. N. Dicken, 'A Classification of the Agricultural Regions
of Europe and North America on a Uniform Statistical Basis', op. cit. (1935).

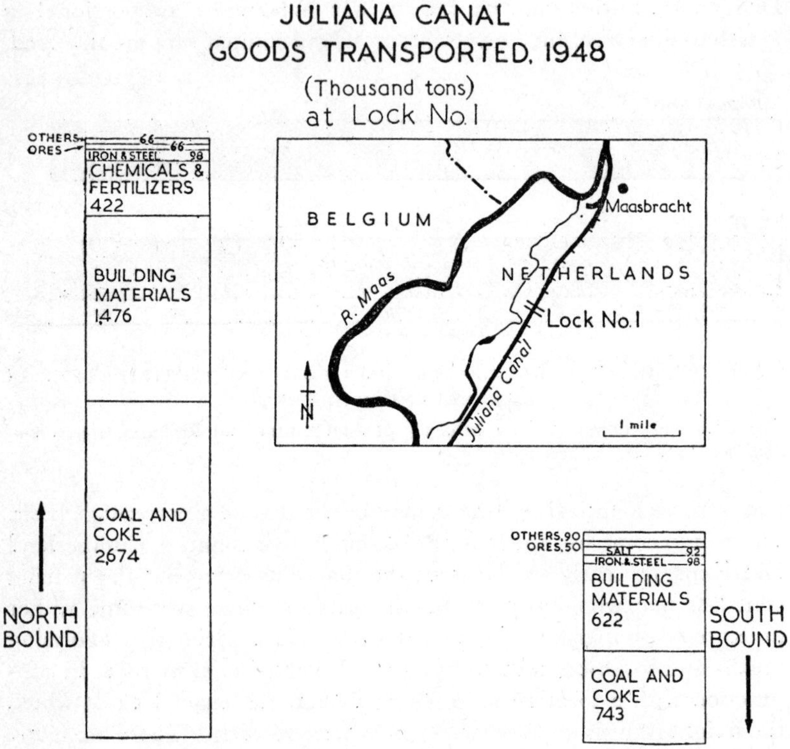

JULIANA CANAL
GOODS TRANSPORTED, 1948
(Thousand tons)
at Lock No.1

OTHERS
ORES
IRON & STEEL 66 66 98
CHEMICALS &
FERTILIZERS
422

BUILDING
MATERIALS
1,476

COAL AND
COKE
2,674

NORTH
BOUND

BELGIUM

R. Maas

Maasbracht

NETHERLANDS

Lock No.1

Juliana Canal

N

1 mile

OTHERS, 90
ORES, 50
SALT 92
IRON & STEEL 98
BUILDING
MATERIALS
622

COAL AND
COKE
743

SOUTH
BOUND

Figure 131. DIVIDED RECTANGLES TO SHOW FREIGHT TRANSPORTED ON
THE JULIANA CANAL, 1948

The data were obtained from the *Centraal Bureau voor de Statistiek* at The Hague.

The Juliana Canal was constructed to by-pass the unnavigable section of the river
Maas, as shown in the inset map, between Maastricht and Maasbracht. The position
of Lock No. 1, at which the freight figures were recorded, is indicated.

The north-bound freight totalled 4·8 million tons in 1948 (cf. 6·8 million tons in
1938); the importance of the canal in the exploitation of the South Limburg coalfield
is shown by the high proportion of coal and coke in the freight figures. The south-
bound freight totalled 1·7 million tons in 1948 (cf. 0·7 million tons in 1938).

basis, while the livestock rectangles were made proportional to the number of animal units per 100 acres of cropland. In the American examples, the same principles were used, except that a triple rectangle was used to indicate the extent of arable, pasture and woodland, the length of which was uniform for every example, but the width of each was made proportional to the actual acreage involved.

An interesting example of this method is by E. S. Simpson,[1] who

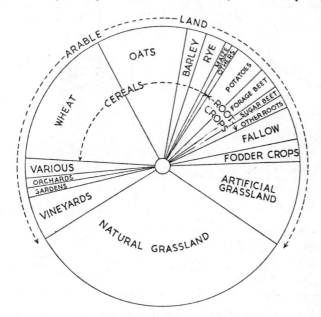

Figure 132. AN INDIVIDUAL DIVIDED CIRCLE TO SHOW THE UTILIZATION OF FRENCH AGRICULTURAL LAND, 1938

The data were obtained from the *Statistique Agricole Annuelle* (Paris, 1939).

analysed the breed-structure of the cattle population of Great Britain for 1908, and for every year from 1935 to 1954, by means of horizontal rectangles divided into thirteen shaded sections according to the percentage representation of each breed.

Divided Circles

The utilization of agricultural land in France is analysed on Fig. 132 by means of an individual circle; no statistics have been added owing

[1] E. S. Simpson, 'The Cattle Population of England and Wales: its Breed Structure and Distribution', *Geographical Studies*, vol. 5, no. 1, p. 47 (London, 1958).

to the crowded nature of the lettering, but even so the diagram is quite revealing. Figure 133 uses divided circles to show the utilization of the arable land in each of the five Australian states; no attempt has been made to draw the circles proportional to the totals, which are lettered

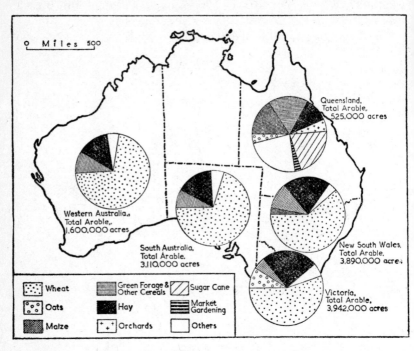

Figure 133. LOCATED COMPARABLE DIVIDED CIRCLES TO SHOW
AGRICULTURAL LAND IN AUSTRALIA, 1947

The data were obtained from the *Official Year-Book of the Commonwealth of Australia* (Canberra, 1948).

The dominance of wheat in every state but Queensland is immediately apparent, while in that state sugar-cane and maize are much more important. One point to note is that the category 'Orchards' appears in the circle for Queensland, but is absent in the others where it actually occupies a greater absolute area; its percentage area was, however, so small in these states that it could not be separately distinguished, and had to be included in the composite category of 'Others'.

below each. The difference between the smallest area involved (i.e. Queensland) and the largest (Victoria) is so great that the eye would find it difficult to compare the areas of the sectors in each, although it is relatively easy to compare the proportions when the circles are of equal size.

A few examples from the very large number available in publications will serve to illustrate the value of divided circles in illustrating distributions. L. S. Wilson uses forty-five individual diagrams to show the nature and direction of Latin American foreign trade.[1] L. D. Stamp employs a series of three-sector divided circles superimposed on a county map of Great Britain, to show the proportions of permanent grassland, tillage and temporary grassland in each county and for Great Britain as a whole between 1937 and 1944.[2] The two maps summarize most effectively the results of war-time ploughing. A series of located divided circles was used to show the output and movement of coal from the Kempen field of north-eastern Belgium, analysing the amounts which moved from each colliery by road, rail and water as a proportion of the total output of each.[3]

H. C. Chew[4] has used very effectively a series of divided circles to illustrate fifteen years of agricultural change in England and Wales (1939–54), for England as a whole and for a number of counties. J. T. Coppock[5] used a series of located divided circles to illustrate forestry developments in Britain; these showed for each region the proportions of State and private planting, of coniferous and deciduous species, etc. Finally, F. J. M. was delighted to see, when he visited the Kongenshus Memorial Park in Jutland, Denmark, that each massive granite boulder in a commemorative avenue relates to a heathland parish; on each are carved two divided circles, one indicating the proportion of heathland in the parish in 1850, the other in 1950.

STAR-DIAGRAMS

Economic distributions may be illustrated by drawing lines, proportional in length to the values they represent, radiating in the approximate direction of movement from the centre of output. Figure 134,

[1] L. S. Wilson, 'Latin-American Foreign Trade', *Geographical Review*, vol. 31, pp. 135–41 (New York, 1931).

[2] L. D. Stamp, op. cit. (1948).

[3] F. J. Monkhouse, 'Coal Movement in Belgium, with Special Reference to the Kempen Field', *Transactions and Papers, 1950: The Institute of British Geographers*, no. 17 (London, 1952). See also H. W. H. King, 'The Canberra–Queanbeyan Symbiosis: A Study of Urban Mutualism', *Geographical Review*, vol. 44, p. 117 (New York, 1954), who used divided circles to analyse traffic flow in and out of Canberra, based on a five-day traffic census.

[4] H. C. Chew, 'Fifteen Years of Agricultural Change', *Geography*, vol. 43, pp. 177–90 (Sheffield, 1958).

[5] J. T. Coppock, 'Britain's Woodlands and Afforestation', *Geography*, vol. 49, pp. 327–33 (London, 1964).

for example, illustrates the movement of coal from two of the large
Kempen collieries, in which the radials indicate the approximate
direction of movement, while the length of each is proportional to the
amount involved. Such star-diagrams are of much less value, however,

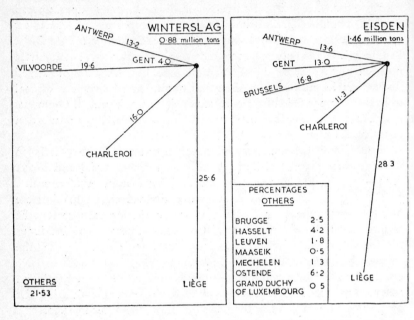

Figure 134. STAR-DIAGRAMS OF COAL MOVEMENT FROM TWO BELGIAN
COLLIERIES, 1948

The data were obtained from the *Association Charbonnière de Bassin de la Campine* in
Hasselt.

The two collieries, Winterslag and Eisden, are in the Kempen coal-field in north-
eastern Belgium.

on a map than is a line of proportional thickness (see Flow-line maps,
below).

Econographs

G. Taylor devised a type of star-diagram based on what he considered
to be the four major controls of white settlement, namely, height above
sea level, rainfall in inches, temperature in ° F and estimated reserves
of coal in tons per square mile.[1] From a central point four axes were

[1] G. Taylor, 'The Distribution of Future White Settlement', *Geographical Review*,
vol. 12, pp. 375–402 (New York, 1922).

drawn, each proportional in length to these four features; the temperature control scale was given double the weight of the rainfall. By joining the ends of the four axes, a rectangular figure was produced, which he called an *econograph*. By comparison with a hypothetical figure representing 'optimum habitability', the shape of the econograph for any particular regions gives an immediate impression of the relative suitability of an area for white settlement, based on these criteria. The principle is akin to that of the climograph (see p. 246).

This type of econograph suffers from the fact that it takes into account only four criteria; in other areas, irrigation, hydro-electric output or oil production may be of greater importance to the habitability. It is possible to plot eight controls instead of four and so obtain representative octagons.

Further, by assuming 1,000 units to occupy the area of an ideal econograph, it is possible to calculate a number to represent proportionally the area of any other econograph. If the value of each is plotted on an outline map in the geographical centre of the area it represents, isopleths can be drawn, to which Taylor gave the name of *isoiketes* (lines of habitability).

FLOW-LINE MAPS

Movement of commodities may be represented by various forms of 'dynamic' map. The uses of proportional circles, bars and star-graphs have already been discussed. Simple movements can be expressed by an arrow linking the point of origin with that of destination. A map on which a whole series of such arrows is plotted may exhibit very revealing patterns, such as those constructed by D. A. Gillmor[1] to show the movements of cattle in Ireland.

The term 'flow-line' map is used of a movement map which shows the direction or route followed by means of a line representing the railway or waterway concerned, while the quantitative impression is conveyed by the width of the line. Thus Fig. 135 depicts the movement of coal along Belgian waterways in 1948, expressed in terms of ton-kilometres (see p. 266). Other values can be used, such as ton-kilometres per kilometre of waterway (Fig. 136), or absolute tonnage in terms of total loadings at points along each waterway or section of each waterway.

[1] D. A. Gillmor, 'Cattle Movements in the Republic of Ireland', *Transactions of the Institute of British Geographers*, no. 46, pp. 143–54 (London, 1969).

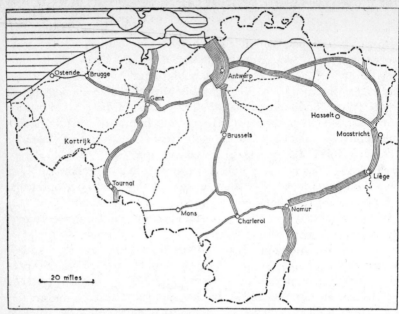

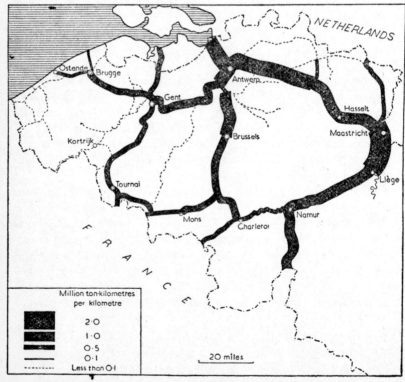

Million ton-kilometres
per kilometre

2·0
1·0
0·5
0·1
Less than 0·1

20 miles

A distinction is sometimes made between a *flow-line*, which shows what actually happens, and a *desire line*, a straight line between points of origin and destination, which may reveal gaps in the existing service.[1]

As a rule, there is rarely sufficiently detailed information available to construct rail or road traffic-flow maps (see pp. 265–6). Occasionally, however, unofficial figures for individual rail-routes can be obtained by patient inquiry; a map was constructed illustrating coal movement in 1949 by rail and water from the Kempen collieries, using information supplied by the seven collieries themselves.[2] In addition, service-frequency maps can be constructed; motor-bus[3] and train time-tables (including unofficial 'working' time-tables) can be analysed, and road-traffic censuses are occasionally available.

Increasing use is being made of the detailed statistics of passenger movement by airlines to construct revealing flow-line maps;[4] some airlines publish such maps as part of their advertisements.

To construct a flow-line map, trace an outline in pencil of the actual routeways. Examine the maximum and minimum quantities involved, and, bearing in mind the complexity of the outline pattern, choose a scale of line thicknesses which can be fitted into the map without confusion. Thus, for example, 1 mm of line-width might represent 1 million ton-km, and values would be represented for each waterway

[1] An example of linkage of ports by ferry services is given by J. H. Bird, 'Seaports and the European Community', *Geographical Journal*, vol. 133, p. 314 (London, 1966); roll-on services are emphasized by heavy lines.

[2] F. J. Monkhouse, op. cit. (1952).

[3] F. W. Green, 'Motor-Bus Centres in South-West England', *Transactions and Papers, 1948: The Institute of British Geographers*, no. 14, p. 60 (London, 1949); and 'Town and Country Services', *Geography*, vol. 34, pp. 89–96 (London, 1949).

[4] See a map of inter-island passenger air-traffic density in the Philippine Islands, which combines flow-lines and proportional circles for handling at airports, in E. L. Ullman, 'Trade Centres and Tributary Areas of the Philippines', *Geographical Review*, vol. 50, p. 209 (New York, 1960). See also an interesting route-map by W. Warntz, 'Transatlantic Flights and Pressure Patterns', *Geographical Review*, vol. 51, p. 205 (New York, 1961), which depicts the optimum route from London to New York for each day in December, 1946; thirty-one lines are superimposed, and an average optimum track and Great Circle tracks are added.

Figures 135, 136. FLOW-LINE MAPS OF BELGIAN COAL MOVEMENT, 1948
The data were obtained from the *Institut National de Statistique* in Brussels.

Figure 135 (*top*) shows coal movement in ton-kilometres. Each single solid line represents 1 million ton-km (to the nearest million), while a pecked line indicates an amount less than this figure.

Figure 136 (*bottom*) shows coal movement in ton-kilometres per kilometre.

by lines of proportional width. Generalizing where necessary the sinuosities of the pattern, and using a pair of dividers, draw parallel lines on either side of the pencil outline, then fill in with black (Fig. 136). Instead of filling in the flow-lines in black, values may be more easily assessed at a glance by drawing a series of parallel lines along each waterway (Fig. 135). Thus, if one line represented 1 million tons, a value of 4·8 million would be shown by five parallel lines; values under $\frac{1}{2}$ million tons could be represented by a pecked or dotted line. Or again, the flow-line can be left in white, which has the advantage that information can be superimposed. Note, for example, the flow-line cartogram of the Mississippi system by J. E. Wrathall.[1]

A more elaborate type of traffic-flow map may be constructed by introducing refinements, such as a differentiation between upstream and downstream traffic[2] (for example, by a compound line in which the upstream value is dotted and the downstream value lined), or even between freight categories.

Another variant can be used for a road-traffic census. If statistics are available for a number of roads in the area under consideration, a single line of dots can be placed along a section of road with, say, 100 vehicles in a period of time, a double line of dots for 200 and so on.[3] A further refinement is to use black dots for one direction traffic, red dots for the other, or different size dots for different vehicles. A variant in less detail is to draw the route-pattern in a single width double-line, and then fill each section with shading or colour corresponding to a quantitative scale.[4]

Special mention should be made of the work of Edward L. Ullman (University of Washington), whose published work is copiously illustrated with various traffic-flow maps. His *American Commodity Flow* (Seattle, Washington, 1957) contains flow-maps of U.S. rail traffic,

---

[1] J. E. Wrathall, 'Recent Developments in the Ohio River Valley', *Geography*, vol. 54, p. 420 (Sheffield, 1969).
[2] See, for example, two maps in H. Ormsby, *France* (2nd edition, London, 1950); one of these (p. 475) differentiates between upstream and downstream traffic in coal on the northern waterways, the other (p. 482) shows coal traffic on the Paris waterways.
[3] G. M. Schultz, 'Using Dots for Traffic Flow Maps', *Professional Geographer*, vol. 13, no. 1, pp. 18–19 (New York, 1961); an example of an area in south-eastern Wisconsin was chosen.
[4] D. E. Christensen, 'A Simplified Traffic Flow Map', *Professional Geographer*, vol. 12, no. 2, pp. 21–2 (New York, 1962), uses this method for a map in the *Florida Reference Atlas* (Tallahassee, 1960). His scale of values was less than 2,500 vehicles, 2,500–7,499, 7,500–12,499, and over 12,499 during a specified period of time; it showed clearly the general pattern and density of movement.

barge and raft traffic on the Mississippi, the movement of coal on the Great Lakes and along the eastern seaboard, and U.S. foreign and domestic seaborne trade in terms of dry cargo, petroleum and total freight.

A series of diagrammatic flow-maps was made by N. J. G. Pounds[1] to illustrate the movement of iron-ore, coal and coke between Lorraine, Luxembourg, the Ruhr, the Saar, etc. Arrows were used pointing broadly in the correct direction, in thickness proportional to the tonnage, and shaded variously to indicate the several commodities.

[1] N. J. G. Pounds, 'Lorraine and the Ruhr', *Economic Geography*, vol. 33, pp. 49–62 (Worcester, Mass., 1957).

5

Population Maps and Diagrams

DATA

Data about population may be grouped for convenience into several major categories. One category comprises pre-census sources such as the Domesday Book, Poll Tax and Hearth Tax records, and occasional individual estimates of population such as that made by Gregory King for England and Wales in 1695.[1]

A second category consists of the various national censuses, which constitute the chief source of information on the distribution and composition of the world's population. In Scandinavian countries and in eastern Canada, censuses were taken in the middle of the eighteenth century, while in Western Europe they originated for the most part about the beginning of the nineteenth century.[2] But there are still some parts of the world, for example, Abyssinia and Arabia, where a full census has never been taken. Also, census returns vary in form, in completeness and in accuracy from one country to another.[3] For the most part, they are held only once in ten years.

A third category includes records of vital statistics which are available for Western Europe and for the United States since the middle of the nineteenth century, and occasionally before. Standards of reliability vary, and only comparatively recently has much vital information, now regarded as essential, begun to be recorded even in countries with well-organized demographic institutions.

Data returned by international organizations such as the United

---

[1] D. V. Glass, 'Gregory King's Estimate of the Population of England and Wales, 1695', *Population Studies*, vol. 3 (Cambridge, 1949–50).

[2] For a review of more recent censuses, see H. J. Dubester, *National Censuses and Vital Statistics in Europe, 1918–39* (Washington, 1947); and H. J. Dubester, *Catalog of the United States Census Publications* (Washington, 1950). See also J. D. Durand, 'Adequacy of Existing Census Statistics for Basic Demographic Research', *Population Studies*, vol. 4 (Cambridge, 1950).

[3] United Nations, *Population Census Methods* (New York, 1951).

Nations,[1] the International Labour Office and l'Institut International de Statistique form a fourth category. These institutions are largely dependent in the first instance upon national census offices for their information, but they play an important part in compiling data, in standardizing them and in making them accessible and so improving their comparability.

Miscellaneous sources may be grouped together in a fifth category. Government inquiries into population problems in connection with planning for war and for peace yield important supplementary data on population. These reports, for example, include those published by the Royal Commissions on Population in Sweden and in Great Britain. Data provided by State Planning Boards and by the National Resources Committee in the United States, and by l'Institut National d'Etudes Démographique in France also come into this category.[2] The British National Register of 1939 and the British Family Census of 1946 deserve special mention. Also of importance are occasional censuses of production and distribution, returns of data based on National Health registrations and on ration-book registrations, returns of migrants made by the British Board of Trade, by the American Immigration Commission and by similar institutions in other countries. Sample surveys of population undertaken by individuals or state-sponsored are becoming increasingly important in providing reasonably accurate data which formerly could only be obtained by means of full-scale censuses.[3]

Mention should also be made here of regional monographs on population which constitute important secondary sources of information; data from many scattered sources are made available in them on a comprehensive basis.[4]

[1] See United Nations Organization, *Demographic Year-Book, 1948* (New York, 1949), and annually. Each volume contains details of the vital statistics of most countries, technical notes in appraisal of the quality of the statistics, and a valuable bibliography of official national censuses and other official demographic publications.

[2] See P. George, *Introduction a l'Etude Géographique de la Population du Monde*, Institut National d'Etudes Démographique (Paris, 1951), for a list of research centres and for bibliographical references.

[3] See F. Yates, *Sampling Methods for Censuses and Surveys* (London, 1949), for a discussion, and for a good example J. R. H. Shaul and C. A. L. Myeburgh, 'Provisional Results of the Sample Survey of the African Population of Southern Rhodesia, 1948', *Population Studies*, vol. 3 (Cambridge, 1948–9).

[4] Outstanding examples of such monographs are: (1) M. Huber, *La Population de la France* (Paris, 1938); (2) J. C. Russell, *British Medieval Population* (Albuquerque, 1948); (3) H. Gille, 'The Demographic History of the Northern European Countries in the Eighteenth Century', *Population Studies*, vol. 3 (Cambridge, 1948–9); (4) D. Kirk,

In spite of all the data available it must be pointed out that our knowledge of the history of the population of the world is still insufficient even for a full appreciation of general trends. Research suffers because the relative novelty of accurate census-taking allows no very deep perspective. Nevertheless, in spite of certain deficiencies, the data available to the geographer are sufficient for him to build up a broad picture of the present population of the world, and more detailed perspective sketches of the population of certain countries like Britain, Belgium, France, Norway, Sweden and the United States. Finally, he must be content to recognize that his finished maps cannot be more accurate than are the data upon which they are based.[1]

Totals and Areas

Population returns may be made on the basis of individual countries, of major administrative units such as states, provinces and counties, and of smaller units such as parishes, townships and communes. The smaller the enumeration area for which population figures are available, the more accurate is the mapping of the population likely to be without recourse to guesswork. The area of each enumeration unit is given in most census returns. This information is necessary in order to calculate population density and also to check any changes which may have taken place in the boundaries of enumeration units between censuses, for both population totals and administrative areas have to be related when trends in population growth are being considered. Some censuses contain base-maps showing the boundaries of the various enumeration units at the date when the census was taken. The lack of such maps in many census returns is a very serious handicap; for example, the absence of parish maps in the successive censuses of England and Wales necessitates much research into changes of parish boundaries before plotting can even begin (see p. 333). It is sometimes necessary to devise base-maps on a diagrammatic basis; this is usually possible when areas of enumeration units are known but

---

Europe's Population in the Inter-War Years (Geneva, 1946); (5) W. S. and E. S. Woytinsky, World Population and Production, Trends and Outlook (New York, 1953); and (6) K. Witthauer, Die Bevölkerung der Erde, Verteilung and Dynamik, Ergänzungsheft Nr. 265 zu Petermanns Geographischen Mitteilungen (Gotha, 1958). For bibliographical purposes, see Population Index, published quarterly by the office of Population Research, School of Public Affairs, Princeton University and the Population Association of America Inc.; and W. Zelinsky, A Bibliographical Guide to Population Geography (Chicago, 1962).

[1] For up-to-date information about population and other data, see Statistical News: Developments in British Official Statistics, H.M.S.O. (London, monthly).

exact details of boundaries are lacking.[1] A grid or network of squares may provide a useful base for plotting data gained by sampling or from field observation, from electoral registers and from the registry of the Factory Inspectorate (see also Fig. 144).[2] The increasing use of statistical methods in population analysis emphasizes the need for a more rational areal enumeration basis than is presently available. Quadrat sampling methods, for example, and multiple regression and correlation analysis would be greatly enhanced by returns of population on a grid square basis.

Social Structure

Rural and Urban Populations. Largely because of environmental differences, rural society presents such a sharp contrast to urban society that in their analyses of population demographers always seek to differentiate between the two. Hence in modern censuses various attempts are made to classify population returns upon this dual basis.

It is necessary to stress the fact that different criteria are used by different census organizations to distinguish between rural and urban populations, and this practice vitiates comparability of the published statistics. In France, the United States and Japan the criterion is numerical; all communities with a population below 2,000, 2,500 and 10,000 persons in those countries respectively are classified as rural. Thus a community of 9,000 in Japan is regarded as rural, whereas in France and the United States it would be classed as urban. Moreover, in the case of each of these countries, the value of the classification is seriously reduced because is makes no allowance for suburban populations living on the fringe of big cities. In England and Wales the criterion for distinguishing between rural and urban populations is based on the arbitrary assumption that all persons enumerated in Rural Districts constitute the rural population.[3] In the U.S.S.R. the economic

---

[1] See, for example, I. Hustich and S. Lindstahl, 'An Area Cartogram of Finland', *Terra*, vol. 68 (Helsinki, 1956).

[2] For an interesting application, see B. L. C. Johnson, 'The Distribution of Factory Population in the West Midlands Conurbation', *Institute of British Geographers, Transactions and Papers*, Publication no. 25, pp. 209–23 (London, 1958).

[3] Moreover, as A. Stevens points out in 'The Distribution of Rural Population in Great Britain', *Institute of British Geographers, Transactions and Papers*, Publication no. 11, pp. 23–53, (London, 1946), within the rural population itself there is a further dichotomy. A distinction may be drawn between the "basic rural population engaged in producing from the land" and a "secondary rural population which is in part ancillary in the exploitation of the land and in part contributes rather to the welfare of the rural community" (p. 28).

function of the community is used as a measure of its rural character, but even this system has its disadvantages. Probably the best solution of the problem to date is to be found in the Brazilian Census of 1940, in which returns of rural, suburban and urban populations have been made upon the basis of regions specially mapped for the purpose.[1]

Education. Until recently the percentage of illiterate persons in the population has been the only index of educational status to be found in most census returns. In parts of Eastern Europe, in Asia, India, Africa and South America, returns of illiteracy are still important as measures of the social character of specific regions.[2] In the United States' census of 1940, a question on the number of years' schooling was included in the schedule, which yielded invaluable information when the returns were cross-tabulated with those of race, age, sex and residence. A similar question was included in the Census schedule for the 1961 Census in Britain.

Housing Conditions. In certain of the censuses of England and Wales, details are to be found of the number of houses in each parish, together with information about the number of rooms per person. Similar data are given in the case of a number of European censuses, for example, that of Belgium. These data have been enlarged in recent censuses, so that these particular sections should prove invaluable in the analysis of inter-regional and sub-areal variations in living conditions.

Sex and Age Structure

Regional differences in the sex and age structure of population are very marked even within comparatively small regions like Lancashire, while in the world at large they are very considerable. A map showing the distribution of population, for example, may depict region X as having precisely the same density as region Y, but in region X the population may consist largely of old people and in region Y of young people. Similarly, region X may have many more women in its population than region Y. These differences are obviously of great significance. They are pointers to the vigour of the population, to its potential labour supply, to its powers of replacement and to its demographic history, and, in fact, affect almost every human activity associated

[1] J. I. Clarke, *Population Geography*, pp. 45–61 (Oxford, 1965).
[2] An attempt to deal systematically with these data was made by J. F. Abel and N. J. Bond, *Illiteracy in the Several Countries of the World* (Washington, 1929).

with the region. Information about these differences is vital also in the sphere of economic and social planning.

It is usual to give general information about the sex and age structure upon a quinquennial rather than upon an individual yearly basis. In national census returns, information about sex and age is often given for larger administrative divisions, but not always for the smallest enumeration areas; for example, British returns provide data for Urban and Rural Districts, but not for parishes.

Ethnic Structure

Race. Data concerning the racial structure of population are to be found in anthropological treatises rather than in census returns. Racial criteria as understood by anthropologists include: (a) certain physical measurements,[1] describing the form of the head, commonly expressed in the form of cephalic indices; (b) measurements of stature; (c) pigmentation; (d) hair form; (e) biometrical measurements, often expressed in the form of indices, for example, Karl Pearson's 'coefficient of racial likeness';[2] and (f) physiological characteristics, in particular blood-group composition.[3]

In some parts of the world where racial admixture has proceeded for thousands of years, constituent 'pure races', if they ever did exist, cannot be readily recognized, and racial characteristics are accordingly ignored in the census returns. This is the case for most of Europe. In other parts of the world where diverse racial stocks have not mixed to the same extent due to later settlement, some attempt is made to classify returns of population totals upon the basis of 'race'. For example, in the first United States census of 1790 a distinction was made between 'free whites' and 'coloured slaves', and in 1850 the classification was expanded to cover 'native whites', 'foreign-born whites', 'Negro' and 'other races'. In 1940 the racial dichotomy 'white', and 'non-white' was introduced, and 'non-whites' were further subdivided into Negro, American Indian, Chinese, Japanese, Filipino, Hindu, Korean, Polynesian and other Asian. Not all of these divisions

[1] For a further discussion, see C. S. Coon, Race of Europe (New York, 1939).

[2] K. Pearson, 'On the Coefficient of Racial Likeness', Biometrika, vol. 17 (Cambridge, 1926), and P. Raymond, Introduction to Medical Biometry and Statistics (Philadelphia and London, 1940), chapter 6 of which is devoted to graphical methods of statistical representation.

[3] See A. Davis, 'The Blood Groups and the Concept of Race', Pts I and II, Sociological Review, vol. 27, pp. 19–342, 183–200 (London, 1935), and A. E. Mourant, 'The Use of Blood Groups in Anthropology', Journal of the Royal Anthropological Institute, 1947, vol. 77, pp. 139–44 (London, 1951).

are truly racial; they might be better described as socio-racial groupings. The tabulation of data under such headings is vitally necessary in countries where problems of fertility, social and economic status, selective migration and so on, are all related to 'racial' structure of the population.

Nationalities. In the many censuses taken in different parts of the world, returns are to be found which deal with the national structure of the population. The data they incorporate are not of equal merit. Because ethnic groupings may in some cases have a particular social or political significance, returns of the strengths of ethnic groups are often weighted in favour of the group which finds itself in power. Moreover, the criteria used to measure national affinities vary from one country to another. In some cases they may be based on 'race' or what are believed to be racial characteristics, for example, skin-colour. In other cases, country of birth, language, religion or 'nationality' may be the test imposed for the purpose of classification. These data afford some measure of group affinities, as apart from any class distinctions which exist in the population. Under certain circumstances people speaking the same language, or of the same religion, or similar in race, may act together or feel themselves bound in allegiance to some common cause. Because ethnic differences cut across class distinctions, they have a regional significance which is of particular interest to the geographer.

The greatest problem in the mapping of such data arises from the unreliability and the complexity of the statistics which have to be used.[1] So much social and political prejudice enters into their compilation, particularly in the case of European countries, that special methods have to be adopted to give an impartial picture of the distributions in any one region. It is often necessary, for example, to produce a series of maps to cover different viewpoints. The geographer can do no more than plot the statistics available, but he must make sure that these are as complete as possible, and that all sources of information have been duly explored.

Occupational and Industrial Structure

Before these statistics can be mapped the categories have to be reduced to a manageable number. The first problem in the geographical

[1] H. R. Wilkinson, 'Ethnographic Maps', *Proceedings, Eighth General Assembly and Seventeenth International Congress, International Geographical Union, 1952*, pp. 547–55 (Washington, n.d.).

analysis of occupational statistics is the wide range of occupations which have to be dealt with, and the variety of classifications which have been adopted in various census returns. The problems of classification and compilation of occupational and industrial population statistics, which are common to all census returns, are best illustrated by reference to the Censuses of England and Wales.

In the first Census of 1801, only three classes of occupations were distinguished. They were (a) persons chiefly employed in agriculture; (b) persons chiefly employed in trade, manufacture or handicraft; and (c) other persons not comprehended in the two preceding classes. In 1841, 877 occupations were listed and grouped into sixteen classes. This system was extended in 1851 to include seventeen classes and ninety-one sub-classes. These were redesignated Orders and Sub-orders in 1861, and this system of Orders has formed the basis of all subsequent Census returns of Occupations.[1]

In the same year, 1861, it was explained in the Appendix to the Report that in future it was by the nature of the products they created that persons were to be classified into occupations. Thus not the kind of work a person did, but the end-product of the industry he worked in was stressed. An example will make this distinction clear. Commercial clerks may be associated with coal-mining, shipping, insurance or with a host of other activities. Under the scheme of 1861, clerks in coal-mining were classified under coal-mining, clerks in shipping under shipping and so forth. This scheme was subsequently modified. In 1881, for example, all commercial clerks were extracted from the various industries and tabulated under one heading – that of *Commercial Clerk*. Such modifications seriously reduced the value of the comparison between one set of census returns and another. By 1891, the 431 occupational divisions distinguished in 1861 had been compressed to 347, but by 1901 the number had risen again to 382, and by 1911 to 472.

The Twofold Classification. Before the Census of 1921, a decision was made to separate the returns formerly designated as occupational into two divisions, occupational and industrial. Thus in both that census and that of 1931, all employed persons were classified twice, first according to the type of work they did, i.e. their occupation, and second, according to the end-product of the industry in which they

[1] H. R. Wilkinson, 'The Mapping of Census Returns of Occupations and Industries', *Geography*, vol. 37, pp. 37–46 (London, 1952).

worked. In the *Occupation Tables* of 1921 about 30,000 occupations
were distinguished, which were classified into 32 orders and some 600
sub-orders, and in the *Industry Tables* about 9,000 industries were distinguished, which were classified into 21 orders and some 400 sub-
orders. The decision to have two independent classifications removed
many of the anomalies to be found in the earlier occupational returns.
Even so, the new Occupation Tables were not entirely satisfactory.
The necessity of reducing 30,000 occupations to a limited number of
orders compelled the classifiers in spite of their professed aims to fall
back in the end on products rather than processes in their classification.
This problem was recognized in the Report of 1931 which stated:
'In the absence of full recognition of the fundamental difference between principles of grouping, classifications have been framed which
although described as occupational, prove on examination to be largely
industrial.'

Place of Work and Place of Enumeration. A certain divergence between
place of work and place of enumeration must always be considered
when census returns of occupations and industries are being mapped.
Unfortunately, nearly all these are based on place of enumeration and
not on place of work. As A. Stevens suggests: 'The census, of course,
is essentially a fiscal provision for the more certain taxing of the lieges.
On sound predatory principle it traces the quarry to its lair. It would
be a more useful demographic instrument in many ways if it located
man at his desk or bench.'[1] In the Census of England and Wales in
1921, an attempt was made to give some regional detail of industries
based on place of work,[2] but in the Report of 1931 it was stated:
'It is to be observed in regard to tables showing the areal distribution
of industry that the areal classification is throughout based on the
individual's place of enumeration and that this may not be the same
as the area in which his place of business is situated. Information
regarding the latter was not obtained in 1931, and its inevitable dis-
regard may for some purposes introduce an element of incongruity.'
Great care is necessary therefore if these tables are to be used to
analyse location of industry. Fortunately, the 1951 and 1961 returns
do have more information on place of work but detailed returns are
still lacking.

[1] A. Stevens, op. cit., p. 34 (1946).   [2] *Industry Tables*, Table 4.

Socio-economic Indices

Demographic Coefficients. In analysing tables of population data, indices which are designed to give a measure of population pressure in different regions are often found useful. Probably the simplest index of this type is obtained from the formula,

$$C = dR$$

where C represents the demographic pressure or demographic coefficient, d the density of population, and R the net reproduction rate.

This is a useful expression because it shows at a glance what future population densities might develop in different regions providing there were no migration, and were the current rates of mortality and fertility to remain unaltered. It helps to explain the building-up of population pressure and migration trends, and is of use too in the study of economic problems and of political relationships between states. In Fig. 147, for example, the actual density of population is compared with the replacement density at a distance of one generation, i.e. about twenty-five years. If replacement densities at two or more generations were employed, the differences in the build-up of the population would be even more striking. The differences at two generations distant are:

$$C = dR^2$$

Mortality and fertility do not, of course, remain constant for any length of time, but the maps may be instructive nevertheless.

A similar coefficient was devised by J. Smolenski as follows:[1]

$$C = dt$$

where t represents the rate of natural increase per thousand inhabitants.

Comparative Density. Population pressure is not purely a function of density and replacement of population, because density of population is obtained by a consideration of the total area of the enumeration unit. This area may vary in its capacity to support population according to its geographical character. Hence more refined indices have been designed to eliminate this anomaly and to illustrate *comparative density.* The simplest index of this type is expressed by the formula:

$$D = \frac{P}{S}$$

where D = comparative density, P = the total population and S = the total cultivable land.

[1] *Congrès International de la Population*, vol. 2, sect. A (Paris, 1938).

But cultivable land in itself varies, according to position, soil and other factors. To consider all the implications and to express them satisfactorily in the form of a simple numerical index is a formidable undertaking, particularly as problems of optimum population remain as yet unsolved. However, it is possible to refine further the above index by weighting the figure for cultivable land.[1]

Other coefficients stress the relationship between the population of an area and all forms of income in that area. In this context

$$Population\ pressure = \frac{Total\ population}{Total\ income}$$

Regional variations can be conveniently mapped by relating local coefficients to the national coefficient as an index, e.g.,

$$Index = \frac{Total\ local\ population}{Total\ local\ income} \bigg/ \frac{Total\ national\ population}{Total\ national\ income} \times 100$$

Such maps are of use in the study of internal migration and regional development.[2] In the field of social area analysis, a wide variety of indices and 'indicators' are in use to measure such things as standard of living, social class, ethnic segregation and 'urbanization'. Many of these have been employed by geographers in regional studies.[3] For urbanization indicators, see the reference to E. Shevky and W. Bell (p. 433). Work in the field of component analysis on British towns suggests that the average number of persons per room affords a good index of social class and housing conditions.

Natural Replacement

The rate at which a population replaces itself – the natural increase – is a central fact in population study. Regional differences in population replacement have an obvious geographical significance, because even within units of the size of an English county marked differences occur. To ascertain rates of natural increase, it is necessary to consult

[1] A method of doing so is outlined by P. Vincent, 'Pression Démographique et Ressources Agricoles', *Population*, vol. 1, pp. 9–19 (Paris, 1946). The problem is also considered in relation to the location of rural population in Great Britain by A. Stevens, op. cit. (1946).

[2] S. Tsubouchi, 'Population Pressure and Rural-Urban Migration in Japan', *Proceedings of the I.G.U. Regional Conference in Japan, 1957*, pp. 512–16 (Tokyo, 1959).

[3] See, for example, P. N. Jones, *The Segregation of Immigrant Communities in the City of Birmingham*, Occasional Papers in Geography, no. 7 (University of Hull, 1967), who makes effective use of K. E. Taueber's index of segregation as described in 'Residential Segregation', *Scientific American*, vol. 213, pp. 2–9 (New York, 1969).

vital statistical records as well as census returns. The keeping of these records is now almost universal, but in relatively few countries are these sufficiently accurate or complete to give more than a superficial view of the natural growth of the population. In England and Wales, civil registration was introduced in 1837, although before this date parish priests had kept records of marriages, births and deaths. Even after registration was introduced, the returns were not always complete. Illegitimate births were often ignored altogether, and such important details as the age of the mother have only recently begun to be recorded.

The *Crude Birth-rate* is usually expressed as the number of children born annually per thousand of the population, obtained from the formula:

$$\frac{Number\ of\ births}{Total\ population} \times 1,000$$

The *Crude Death-rate* is also expressed as an index – the number of deaths per thousand of the population obtained from the formula:

$$\frac{Number\ of\ deaths}{Total\ population} \times 1,000$$

This is not a very refined measure of mortality because the age and sex structure of the population is not taken into account in its calculation.

Natural Increase. Crude birth- and death-rates may be used to find the natural increase in the population. For example, the birth-rate for 1948 in England and Wales was 17·8, the death-rate was 11; therefore the rate of natural increase was:

$$17·8 - 11 = 6·8\ per\ thousand\ of\ the\ population$$

The estimated population of England and Wales in 1948 was 42,750,000 and therefore the natural increase was:

$$42,750 \times 6·8 = 290,700$$

Standardized Rates. The crude birth-rate is not a very good index of fertility because variations in the sex and age structure of the population are not taken into account in its calculation. A standardized birth-rate, in which age and sex anomalies are smoothed out by comparison with a hypothetical standard population, is a better gauge.[1] For

[1] Procedures for standardization are discussed in, for example, T. L. Smith, *Population Analysis* (New York, 1948), and in P. R. Cox, *Demography* (Cambridge, 1950).

example, the crude birth-rates in the United States for urban and rural population in 1940 are inaccurate measures of fertility, because the urban population contained a greater proportion of women than

Population of the United States, 1940		
	Crude birth-rates	*Standardized birth-rates*
Urban	16·8	15·8
Rural	18·3	19·5

did the rural population. It may be seen from the above table that when allowances are made for this, the figures are very different.

Fertility Ratios. Standardized rates are difficult to compute, however, and there is a more simple method of arriving at a good estimate of fertility by relating the number of young children in the population to the number of women of child-bearing age. This fertility ratio is usually expressed as:

$$\frac{\textit{Number of children under 5 years old}}{\textit{Number of women aged 15–50}} \times 1,000$$

It provides a simple index to fertility which can easily be mapped. The information necessary for the calculation is obtained directly from the tabular data recorded in various census returns, and in the case of Britain, from the *Registrar General's Estimates of Ages and Sexes in the Population*. It can moreover be extracted from quinquennial tables, which makes the calculation less laborious. Information about age and sex groups dating back to about the middle of the nineteenth century are to be found in most European census returns – in the case of England and Wales from 1841. Hence it is possible to produce maps showing distribution of fertility ratio for over a hundred years.

Net Reproduction Rate. A useful index to the rate at which a population is replacing itself is provided by the net reproduction rate, first demonstrated by R. R. Kuczynski.[1] He pointed out that excess of births over deaths was a deceptive measure of replacement, because many countries with a surplus of births over deaths were not actually replacing their populations. In other cases, the population increase was being

[1] R. R. Kuczynski, *The Measurement of Population Growth* (London, 1935).

maintained merely because older people were surviving longer. Kuczynski maintained that population replacement could only be effectively measured if a whole generation were put to the following test. Was the fertility and mortality in a population such that a generation, permanently subject to them, could during its own lifetime produce enough children to replace itself? Such a test could be devised by considering only female births. If a generation of mothers is successful in producing exactly the same number of mothers in the second generation, then it could be said to be exactly replacing itself. To take an example, the fertility tables for Australia, 1920–2, show that on the average a thousand mothers gave birth, during the whole span of child-bearing, to 1,517 female children. A reference to the mortality tables indicates that of these, 1,318 could be expected to become mothers. Hence 1,000 mothers had given birth to 1,318 future mothers. The population was replacing itself in the ratio of:

$$\frac{1,318}{1,000} = 1\cdot318$$

$1\cdot318$ *being in this case the net reproduction rate.*

The beauty of the net reproduction rate is its simplicity as a measure of replacement. If it is unity, the population is exactly replacing itself; below unity it is failing to do so, above unity it is increasing.

Male and Female Reproduction Rates. R. R. Kuczynski's concept of replacement of population as measured by female net reproduction rate only has been criticized recently, notably by P. H. Karnul.[1] He points out that in actual population, two measures are available – female and male net reproduction rates respectively. An alteration of sex-age distribution caused, for example, by war will cause fluctuations in the male and female rates respectively, so that neither in itself may be considered as a sufficiently reliable index to population replacement, but both must be considered together.

Marriage Standardized Reproduction Rates. It may be useful occasionally to consider replacement rates as measured in the married population

[1] P. H. Karnul, 'The Relations between Male and Female Reproduction Rates', *Population Studies*, vol. 1 (London, 1947); and 'An Analysis of the Sources and Magnitudes of Inconsistencies between Male and Female Net Reproduction Rates in Actual Populations', ibid., vol. 2 (London, 1948–9).

without reference to the unmarried females.[1] The necessary calculations are somewhat laborious, but they are of value in the analysis of the causes of differential fertility rates.

The Family Census of 1946. Regional variations in the size of the family, where such information is available over a long period, affords most important evidence on the replacement of population. In the *Royal Commission on Population* the results of the Family Census of 1946 were used as the basis of population projections rather than conventional analysis by means of age–specific fertility rates. Unfortunately, not much regional data on family formation and family building are at present available, but such data will probably be compiled in future censuses.

Migration and Movement

Growth of population is affected not only by natural replacement but also by migration of population. Not only permanent migration of population has to be considered in the measurement of population characteristics, but movement of the population also. Some populations are permanently on the move – the nomadic peoples of desert, steppe and forest; itinerant groups, as the Vlachs of the Balkan Peninsula whose movements are associated with the practice of transhumance; seasonal workers, as the Italians who seek work in France during the harvest period; daily workers who travel long distances between their place of residence and their place of work. Mobility is a population characteristic no less than age structure, although considerably more difficult to classify and measure.[2]

Because there are so many types of movement of population, only some of which are recorded by national authorities, information about these movements is difficult to compile. These difficulties are discussed at length in various demographic monographs, and it is not intended to do more than summarize a few of the chief ones here, in order to understand the nature of the problems involved in mapping the available statistics.

Comparison of Vital Statistics and Population Totals. By adding the surplus of births over deaths occurring during a particular decade to the total population of a region at the beginning of the decade, and com-

[1] In the *Report of the Royal Commission on Population* (London, 1949), a method of calculating standardized net reproduction rates was devised for this purpose.
[2] See J. Beaujeu-Garnier, *Geography of Population*, pt. 4 (London, 1966).

paring this sum total with the actual total of enumerated population at the end of the decade, the net migration which has taken place during the period under review may be calculated, by subtracting one total from another. A possible error in the calculation may arise because the vital statistics themselves may not have been recorded accurately and deficiencies in them may lead to the assumption of a higher or lower migration than has actually occurred. Generally, however, the calculation of net migration by this method is sufficiently accurate for most purposes. But this information does not indicate how much immigration and emigration have taken place. The net migration may be nil, although great movements of the population have taken place which counter-balance one another.[1] Net migration data are useful for showing general trends in the movement of population, for example, movement from country to town, or the wider movement from Eastern to Western Europe which took place during the inter-war years.

Net migration data are available not only on a national basis, but also sometimes for quite small local regions, so that they are of use in determining internal trends in population as well as international and intercontinental movements.

Place of Birth and Place of Residence. This information is largely derived from census returns and may be used to indicate both local and international movements. For example, place-of-birth statistics can be used to illustrate overseas' migration to the United States. They may also be used to show inter-county movements from decade to decade in England. It is not, of course, possible to measure population migration accurately by place-of-birth statistics. They reveal little of the nature of the migrants, nor do they distinguish between permanent and temporary migrants. For example, the Census of 1921 for England and Wales was taken in June when large numbers of people were on holiday; place-of-birth statistics for Blackpool in 1921 indicate a great migration there which was naturally very temporary. Since place-of-birth statistics are recorded in most cases only when a census is taken, it is impossible to find out the precise date when migration took place.

Another deficiency, in the case of certain British censuses, arises from the divergence between 'civil counties' and 'registration counties',

---

[1] This point is stressed by M. P. Newton and J. R. Jefferey, *Internal Migration: Some Aspects of Population Movements within England and Wales*, General Register Office (London, 1951).

for only occasionally were the two coincident.[1] In the returns of the
censuses of England and Wales in 1851 and in 1861, the figures tabulated show how persons born in geographical or civil counties were
distributed throughout the registration districts and counties. Apart
from these 'inconveniences and perplexities' arising out of the complex regional basis of the returns, the place-of-birth statistics provide
valuable information on general population movement. They are of
great value, particularly to the historical geographer, who has no other
data available to provide that perspective so necessary for an understanding of long-term trends in migration.

Records of Passenger Statistics. Another source of information on migration and movement of population is to be found in a variety of records
of passenger movements. These records are, for the most part, relatively
recent, and they vary immensely in their character, accuracy and completeness from one country to another. Largely owing to the efforts of
international organizations, such as the International Labour Office,
these records are being brought into uniformity.[2] They include port
statistics, frontier-control statistics, data compiled from registration
coupons which are issued by home and consular authorities, information
based on tickets sold by transport agencies and passport statistics. In
some cases, local registration of population movements is customary,
and such records are available. These data are naturally of most use in
the measurement of international and intercontinental movement of
population.

There are numerous difficulties in compiling data relating to internal
movements of population. In some countries in Europe an attempt is
made to keep local registers of persons making a permanent change
of address involving movement out of a given administrative district.

Daily Movement of Population. In the Census of England and Wales
for 1921, the place of work of the enumerated population, as well
as place of enumeration, was recorded. The returns were published
in a separate volume entitled *Workplaces*, and provide useful data
concerning the daily movement of population between work and residence.

---

[1] For a full discussion, see H. C. Darby, 'The Movement of Population to and from
Cambridgeshire between 1851 and 1861', *Geographical Journal*, vol. 101, pp. 118–25
(London, 1943).

[2] See United Nations, *Problems of Migration Statistics* (New York, 1951).

Sex and age structure of the population may also be used to reveal information on the movement of populations. If, for example, the balance between the sexes becomes increasingly disproportionate with increasing age, sex-selected migration has probably been taking place, either into or from the region under consideration.

Miscellaneous sources of information about population movement are many. Local passenger services often keep records, road censuses are held from time to time, and the railway companies issue returns of passengers carried. But these miscellaneous statistics are difficult to compile and often only give an incomplete picture of movement at any one time. An interesting source for historical studies of population movement exists in parish marriage registers, which provide a reasonably random selection of movements between parishes about which census returns give no information at all.[1]

Population Growth

The growth or decline of the population of a region over a certain length of time amounts to the sum of the natural replacement which has taken place plus the net migration into the region during the period under consideration. Regions which enjoy a high natural replacement may experience an actual decline in population because emigration is heavy, and vice versa. The growth of population is often, but not always, related to economic circumstances. Malthus's original theory of the cycle of population growth stated this relationship. The theory has aroused more controversy than any other in the field of demographic studies. The growth of population in any region, whether it is positive or negative, undoubtedly does reflect the history of man's response to the environmental possibilities present in the region. Moreover, future trends in population growth may only be estimated from the evidence provided by the past behaviour of population growth.

The dynamics of population growth vary considerably from one region to another. The growth may take the form of a steady, long-continued increase, it may fluctuate widely from one decade to another, it may exhibit an accelerated or a decelerated rate, it may reach a peak and then decline. Providing that sufficient data are available, and that the boundaries of enumeration units on which the returns are based have not varied too widely from one census to another, the

[1] A. Constant, 'The Geographical Background of Inter-Village Population Movements in Northamptonshire and Huntingdonshire, 1754–1943', *Geography*, vol. 33, pp. 78–88 (London, 1948).

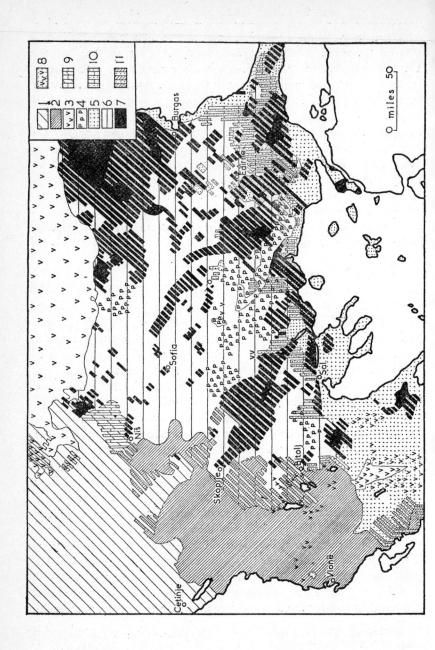

growth and rates of growth of population can be plotted with tolerable accuracy.

Predictable models of population growth have considerable importance in relation to future physical and economic planning needs. Not only is there a need for forecasting future population totals but also how the growth will be geographically distributed. Hence there is a growing literature concerned with regional model-making.[1]

NON-QUANTITATIVE MAPS

Very few population maps are non-quantitative, but certain features of the occupational and ethnic structure of population can be shown in a non-quantitative manner. Two examples of such techniques of mapping will be considered, with reference to the distribution of ethnic structure.[2]

The Chorochromatic Technique

The most popular type of ethnographic map found, for example, in most atlases, is that which shows territory shaded according to the major ethnic affinity of its population. The popularity of these maps is the result of their application in the field of political geography. Such maps do not usually convey population density, although attempts

[1] For a general discussion, see W. Isard, *Methods of Regional Analysis: An Introduction to the Regional Science*, pp. 1–79, 493–568 (New York, 1960).

[2] For fuller discussions of the methodology of ethnographic mapping, see (1) H. R. Wilkinson, op. cit. (1952); (2) W. Krallert, 'Methodische Probleme der Völker- und Sprachen-Karten', *International Yearbook of Cartography*, vol. 1, pp. 99–120 (London, 1961); (3) S. I. Bruk, 'Basic Methodological Problems in Ethnic Mapping', *Soviet Geography, Review and Translation*, vol. 3, pp. 32–4 (Washington, D.C., April 1962).

Figure 137. A CHOROCHROMATIC MAP OF ETHNOGRAPHIC DISTRIBUTIONS

Based on H. R. Wilkinson, *Maps and Politics: A Review of the Ethnographic Cartography of Macedonia* (Liverpool, 1951).

The map shows the distribution of ethnic groups in the Balkan Peninsula according to a map compiled by K. Sax in 1878.

The references in the key are as follows: **1.** Serbo-Croats; **2.** Albanians; **3.** Rumanians; **4.** Pomaks; **5.** Greeks; **6.** Exarch Bulgarians; **7.** Circassians, Tartars, Turks and Turcomans; **8.** Graeco-Vlachs; **9.** Serbo-Bulgarians; **10.** Graeco-Bulgarians; **11.** Graeco-Albanians.

Note the use of 'interdigitations' to show both admixture of ethnic groups and transitional types. The original map was in colour but by choosing distinctive shading it has been interpreted in black and white.

have been made to overprint distributions on density maps.[1] Nevertheless, they are clear, and the delimitation of ethnic divides is distinctive, indeed much more so than it is in the field. But on such maps
admixture of population is difficult to depict, even with the aid of
'interdigitation' (Fig. 137) and even with additional symbols to show
the population of towns. Moreover, these maps generally over-
simplify distributions; huge areas may be shaded to represent an ethnic
group with a population of less than one person per square kilometre,
while densely populated districts shaded to represent another group
may be so small as to escape notice. It is also particularly difficult for
the eye to give equal weight to different types of shading, especially
if colour is used; red always gives an impression of a more intensive
distribution than would be the case if purple or yellow were used.

Inscriptions

In the earliest ethnographic maps the problem of showing distributions
was solved in a crude but effective manner. The names of the ethnic
groups were inscribed over the territory with which they were associated; majority populations were shown by capital letters and minorities
by lower-case. This method is still in use today, for example, to show
tribal distributions in African territories. It does not convey density of
population, although the names may be inscribed over population
density maps, nor does it give a sufficiently accurate impression of
complex ethnic structure when an inter-mixture of groups has to be
plotted. It does not allow comparative strengths of the component
groups to be portrayed, and it fails to convey any accurate impression
of the limits of specific groups.

CHOROPLETH MAPS

If the isopleth is the chief tool of the climatologist, the choropleth may
be said to be the chief tool of the human geographer in his quantitative treatment of the distributional aspects of population. Therefore
choropleth maps receive priority of treatment in this section, and they
are dealt with in more detail than other quantitative methods.[2]

---

[1] A. Haberlandt, 'Karte der Völker Europas nach Sprache und Volksdichte',
1 : 3,000,000 (Vienna, 1927).

[2] Reference should be made to a Special Number of the Transactions of the *Institute
of British Geographers*, no. 43, which contains a series of thirteen detailed population
maps of the British Isles, ed. A. J. Hunt (London, 1968). See also R. J. Johnston,
'Population Movements and Metropolitan Expansion: London, 1960–1', *Transactions
of the Institute of British Geographers*, no. 46, pp. 69–91 (London, 1969); and (for world
maps) *A Geography of Population: World Patterns* (New York, 1969).

Population Density

To create an accurate impression of density, it is necessary to consider the areas of enumeration districts as well as the totals of population. Density is usually expressed as the ratio of a specific number of persons to a given area of land, as twenty persons per square mile. It may for certain purposes be represented as the ratio of a specific area to a given number of persons, as 4 acres to one person (Fig. 142).

The problem of constructing density maps may be clarified by considering, as a specific example, the distribution of population in Southwest Lancashire in 1891 (Figs. 138–43), based on the Census of England and Wales for that year. Areas present something of a problem in that returns of parish areas are not accurate in early censuses, because detailed cadastral survey was often lacking; frequent changes of parish boundaries give rise to anomalies between recorded and actual areas of parishes; and coastal parish acreages often included areas of sea and foreshore. By a careful scrutiny of the census acreages and a comparison with base-map data, major errors in the calculation of the density of population in each parish may be avoided. Since the census returns give the totals of population and acreage for each parish, it was necessary to reduce these figures to persons per unit area in each parish, and a nomograph was used (see p. 45).

When the densities had been calculated, it became obvious that because only a limited range of shading was practicable, only certain categories of population density could be shown on the map. As they were to be divided into some half-dozen categories, care had to be exercised in the selection in order to ensure that the major density values were given full weight (Fig. 142). A dispersion diagram was helpful in ensuring that important variations in density were not overlooked. It should be noted that certain aspects of the distributions of population are better illustrated by expressing regional differences in terms of deviations from average national density. This method does not altogether solve the problem presented by great ranges of density, but has a special value in the portrayal of population concentrations.[1]

R. M. Prothero has suggested an ingenious method for converting a dot distribution map to a population density map by superimposing a grid over the dot map. The dots in each square can then be counted

[1] See A. Hoffman, 'India: Main Population Concentrations', *Geographical Journal*, vol. 111, pp. 89–100 (London, 1948).

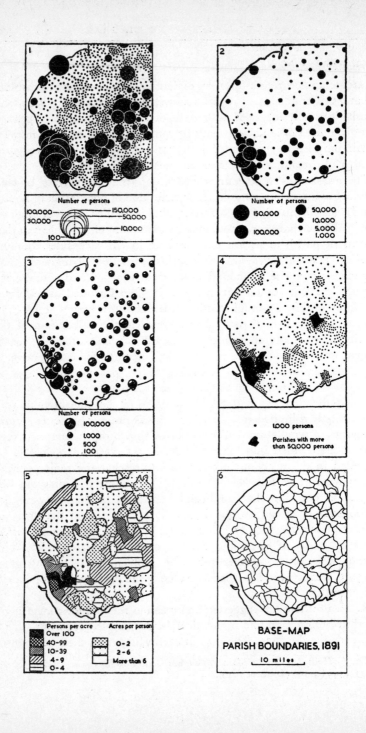

1

Number of persons

100,000 — 150,000
30,000 — 50,000
— 10,000
100 —

2

Number of persons

● 150,000   ● 50,000
              ● 10,000
● 100,000   ● 5,000
              · 1,000

3

Number of persons

● 100,000
● 1,000
● 500
· 100

4

· 1,000 persons

◆ Parishes with more
   than 50,000 persons

5

Persons per acre   Acres per person
■ Over 100
▨ 40-99   ▒ 0-2
▤ 10-39   ▥ 2-6
▧ 4-9   □ More than 6
▦ 0-4

6

BASE-MAP

PARISH BOUNDARIES, 1891

—— 10 miles ——

and the map shaded accordingly (Fig. 144). This method is likely to be of value only when dots have been carefully located in the first instance. It could well be applied to show density of urban population.

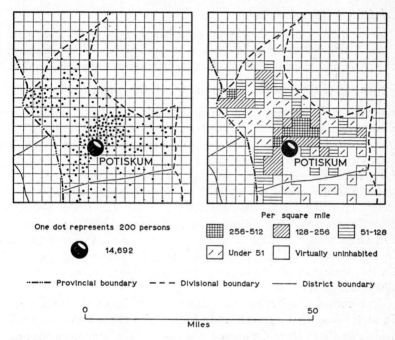

One dot represents 200 persons

14,692

Per square mile

256-512   128-256   51-128

Under 51   Virtually uninhabited

··—·— Provincial boundary   — — — Divisional boundary   ——— District boundary

0   50
Miles

Figure 144. DISTRIBUTION OF POPULATION NEAR POTISKUM, NIGERIA
Based on R. M. Prothero, 'Problems of Population Mapping in an Under-developed Territory (Northern Nigeria)', *Nigerian Geographical Journal*, vol. 3 (Ibadan, 1960). Each square in the grid represents 4 square miles (10 sq. km).

Line-shading or colour-wash are the most effective methods of depicting the densities once they have been decided; the problem of shading is discussed above (p. 47). In the case of population distribu-

---

Figures 138–43. CONTRASTING TECHNIQUES OF SHOWING POPULATION DISTRIBUTION
Source of data: *Census of England and Wales, 1891*, vol. 1 (London, 1893).
The maps show the distribution of population in south-western Lancashire only. In **1** dots and proportional circles are employed; in **2** proportional circles only are employed and the form of key and proportional scale have been changed; in **3** proportional spheres are employed; **4** is a modified dot map; and **5** is a simple choropleth map. The data used for each map are identical.

tions, the conventional impression of increasing density conveyed by the graduation from white to black is so firmly established that there is little to be gained by using colour.

Mapping by 'Standard Scores'

This method, while it does require further processing of the original data, is relatively simple. It involves the calculation of the standard deviation and subsequently the standardization of the original values by the use of the formula $x - \frac{\bar{x}}{\sigma}$, where x is the original value for the unit; $\bar{x}$ is the arithmetic mean of the original values for all units in the distribution, and σ is the standard deviation of the distribution. The standardized distribution thus produced has zero mean and unit S.D. (standard deviation) and the standard scores, being 'pure' numbers, facilitate direct comparisons among local authorities and among variables. Effective use of this method, however, depends to a very large extent upon the normal frequency distribution. Only where distributions do not significantly differ from normal (or at least from each other) can direct comparisons be made between the standard scores attained by a particular local authority area.[1] The Chi-square test (p. 502) can be used to establish the normality of frequency distributions. Figure 145 illustrates the method.

In a normal distribution the variates are distributed about the mean in such a way that approximately 68 per cent of the frequencies will fall under that part of the curve within ± 1 S.D. of the mean (see key, Fig. 145) and it was therefore considered useful to divide this central part of the distribution into cell intervals of $\frac{1}{2}$ S.D. There is no statistically significant difference between adjacent cells in this part of the distribution but valid comparisons can be made between non-adjacent cells. Towards the tails of the distribution, as the frequency within the S.D. units decreases, the width of the cell has been increased; adjacent cells within the tails of the distribution (S.D. greater than 1) *are* therefore significantly different. The frequency distribution for the standardized data is shown on the map by means of (i) arithmetic mean and (ii) standard deviation which, when n (the number of cases) is known, describe the distribution. The parameters are those of the observed distribution plotted as a polygonic diagram and fitted with a

[1] D. G. Symes and E. G. Thomas with R. R. Dean, *The Yorkshire and Humberside Planning Region: An Atlas of Population Change, 1951-66*, Department of Geography, University of Hull, Miscellaneous Series no. 8 (Hull, 1968).

Figure 145. PERCENTAGE CHANGE BY MIGRATION IN PART OF EAST YORKSHIRE AND LINCOLNSHIRE, 1951–61

The method of mapping is by 'standard scores' based on D. G. Symes, E. G. Thomas and R. R. Dean, *The Yorkshire and Humberside Planning Region: An Atlas of Population Change 1951–66*, Department of Geography, University of Hull, Miscellaneous Series no. 8 (Hull, 1968).

The data source is the Census of England and Wales, 1961 and the Registrar General's Statistical Reviews, 1951–60. Chi-square value is 15·22.

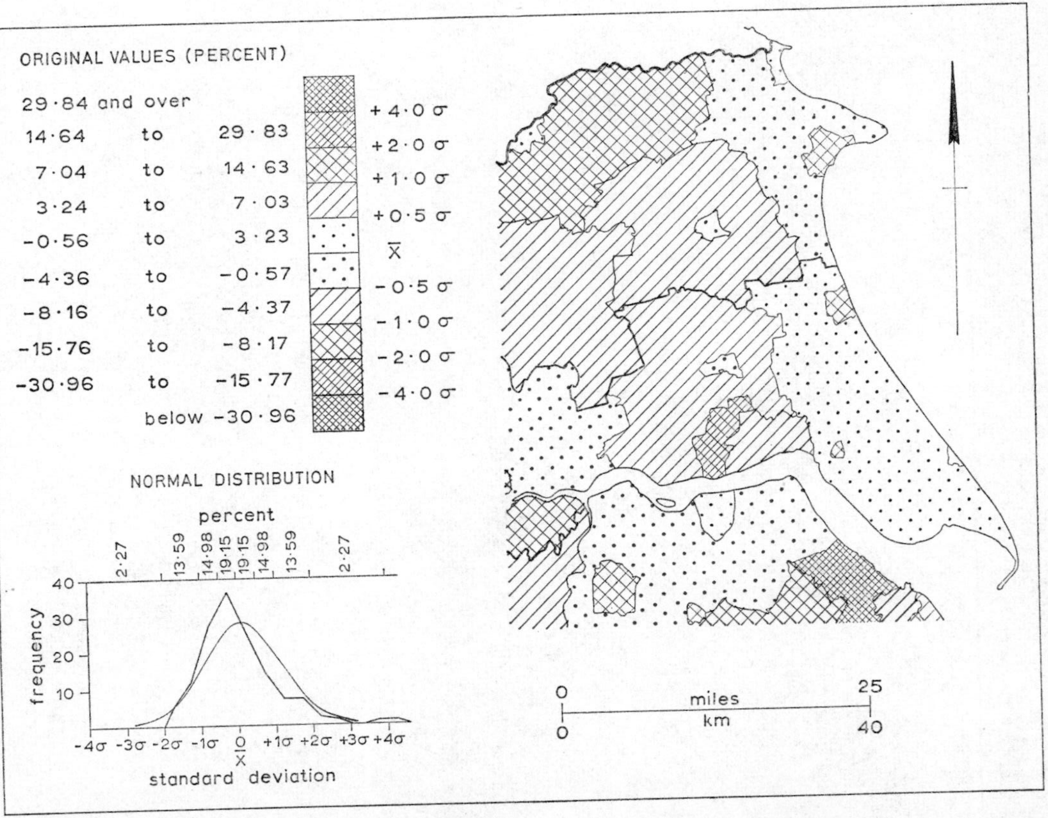

ORIGINAL VALUES (PERCENT)
29·84 and over
14·64 to 29·83
7·04 to 14·63
3·24 to 7·03
-0·56 to 3·23
-4·36 to -0·57
-8·16 to -4·37
-15·76 to -8·17
-30·96 to -15·77
below -30·96
+4·0 σ
+2·0 σ
+1·0 σ
+0·5 σ
X̄
-0·5 σ
-1·0 σ
-2·0 σ
-4·0 σ
NORMAL DISTRIBUTION
percent
2·27
13·59
14·98
19·15
19·15
14·98
13·59
2·27
frequency
40
30
20
10
-4σ -3σ -2σ -1σ 0 X̄ +1σ +2σ +3σ +4σ
standard deviation
0
miles
25
0
km
40

smoothed normal curve having the same $\bar{x}$, S.D. and N as the observed distribution. The differences between the observed and the expected frequencies have been subjected to the Chi-square test. The values which Chi-square must reach to be significant at the various accepted levels, given the appropriate number of degrees of freedom, are tabulated. Where distributions do significantly differ from the normal but are demonstrably more nearly normal than of any other type, it may be possible to find a more normally distributed function of the variable, e.g. the logarithms of the numbers.

The Dasymetric Technique

It may so happen that the data for any given region may not be sufficiently complete to give a detailed picture of the variations in density from one place to another, as is the case when the returns are made for very large administrative divisions only. In this case, it may be necessary occasionally to ascertain details of distribution on the map which are not actually given in the returns.[1] Or, even if returns are made on a parish basis, it may be desirable for certain purposes 'to emancipate ourselves from undue restraint of administrative boundaries . . . and so avoid the bizarre sort of tartan pattern frequent in population distribution maps . . .'[2] For such work it is necessary first, to consult topographical and land-utilization maps of the region under consideration, and then to pick out from these maps districts such as moorland, marsh, heath, etc., on which no settlement occurs, and second, to delimit settlement zones as denoted by the settlement patterns. These zones have to be superimposed upon the base-map. Their areas must also be calculated. Suppose, for example, that for each administrative division the information available were as follows:

	Population	Total area
Districts without settlement	Unknown	80 sq. km
Intermediate districts	Unknown	60 sq. km
Closely settled districts	Unknown	10 sq. km
Administrative division	60,000	150 sq. km

The mean density of population for the whole unit is 400 persons per square kilometre. If, however, the districts without settlements are

---

[1] J. K. Wright, 'A Method of Mapping Densities of Population with Cape Cod as an Example,' *Geographical Review*, vol. 26, pp. 103–10 (New York, 1936).

[2] A. Stevens, op. cit., p. 49 (1946).

regarded as unpopulated, then the intermediate and closely settled districts may be assumed to have a population density of:

$$\frac{60,000}{60 + 10} = 857 \ persons \ per \ square \ kilometre$$

These new density values, o and 857 respectively, will give a reasonably accurate and more detailed picture of distribution of population than the one uniform density of 400. The calculation may be carried a stage further and reasoned guesses made of the population densities for the intermediate and closely settled districts.

A. Geddes used these methods in his analysis of the population of Bengal,[1] in his case not because data for smaller units of enumeration were not available, but because the labour involved in utilizing them would have been enormous. He was also concerned with showing the 'geographical' distribution of population in relation to land utilization rather than in relation to administrative boundaries. Hence he correlated statistical and topographical data to produce densities of population for different categories of land, such as forest and cultivated land. His final densities were directly related to the 'unit area of daily life' of the persons making up the population. The formula used for obtaining these densities is worth noting. If for an administrative unit of area A the population returned by the census is P, the statistical density is P/A. If the population is to be redistributed between forest and cultivated land, then a reasoned guess must be made of the number of forest dwellers (P_1) and the area of forest calculated from land-utilization maps (A_1). The remaining population living in the cultivated lands will then be $P - P_1$, and the area of cultivated land $A - A_1$. The density of population in the cultivated land will be:

$$\frac{(P - P_1)}{(A - A_1)}$$

This method has many applications, for example, it might be applied to the distribution of population as between moorland and vale, or between built-up areas of towns and scattered rural settlement.[2]

The dasymetric technique can also be applied to the mapping of indices of 'comparative density' (see p. 321), by the choropleth method, where it is desired to show population pressure in terms of availability

[1] A. Geddes, 'The Population of Bengal, its Distribution and Changes: A Contribution to Geographical Method', *Geographical Journal*, vol. 89, pp. 344–68 (London, 1937).

[2] See *Great Britain, Population Density, 1931*, compiled by the Ministry of Town and Country Planning, sheet 2, 1 : 625,000 (Ordnance Survey, Southampton, 1944).

of agricultural land. Referring to maps of this type, A. Stevens points out that in Scotland the head-dyke might be used to limit 'the "field" of pressure on the land'.[1]

Urban Population

Especial difficulties arise in the mapping of populations in towns, which are by their nature areas of population concentration. Yet within towns notable variations in population density occur. The extent to which these variations can be accurately portrayed does depend on the amount of detailed information available on a geographical basis. The breakdown available in published census returns may be on a ward basis only, which is too large and too heterogeneous a unit to provide a satisfactory basis. Enumeration districts provide a more satisfactory basis for most purposes. In Britain returns are not published on this basis, but it is possible to consult unpublished data, and the value of maps based on enumeration districts as compared with wards has been nicely illustrated by Emrys Jones.[2]

Census returns for different countries differ widely in their methods of compiling returns of population for urban areas. In the United States experiments have been made on the basis of 'census tracts', which are areas specifically delimited on a social basis in an attempt to gain some measure of homogeneity. All such data can be effectively mapped by choropleths. Where even greater detail is required, it may be necessary to adopt special measures involving large-scale plans and the use of located symbols or dots.

Sex and Age Distribution

A series of choropleth maps can be used to show the detailed distribution of population by ages. Anomalies are more effectively illustrated if the percentages of persons in each group is plotted and not the absolute totals. Regional deviations from the national norm may also be plotted with effect (Fig. 146). The regional deviations are obtained by noting the difference between local age-group percentages and national age-group percentages. A cartographical analysis of population structure along these lines has many uses. It affords information about the availability of male and female labour in different localities; it is of help in the interpretation of differential fertility rates and trends;

[1] A. Stevens, op. cit., p. 51 (1946).
[2] Emrys Jones, 'Sociological Aspects of Population Mapping in Urban Areas', *Geography*, vol. 46, pp. 9–17 (London, 1961). See also B. T. Robson, *Urban Analysis* (Cambridge, 1969).

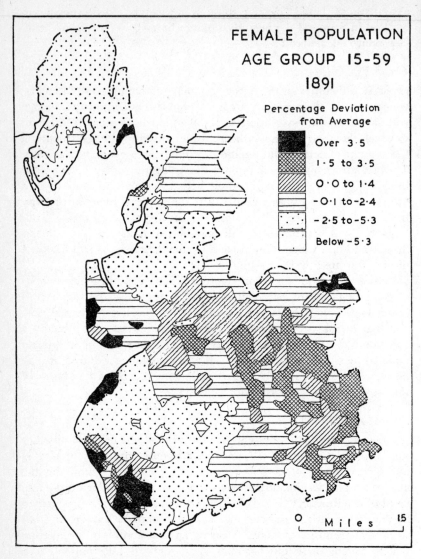

FEMALE POPULATION
AGE GROUP 15-59
1891

Percentage Deviation
from Average

■	Over 3·5
▨	1·5 to 3·5
▨	0·0 to 1·4
▤	-0·1 to-2·4
▦	-2·5 to-5·3
□	Below -5·3

0   Miles   15

Figure 146. A DEVIATIONAL CHOROPLETH MAP

Source of data: *Census of England and Wales, 1881*, vol. 3 (London, 1883).

The map is intended to reveal regional variations in the age structure of the female population in Lancashire aged over 15 and under 60. By showing deviations from the national average, the variations are given sufficient emphasis to stand out clearly.

The map was compiled on the basis of administrative divisions.

it gives many clues about long-term migration. Maps of this type are indispensable to planners in economic and social spheres.[1]

Ethnic and Occupational Structure

The choropleth technique is not eminently suited to depict complex distributional features of the ethnic and occupational structure of population but in certain cases it proves useful. If two elements in the ethnic structure are to be plotted, for example, Czechs and Germans in Bohemia, a proportional scale of values to show the ratio of one nationality to another can be devised and the distribution of ratios shown by choropleths.[2] To show more than two elements, for example, a number of nationalities, or a number of industrial employments, colour has to be employed.[3]

Replacement Rates

Replacement of population, whether indicated in terms of gross and net reproduction rates (Fig. 147), or by fertility ratios, or by crude birth (Fig. 149), death and natural replacement rates, may be most conveniently mapped by means of choropleths. Changes in replacement rates may also be effectively mapped in this manner, but in this case a problem occurs; to show positive and negative rates or positive and negative changes, it is necessary to use special systems of shading (see p. 50). Dudley Kirk (op. cit., 1946), in his map of the distribution of net reproduction rates in Europe, used two colours to overcome this difficulty, black to represent positive and red negative qualities.

Mortality and Morbidity Rates

Most countries of the world now publish data about causes of death and these can be mapped on a geographical basis to reveal marked

[1] The economic significance of age differences in the population has been analysed in A. Sauvy, *Richesse et Population* (Paris, 1943).

[2] A pioneer of this method was E. Hochreiter, 'Nationalitätenkarte von Böhmen', 1 : 1,850,000, *Petermanns Mitteilungen*, vol. 29 (Gotha, 1883). See also a map in E. Jones, 'The Distribution and Segregation of Roman Catholics in Belfast', *Sociological Review*, vol. 4, pp. 167–89 (Keele, 1956) based on a 'segregation index', i.e. the degree to which the proportion of Roman Catholics in any part of the city departs from the proportion for the city as a whole.

[3] This method was adopted in '*Langues Maternelles dans le Royaume S.H.S. par Communes*', 1 : 1,500,000, Publié par la Direction de la Statistique d'Etat (Sarajevo, 1924), which showed the distribution of languages in Jugoslavia according to the Census of 1921. See also B. v. Semenow-Tian-Schanskÿ, 'Handel und Industrie im Europäische Russland', 1 : 750,000, *Petermanns Mitteilungen*, vol. 59, Tafel 36 facing p. 236 (Gotha, 1913), in which 8 occupational groupings are shown according to the percentage of persons employed (0–20 per cent, 20–50 per cent and over 50 per cent).

regional variations not only, as is to be expected, from one part of the world to another, but within particular countries. The plotting of the geography of disease calls for techniques which show not only distribution but also intensity and variability in time and place. Choropleths can be used for this purpose within certain limits. A simple

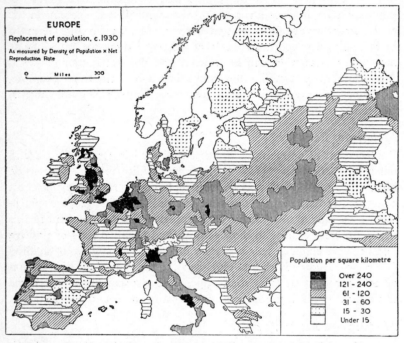

Figure 147. REPLACEMENT OF POPULATION IN EUROPE, 1930

Data and base-map from D. Kirk, *Europe's Population in the Inter-War Years* (Geneva, 1946). Replacement of population on this map is expressed in terms of projected population density per sq. km.

illustration is a map showing regional variations in the average death rates for England and Wales, as compiled by M. A. Murray. This shows the ratio of local age and sex adjusted death rate to the national rate for different localities.[1] Rather more elaborate is the elegant system

[1] M. A. Murray, 'The Geography of Death in England and Wales', *Annals of the Association of American Geographers*, vol. 52, pp. 130–49 (Lancaster, Pa., 1962). See also M. A. Murray, 'The Geography of Death in the U.S.A. and the U.K.', *Annals of the Association of American Geographers*, vol. 57, pp. 301–14 (Lancaster, Pa., 1967); and *National Atlas of Disease Mortality in the United Kingdom*, prepared by G. Melvyn Howe, on behalf of the Royal Geographical Society (London, 1963).

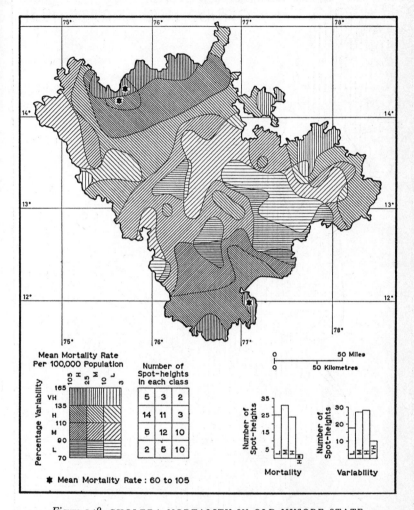

Figure 148. CHOLERA MORTALITY IN OLD MYSORE STATE

Based on A. T. A. Learmonth and Manindra Nath Pal, 'A Method of Plotting Two Variables (such as Incidence and Variability from Year to Year) on the Same Map using Isopleths', *Erdkunde*, vol. 13, pp. 145–50 (Bonn, 1959).

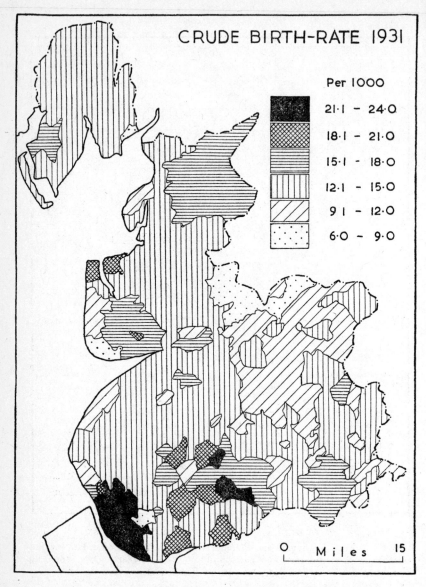

Figure 149. A CHOROPLETH MAP OF VITAL STATISTICS

Source of data: *The Registrar-General's Statistical Review of England and Wales for the Year 1931*, Tables, pt 2, Civil (London, 1933).

The distribution is by administrative divisions.

of cross-tabulation devised by A. T. A. Learmonth to incorporate both degree of intensity and amount of variability of the incidence of disease in one choropleth map (see also p. 358 and Fig. 148).[1]

Migrations

Out-migration, in-migration and balance of migration may be shown by means of choropleth maps. The methods are best illustrated by reference to a specific example, in this case out-migration from Worcestershire in 1861 (Figs. 150–3). The data are based on birthplaces given in the census returns of England and Wales for 1861 for over 600 registration districts. The number of persons born in Worcestershire and enumerated outside the county for each registration district was first of all calculated. These totals were mapped in four ways; (a) each total was expressed as a percentage of the total number of migrants from Worcestershire until that year; (b) each total was related to the area of its particular registration district and expressed as a density in terms of persons per 10,000 acres; (c) each total was expressed as a percentage of the total population in its enumeration area; and (d) absolute totals were considered *per se*.

Growth of Population

Probably the most obvious method of showing population growth is to prepare a series of maps in chronological order showing the past and present distributions. Such a series gives a broad picture, but it is a difficult task to measure and compare the growth of population from one district to another without recourse to a very close examination of the map.

Intercensal Changes in Population. To overcome this difficulty, intercensal changes in population can be shown directly, usually in the form of the percentage change in the increase or decrease of population which has occurred between two censuses. These changes are best plotted in the form of a choropleth map. Some further indication of absolute numerical changes may be given by super-imposing proportional symbols on the choropleths.[2]

There are three main difficulties in the preparation of choropleth

[1] A. T. A. Learmonth, 'A Method of Plotting on the Same Map Health Data of both Intensity and Variability of Incidence', *Annals of the Association of Tropical Medicine and Parasitology*, vol. 48, pp. 345–8 (Liverpool, 1954),

[2] See *Great Britain, Population. Total Changes, 1931–39*, compiled by the Ministry of Town and Country Planning, sheet 2, 1 : 625,000 (Ordnance Survey, Southampton, 1942) and subsequent publications.

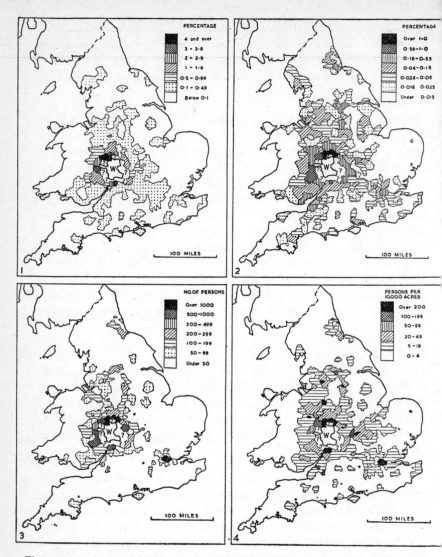

Figures 150–3. FOUR METHODS OF MAPPING MIGRATION BY CHOROPLETHS

Source of data: *Census of England and Wales, 1861*, vol. 2 (London, 1863).

The number of natives of Worcestershire living outside the county in 1861 has been mapped to give an indication of migrants from Worcestershire on the basis of registration districts. **1** shows migrants from Worcestershire expressed as a percentage of the population of each registration district into which they have moved; **2** shows migrants for each registration district expressed as a percentage of the total number of emigrants from Worcestershire; **3** simply shows registration districts shaded to show the absolute number of migrants they each received; and **4** shows the number of migrants from Worcestershire in each registration district expressed in terms of number per unit of area. No separate data were available for the Isle of Man, and Scotland has been excluded.

maps of this nature. In the first place, the dates of the two censuses have to be chosen with care. For example, if the years 1801 and 1931 were chosen for Ireland, these would give no indication of the real nature of the growth and decline of Irish population in the interim decades. In the second place, the map must be based on the enumeration units given in the latest census to be used. Almost certainly changes of unit boundaries will have taken place between the two censuses, and amalgamation or partition of various units will also obscure the issue. In some parts of Britain, for example, in the Manchester and London areas, it is a most formidable task to equate the regional bases of the returns between a census of the early nineteenth century and another of more recent date. Intricate research into changes in the boundaries of parishes, wards and boroughs is a necessary prelude to the preparation of the base-map. Since many changes are not even recorded, this task involves laborious comparisons of acreages. The third problem is one of presentation. Almost invariably some parishes will have declined in population, others will have gained. A two-fold division of shading is therefore necessary to distinguish between each category; this problem has been discussed in Chapter 1 (see p. 50).

Intercensal changes may be plotted also by means of indices showing the rate of change as compared with the corresponding national rate. An index of the national change is arrived at by dividing the national total of population for, say, 1931, by that for 1831, if these two censuses are being considered. In the example given below:

$$Index\ of\ change = \frac{National\ total,\ 1931}{National\ total,\ 1831} = \frac{5,000,000}{2,000,000} = 2 \cdot 5$$

Local indices are similarly calculated, and then plotted using the choropleth method. Some distinction should be made between parishes with an index of less than 2·5, i.e. below the national rate of growth, and those above that figure.

Population Peaks. Line-graphs may show the behaviour of the growth of population for individual localities fairly well, but they are hard to use when hundreds of localities are under consideration. To analyse geographically the evidence provided by the graphs some of the essential features have to be transferred to maps. For example, the dates at which population peaks are attained are easily plotted on a map.[1]

[1] S. D. Dodge, 'A Study of Population Regions in New England on a New Basis', *Annals of the Association of American Geographers*, vol. 25, pp. 197–210 (Lancaster, Pa., 1935).

The distribution of the dates of peaks of population plotted in this manner indicates those regions the procedure of which is in decline; the earlier the peak, the longer the period of decline. When contiguous districts have roughly the same dates of peak population, some underlying regional economic cause, such as mineral exploitation, may be judged to be responsible.

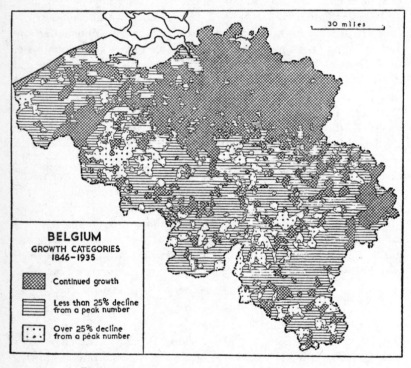

Figure 154. GROWTH CATEGORIES IN BELGIUM

Based on a map in H. M. Kendall, 'A Survey of Population Changes in Belgium', *Annals of the Association of American Geographers*, vol. 28, p. 152 (Lancaster, Pa., 1938).

Growth Categories. It is possible to show the distribution of peak populations according to the time when they occurred. S. D. Dodge distinguished four categories of population trend based on a careful analysis of patterns of growth of population for parts of New England. The categories were (*a*) continued growth, (*b*) decline of 25 per cent from a peak, (*c*) decline of 25–50 per cent from a peak, and (*d*) decline of over 50 per cent from a peak (see reference on p. 347).

These categories may, of course, be expanded to include various degrees of irregularity in growth and decline. They may be effectively illustrated by the choropleth technique, in which case they reveal long-term changes in the population balance of various regions. They are particularly useful in illustrating relative declines of population. H. M. Kendall used this method of analysis in his treatment of the population of Belgium (Fig. 154).

Other types of growth categories can be used. For example, C. F. Kohn in his work on population trends in the United States employed

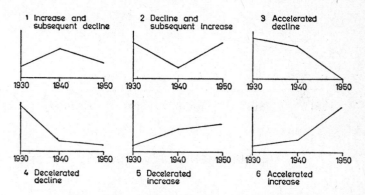

Figure 155. KOHN'S GROWTH CATEGORIES

the six categories given in Fig. 155.[1] These categories are best shown by a series of maps, but they may be incorporated in one map with the aid of an ungraded shading system (see pp. 47–50). Other population-growth categories may be distinguished upon inspection of the data relating to the growth of particular populations.[2] (In this connection, see also p. 392.)

Variability. In a paper on population trends in India, A. Geddes demonstrated that variability in the growth of population has a distinctive regional significance.[3] The normal type of intercensal population-growth map does not show variability. Two enumeration districts,

[1] C. F. Kohn, 'Population Trends in the United States since 1940', *Geographical Review*, vol. 25, pp. 98–106 (New York, 1945).
[2] See, for example, 'Carte dynamique de la région parisienne', in A. Demangeon, 'France Economique et Humaine', pt 2, *Géographie Universelle*, vol. 6, p. 835 (Paris, 1948).
[3] A. Geddes, 'The Population of India. Variability of Change as a Regional Demographic Index', *Geographical Review*, vol. 32, pp. 562–73 (New York, 1942).

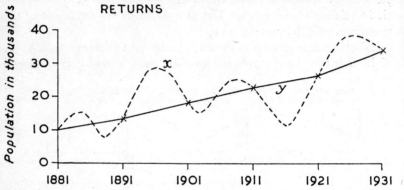

1. A HYPOTHETICAL CASE OF HOW VARIABILITY IN POPULATION GROWTH MAY BE CONCEALED IN CENSUS RETURNS

2. MEASUREMENT OF VARIABILITY INDEX

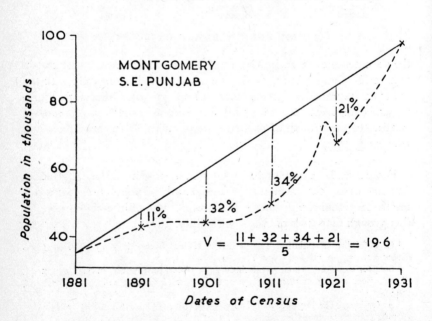

MONTGOMERY
S.E. PUNJAB

$$V = \frac{11 + 32 + 34 + 21}{5} = 19 \cdot 6$$

Dates of Census

for example, might exhibit similar percentage growths of population, but whereas one may have had a steady increase of population, the other may have experienced all kinds of population vicissitudes. The census dates are purely arbitrary and may tend to cloak any variability in the population growth (Fig. 156, *top*). Geddes showed distribution of variability by means of a variability index. This was calculated by comparing the actual curves of growth of population, in the case of each enumeration district, with a theoretical normal curve of growth; the mean percentage deviation between the two curves was taken as an index (Fig. 156, *bottom*).

In practice it is not necessary to calculate a hypothetical curve of growth. A straight line drawn between the two intercensal plottings will provide a sufficiently accurate curve for the measurement of variations. Alternatively, a curve may be sketched by reference to the national figures. If 1801 and 1921 are taken as the intercensal dates, an index of population variability may be calculated in the following way. Superimpose the percentage growth of population, decade by decade, for the enumeration district under consideration, upon the hypothetical curve of growth represented by a straight line drawn between the plottings for 1801 and 1921. The two curves will deviate from one another at as many as ten points. These deviations may be measured in terms of percentages by reference to the scale used on the graph. All ten variations should be added, and the mean percentage deviation found by dividing by eleven; the variability index so obtained can be plotted conventionally by means of a choropleth map. A. Geddes found, for example, that a range of six variability indices was practicable in the case of India.

Daily Movement of Population

Where the scale of the map permits, daily movements of population can be accurately shown by flow diagrams and other techniques (see

Figure 156. MEASUREMENT OF POPULATION GROWTH VARIABILITY

In the top diagram, the pecked line (*x*) indicates the true curve of the growth of population and the solid line (*y*) the curve as adjudged by reference to census returns.

The bottom diagram is based largely on A. Geddes, 'The Population of India', *Geographical Review*, vol. 32, p. 569 (New York, 1942).

The pecked line shows the population curve sketched from census returns and from known dates of crises. The solid line represents the theoretical curve of growth. The two curves are known to deviate at four points, each marked by a cross. The amounts of deviation are calculated on a percentage basis and the variability coefficient (V) found by Geddes' formula.

pp. 307 and 375–6). Where, however, such movements are to be mapped on a broad basis the choropleth technique proves useful and economical. R. Lawton has used it to give an overall picture of the daily movement into and out of local authority areas in England and Wales as well as the net daily balance of movement into and out of such areas. He was able to plot the number of people travelling daily from one authority to another as a percentage of the total resident population of the administrative area into which they came to work. Lawton also used J. Westergaard's concept of the 'job ratio' to compile a complementary map showing the ratio of the working resident population and the total occupied population in each authority.[1]

QUANTITATIVE SYMBOLS

Dots

The dot technique may be employed to show the absolute total distribution of population or the distribution of individual elements within the population, such as Poles in Belgium or France. Technical problems in the construction of dot maps have been referred to in Chapter 1 (see p. 25). In the case of population maps, if the value of each dot is too big sparsely populated districts will not be represented at all, if too small the dots will coalesce in densely populated districts. The stratagem of dispensing with dots altogether in densely populated parishes and shading them black has been adopted in Fig. 141, but this is hardly a true solution of the problem. Occasionally it may be useful to employ two sizes of dot for special purposes (Fig. 157).

As a general rule, the dots have to be distributed evenly within the boundaries of the area of the population which they represent, but it is possible in the case of dots of small value to place them according to the distribution of dwellings. This involves, of course, the consultation of topographical maps and plans. Dot maps of town populations are often shown in this way, one dot with a value of five or more persons being placed over each house or each group of houses.[2]

---

[1] R. Lawton, 'The Daily Journey to Work', *The Town Planning Review*, vol. 29, pp. 241–57 (Liverpool, 1959).

[2] See such maps in W. William-Olsen, 'Stockholm', *Geographical Review*, vol. 30, pp. 420–38 (New York, 1940). An interesting technique was used by W. Applebaum, 'Technique for Constructing a Population and Urban Land Use Map', *Economic Geography*, vol. 28, pp. 240–3 (Worcester, Mass., 1952). He drew an urban land-use map, with industrial, commercial and other categories shaded, and residential areas in white. On the last he placed dots (1 dot = 20 people), located by the blocks of apartments in which they lived.

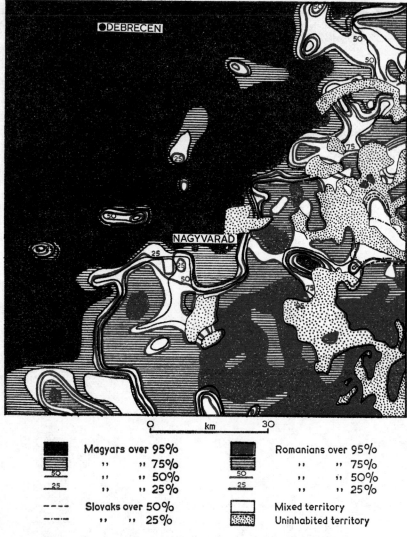

Figure 160. ETHNOGRAPHIC DISTRIBUTIONS BY ISOPLETHS

Based on 'Hungary, Atlas', *The peoples of Austria-Hungary, I* (N.I.S.D. 1919). Data from the Austro-Hungarian census of 1910.

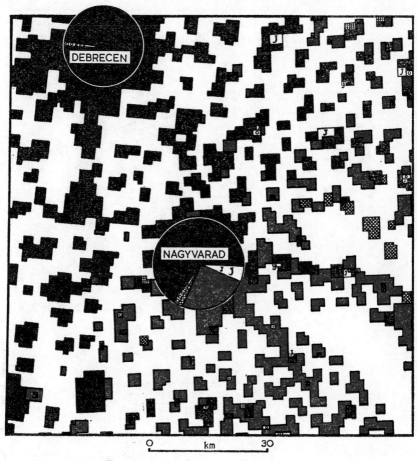

One square millimetre = 200 persons

Towns over 50,000 represented by a circle

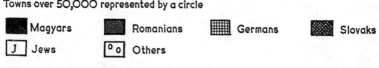

| Magyars | Romanians | Germans | Slovaks |
| J Jews | ⁰₀ Others | | |

Figure 159. ETHNOGRAPHIC DISTRIBUTIONS BY GROUPED SQUARES
Based on *Ethnographical Map of Central Europe,* Institute of Political Sciences (Budapest, 1942). Data from various inter-war censuses.

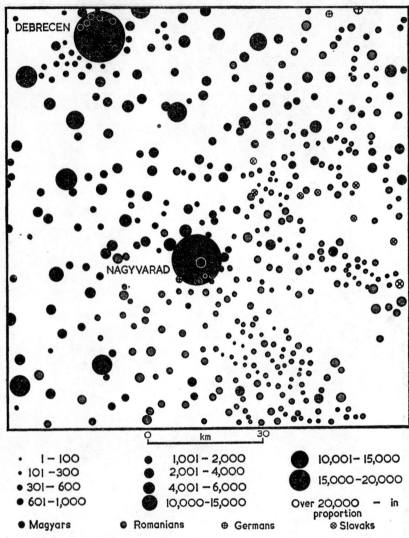

Figure 158. ETHNOGRAPHIC DISTRIBUTIONS BY PROPORTIONAL CIRCLES

Based on *Roumanian Ethnographical Maps and their Value*, facing p. 34, Institute of Political Science (Budapest, 1942). Data from various inter-war censuses.

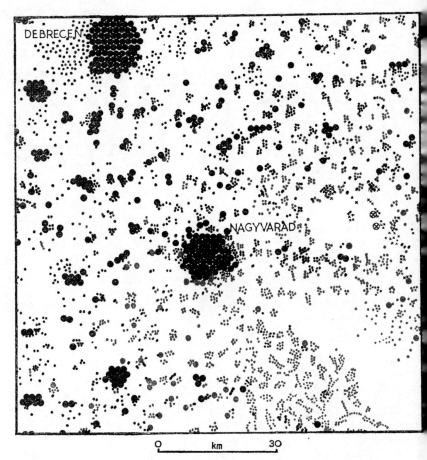

DEBRECEN

NAGYVARAD

0   km   30

	Magyars	Romanians	Germans	Slovaks
1,000 persons	●	●	⊕	⊗
100 persons	·	·	+	×

Figure 157. ETHNOGRAPHIC DISTRIBUTIONS BY DOTS OF TWO SIZES

Based on Z. S. Bátky *et al, Magyarország néprajzi térképe település és lélekszam szerint,* 1:300,000 (Budapest, 1919). Data from Austro-Hungarian Census of 1910.

In common with Figs. 155-7, this map shows part of the border country of Transylvania lying between Hungary and Rumania.

In Britain a scale of one dot for five persons has been found practicable, using 1 : 2,500 town plans. The data in this case may be drawn from electoral registers. Because persons 18 (until recently under 21) are not listed, it is necessary to make use of a multiplier calculated from the ratio of total local population to total electoral population in order to correct the data. Field-work is also necessary to locate residential properties and house numbers on the plans.[1]

The dot method comes into its own, so to speak, in the mapping of widely scattered rural populations where information about local village areas or details of local boundaries may be lacking. R. M. Prothero has drawn attention to the problems of mapping population distribution in such territories as Northern Nigeria, and has suggested possible solutions.[2] In his opinion, 'a dot distribution map would seem to be the first essential in any attempt to delineate population'.

Proportional Squares, Circles and Shaped Symbols

The total population of each enumeration unit may be represented by a symbol of proportional size, either by a square or a circle. These symbols are particularly suitable for showing distribution of population by countries to demonstrate the political and economic significance of varying population resources. Rectangles proportional in size to populations of different regions, countries or states, for example, may be grouped in their approximate geographical position for purposes of comparison of populations and land areas. They can also be used to show detailed distribution, but in this case the problem arises of choosing a scale of symbols so that small populations may be distinguished, and yet large populations not obscure neighbouring small populations (Figs. 139–40). The disadvantages attached to this method are first, that the extreme range of population density to be found in many regions often leads to an overlap of symbols and causes difficulty in the drawing and in the interpretation of the map, and second, that the eye has difficulty in equating the symbol to the area it represents and the method does not readily convey density of population. Populations of countries may also be depicted proportionally by symbols associated with their respective shapes, thus affording an instantaneous visual impression, as in Fig. 183.

[1] A. J. Hunt and H. A. Moisley, 'Population Mapping in Urban Areas', *Geography*, vol. 45, pp. 78–89 (London, 1960).
[2] R. M. Prothero, 'Problems of Population Mapping in an Under-developed Territory', *Nigerian Geographical Journal*, vol. 3 (Ibadan, 1960).

Proportional squares and circles may be effectively employed to show ethnic structure of population (Fig. 158),[1] occupational structure (Fig. 122), and also to show migration, especially in-migration to large towns,[2] and time changes in the totals of populations.[3]

Proportional Spheres and Cubes

To overcome the disadvantage of overlap, proportional spheres or cubes may be used instead of circles and squares. The introduction of the third dimension enables a greater range of densities to be more effectively shown (see pp. 29–32). The symbols have to be shaded to create the illusion of sphericity (Fig. 140). A further refinement may be introduced by tinting the spheres according to the major industrial activity of the population in each centre; thus an additional indication of economic structure is given.[4] Distributions of occupations may also be shown by spherical symbols, because they can be grouped together without difficulty, providing that the classification is not too complex.

Dots and Circles

Two or more techniques in conjunction may profitably be employed to overcome some of the disadvantages commonly associated with a single technique. Dots and circles can be used, for example, to depict the distribution of population, particularly on a continental scale. Thus the dots may show the scatter of rural population, the circles may represent the population of towns (Fig. 138).[5]

In a composite map of this type, it is desirable that the symbols should be related to one another, so that the size of the dot is com-

[1] See an excellent map of nationalities by proportional squares method in l'Académie Tchéque, *Atlas de la République Tchécoslovaque* (Prague, 1935).

[2] A variety of population maps can be seen in G. T. Trewartha and W. Zelinsky, 'Population Distribution and Change in Korea, 1925–1949', *Geographical Review*, vol. 45, pp. 1–26 (New York, 1955). A map of absolute change (facing p. 24) uses solid black proportional discs for increase in rural units, stippled discs for decrease in rural units, and diagonally shaded discs for increase in urban units. An interesting example of the use of proportional circles to show time changes in the totals of populations was used by H. W. H. King, 'The Canberra–Queanbeyan Symbiosis: A Study of Urban Mutualism', *Geographical Review*, vol. 44, p. 114 (New York, 1954). Using proportional discs placed at the ends of radii for each town, he showed the population totals at each intercensal period, 1841–1947; this diagram was called a *chronograph*.

[3] See, for example, Ministry of Town and Country Planning, Population Maps.

[4] See, for example, *Ekonomisk-Geografisk Karta över Sverige*, 1 : 1,000,000 (based on the 1940 Census and prepared by W. William-Olsson, Stockholm).

[5] W. Coulter, 'A Dot Map of Distribution of Population in Japan', *Geographical Review*, vol. 16, pp. 283–4 (New York, 1926).

mensurate with the size of the circle. If a dot $\frac{1}{10}$-inch in diameter represents, say, 10,000 persons, a city of a million should be represented by a circle 1 inch in diameter. Expressed mathematically, the relationship is as follows:

$$\frac{\textit{Diameter of circle}}{\textit{Diameter of dot}} = \frac{\sqrt{1,000,000}}{\sqrt{10,000}} = \frac{10}{1}$$

A dot value of 10,000 is often too high to portray rural population distribution, even on a continental scale. If a smaller value is used, the circles representing urban population centres become correspondingly bigger. Even if they are drawn as rings, in order not to obscure neighbouring rural distributions, their size becomes impracticable. For example, a dot value of 100 could be usefully employed to show the distribution of the rural population in Australia, but as some of the great Australian cities have populations exceeding the million mark, these would have to be represented by circles 10 inches in diameter.

Dots and Spheres

To make the symbols for rural and urban population commensurate and yet practicable to draw, the urban populations may be shown in three dimensions by means of cubes or spheres. In the case of the Australian population distribution quoted above, the 'million' cities could each be represented by a sphere, the diameter of which would be $\sqrt[3]{\frac{3}{2} \times 10^2} = 5 \cdot 3$ in., a much more practicable size to draw. This method of showing population distribution has many advantages, and conveys a good impression of density. It was first used with effect by Sten de Geer, the Swedish geographer.[1]

Grouped Squares

Probably the most satisfactory manner of depicting a complex ethnic structure is by means of symbols grouped in that locality where the population is known to be densest. Information has to be obtained by reference to the settlement pattern from large-scale topographical maps.

Some care is needed in the choice of size of symbol for this type of

---

[1] S. de Geer, 'A Map of the Distribution of Population in Sweden: Method of Preparation and General Results', *Geographical Review*, vol. 12, pp. 72–83 (New York, 1922). The maps themselves were published in the form of an atlas – *Karta över Befolkningens Fördelning i Sverige den 1 Januari, 1917*, 1 : 500,000, 12 plates (Stockholm, 1919).

map. It depends on the range of density of population which occurs within the area under consideration. It is not necessary to calculate the scale of symbol from the situation as it exists on the ground; it is more simple to work directly from the base-map selected. If the area on the map of the administrative unit which has the greatest density is calculated to be 2·5 sq. cm and the population of the unit is, say, 40,000, then the density on the map, i.e. the scale of symbol to be adopted, must not be more than:

$$\frac{40,000}{2·5} \text{ persons per square centimetre} = 160 \text{ persons per square millimetre}$$

If this scale of symbol were adopted, every 160 persons in each ethnic group would be represented on the map by a shaded area 1 mm square. The millimetre squares are built up into shapes, or in the case of towns incorporated into circles, and placed in position on the map according to information gleaned from an examination of topographical maps (Fig. 159). Thus the administrative unit with a population

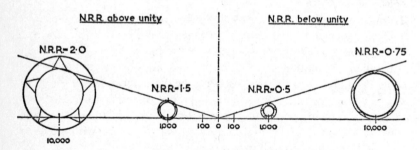

Figure 161. NET REPRODUCTION RATE SYMBOLS

Circles with thin lines indicate sizes of actual populations and circles with thick lines indicate the product of population total and by current net reproduction rate. Outward-facing points denote potential increase and inward-facing points denote potential decrease in the respective sizes of populations.

of 40,000 mentioned above would be fairly well covered with squares arranged in a pattern corresponding to its shape. A unit with a population of only 160, however, would have only 1 mm square placed in the vicinity of the largest village or habitation. Any population too small to show is added to the population of a neighbouring unit, so that every person in each ethnic group is represented on the final map.

One difficulty in the choice of scale of symbol is likely to be encountered. Where very large cities have to be considered they will inevitably dictate the scale to be adopted, with disastrous consequences or the representation of rural populations. Accordingly, these cities should be treated separately – represented perhaps by circular diagrams on the margins of the map, or by proportional spheres grouped *in situ*. If the former are used, care should be taken to adopt the same symbol scale as for the rest of the map, and to show the position of the towns by rings on the map, the rings being proportional in size to the total population.

Special Proportional Symbols

Special symbols may be designed to show distribution of particular features of population structure. For example, H. J. Fleure devised special symbols to show the distribution of net reproduction rates in Europe (Fig. 161).[1] The symbols in this case were devised to show population totals and negative and positive net reproduction rates.

ISOPLETHS

The use of isopleths has only a limited application in the case of population maps. They are used mainly to depict population distributions and characteristics on a continental scale, where it is desired to smooth out local variations in order to get a broad picture.[2] They are, however, increasingly important in the depiction of statistical 'surfaces', e.g. trend surfaces.

Population Density

The administrative unit has to be chosen with care for the construction of maps showing density of population.[3] To use communes, for example, in the case of France, or parishes in the case of Great Britain, would lead to great difficulty in the interpolation of isopleths because of the detail involved. Points for interpolation are arbitrarily fixed in the geographical centre of the administrative unit, or over the biggest centre of population within that unit, their value being the mean density of the population in the unit concerned. Skill in such interpolation only comes with practice. The spotting of key isopleths enables the others to

[1] H. J. Fleure, *Problems of Population in Europe* (Manchester, 1942).
[2] In attempting correlations between population density and rainfall, for example. See Arthur H. Robinson, 'Mapping the Correspondence of Isarithmic maps', *Annals of the Association of American Geographers*, vol. 52, pp. 414–25 (Lancaster, Pa., 1962).
[3] Techniques of construction are discussed in J. W. Alexander and G. A. Zahorchak, 'Population Density Maps of the United States; Techniques and Patterns', *Geographical Review*, vol. 33, pp. 457–66 (New York, 1943).

be quickly drawn. It is often necessary to construct a dispersion graph to help decide the isopleth intervals to be selected.[1]

It has been noted above that the chief disadvantage of the isopleth technique is its unsuitability for showing details of distribution. In areas where great variations in density occur, key isopleths are impossible to discern and the whole map becomes a series of concentric circles centred on the towns. This disadvantage can be offset if the town populations are shown by spheres, and the rural populations are shown by isopleths; this method has been used effectively to depict distribution of populations in, for example, central Oklahoma.[2]

Not so successful but nevertheless interesting is the application of relief shading to isopleths of population density, as suggested by J. C. Sherman.[3] T. Hägerstrand makes very effective use of isopleths in his work on innovation and diffusion processes among the rural population of a Swedish province, including their use in predictive models (see p. 372).

Expectancy of Life and Mortality

Reference has already been made to the plotting of various categories of demographic data by choropleths. Isopleths, however, do serve a useful purpose when variations in such data have to be plotted on a continental or global scale. A good example of their use is to be found in an inaugural lecture by D. L. Linton on the Tropical World.[4] He makes use of maps showing expectancy of life over the world at different dates by isopleths, and summarizes the progress of death control by isochronic lines on the basis of ten-year intervals of life expectancy for males at birth first exceeding fifty years. The use of isopleths can plainly be applied to a wide variety of data relating to birth-rates, fertility and so on, where broad patterns need to be presented. It may even be practicable to show two sets of related data on the same map, as A. T. A. Learmonth has shown in an analysis of cholera mortality in Mysore (Fig. 148). The data in this case were mean mortality and

[1] The case for isopleths is strongly argued and exemplified by M. Pinna, *Carta della Densità della Populazione in Italia* (*Cens. 1951*) (Consiglio Nazionale della Richerche, 1960).

[2] C. J. Bollinger, 'A Population Map of Central Oklahoma', *Geographical Review*, vol. 20, pp. 283–7 (New York, 1930).

[3] J. C. Sherman, 'New Horizons in Cartography', *International Yearbook of Cartography* (ed. E. Imhof), vol. 1, pp. 13–19 (London, 1961).

[4] D. L. Linton, *The Tropical World. An Inaugural Lecture delivered in the University of Birmingham* (Birmingham, 1961).

variability, as measured by the percentage of mean absolute deviation from the mean mortality rates.

Population Potentials

Isopleths may also be used with effect in the plotting of *potentials* of population.[1] The word potential here is used in a similar but not the same sense as in physics; the gravitational influence of a planet on the earth, for example, may be expressed in terms of its mass and its distance away from the earth. The influence of any concentration of population from one part of the earth on another may be expressed as the number of people in that concentration divided by its distance to the spot at which its potential is being assessed. This potential has obviously great cultural, social, economic and even political significance. It may be mapped by means of isopleths with an interval expressed in terms of persons per mile. In the case of Europe, for example, the points for interpolation may be located in the centre of administrative units of a size approximating to that of a county. The potential of each of these points has then to be calculated by the formula:

$$\frac{n}{d} = P$$

where n is the number of persons, d is the distance and P is the potential.

The population potential at each county centre will be the sum of the influence of all other centres upon it, plus its own influence upon itself. An example will make this clear. Consider Fig. 162, which incorporates four concentrations of population at A, B, C and Z. The potential in terms of persons per mile at Z will be as follows:

From A to Z: $\frac{50}{10} = 5$ *persons per mile*

From B to Z: $\frac{100}{12} = 8{\cdot}33$ *persons per mile*

From C to Z: $\frac{10}{5} = 2$ *persons per mile*

Own (Z) $= 80$ *persons per mile*
Potential at Z $= 95{\cdot}33$ *persons per mile*

[1] J. Q. Stewart, 'Empirical Mathematical Rules concerning the Distribution and Equilibrium of Population,' *Geographical Review*, vol. 37, pp. 461–85 (New York, 1947).

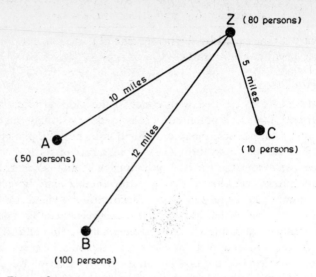

Figure 162. CALCULATION OF POPULATION POTENTIAL

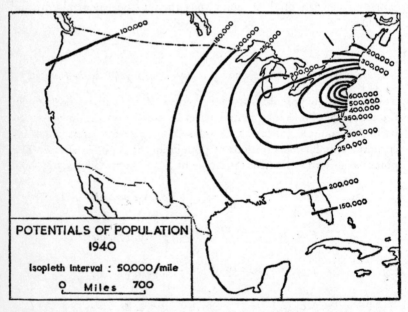

Figure 163. POPULATION POTENTIALS BY ISOPLETHS FOR THE UNITED STATES

Based on J. Q. Stewart, 'Empirical Mathematical Rules concerning the Distribution and Equilibrium of Population', *Geographical Review*, vol. 37, p. 476 (New York, 1947).

It may be appreciated that a considerable amount of labour is necessary to calculate potentials at each point when a large number have to be considered. When the potentials have been computed and plotted, the isopleths are interpolated in the usual manner. The completed maps often reveal extraordinary concentrations of potential at one point within a land mass of continental size (Fig. 163).[1]

Ethnographic Distributions

Maps depicting the distribution of nationalities by isopleths have certain advantages, for they show comparative concentrations of the various ethnic groups in zones, they indicate sharp divides between one group and another, and they delimit mixed areas.[2] Data for the interpolation of isopleths are worked out separately for each ethnic group. For example, the ratio of group X of the total population is obtained for all the administrative units under consideration. The percentages are plotted in the centre of the unit concerned, and then the isopleths are drawn at suitable intervals of 10, 20 or 25 per cent, according to the detail required. Where a number of groups has to be considered, isopleths must be drawn in colour in order to be distinguishable (Fig. 160).

DIVIDED CIRCLES

Divided circles ('pie-graphs') are widely used to illustrate features of the population structure. The circles are divided to show constituent elements in the population of any area. There are two varieties: (a) proportional divided circles, which vary in size according to the respective totals of population under consideration; and (b) comparable divided circles, which are of standard size, and which are usually employed where the population totals have such a wide range that proportional circles would vary in size from a dot to a large circle. Both varieties may be used to show the ethnic, racial, social, occupational, industrial and age structure of the population. Most useful are two-element divided circles to show such fundamental proportions as rural versus urban population, migrants versus native-born,

---

[1] For a detailed consideration of population potential mapping, see W. Warntz, 'A new Map of the Surface of Population Potentials for the United States, 1960', *Geographical Review*, vol. 54, pp. 170–84 (New York, 1964). This paper includes a map showing potentials of population in the United States in 1960 in some detail.

[2] B. C. Wallis, 'The Distribution of Nationalities in Hungary', *Geographical Journal*, vol. 47, pp. 177–88 (London, 1916).

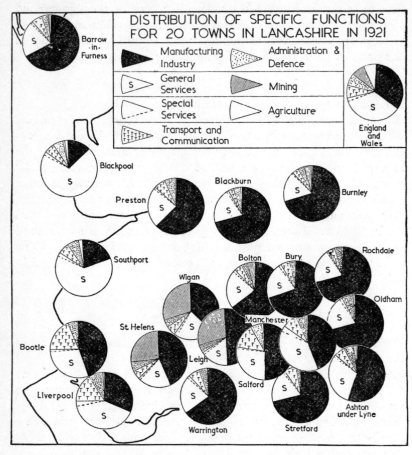

Figure 164. LOCATED COMPARABLE DIVIDED CIRCLES

Source of data: *Census of England and Wales, 1921, Industry Tables* (London, 1925).

By comparing the diagram for each town with that for England and Wales as a whole, some idea of the amount of specialization of function for each of the towns can be estimated. The majority of Lancashire towns fall into the manufacturing category. Their crowded distribution leaves but a limited urban field for each town and services are under-developed. The extraordinary concentration of services in Southport and Blackpool is immediately apparent. Manchester approximates most closely to the average distribution of functions, thus its metropolitan status is reflected. In contrast Liverpool emerges as a transport town but with services above the average.

residential versus working population and ethnic minority versus
majority (see pp. 317–18). Another useful function of these diagrams is to
show time-changes from one decade to another, for example, in population totals or in the numbers of migrants. It is not intended to discuss all these applications here, but the principles of construction of
divided circles may be illustrated with reference to the mapping of the
industrial structure of population in some twenty Lancashire towns.

Located Divided Circles

The data for this example are drawn from Table 4 of the *Industry
Tables* of the Census of England and Wales for 1921, in which details
are listed of the number of persons engaged in some 400 industries,
based on place of work. There are many ways in which these industrial data may be classified for the purpose of cartographical illustration. In this case they have been grouped into seven categories on the
basis of 'function', and each category may be said to represent a
functional activity of the population. The total number of persons
engaged in each category has been calculated as a percentage of the
total engaged in all seven and the results have been plotted (Fig.
164).

COLUMNAR DIAGRAMS

There are essentially three forms of columnar diagram which may
be used to illustrate population data: (*a*) simple columns to depict
quantitative data; (*b*) compound, in which each column is subdivided
to show structural aspects; and (*c*) superimposed, in which departures
from normal distributions are shown to illustrate regional variations.[1]

Located Superimposed Columnar Diagrams

The location and superimposition of columnar diagrams deserves
fuller treatment, both as an example of the principle of columnar
representation of population data, and as an example of how columnar
diagrams may be adapted to show regional variations. The example
taken illustrates data drawn from the *Industry Tables* of the Census of
England and Wales in 1921. The data are based on place of work
and not on place of enumeration, so that they can be used to indicate

[1] A number of well-executed columnar diagrams is to be found in United States
Department of Commerce, Bureau of Census, *Statistical Atlas of the United States*
(Washington, 1935).

location of industry (see p. 318–19). For the purposes of the illustration, the total number of persons engaged in manufacturing of all types in each of the ten biggest towns in Cheshire in 1921 has been considered, and these totals have been broken down into six major categories. The number of workers in each category in each town has been expressed as a percentage of the total workers in all six categories. These percentages have been expressed in the form of columns arranged

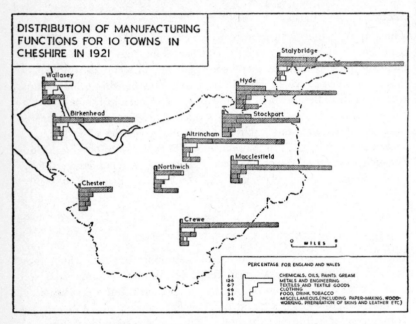

Figure 165. LOCATED SUPERIMPOSED COLUMNAR DIAGRAMS
Source of data: *Census of England and Wales, 1921, Industry Tables* (London, 1925).

in a horizontal series to facilitate mapping. These columnar diagrams have been superimposed, in each case, on another which depicts the percentages for England and Wales as a whole (Fig. 165). By this method the manufacturing structure of the population in each town is boldly depicted and the degree of localization of the six categories of industrial occupations may be quickly appreciated by inspection.

Another interesting example of this versatile technique is its application to peak-period travel in London. The number of passengers arriving in the morning and departing in the evening, at intervals of fifteen minutes during the peak periods, can conveniently be shown by columnar diagrams superimposed at the places of arrival and departure, and a graphic picture of the volume and distribution of traffic emerges.[1]

PYRAMIDS

Columns constructed to represent specific quantitative population data may be arranged in the form of a pyramid. Pyramids of this type are employed primarily in the analysis of population growth, and of population composition.

Age and Sex Pyramids

The simplest method of representing the sex and age structure of a population is to represent it in the form of a pyramid built up in age groups, males on one side, females on the other, and with the base representing the youngest group, the apex the oldest (Fig. 166). These pyramids may take the form of groups for individual years, in which case there may be as many as ninety or more tiers in the pyramids, or, more usually, of five-year or quinquennial groups. Pyramids based on quinquennial groups provide the geographer with as much information about sex and age structure as he is likely to need for most purposes. In any case, detailed local information for one-year groups is difficult to obtain, whereas quinquennial data are published in international returns, in individual census returns, and occasionally in special returns of the Registrar General. There are slight variations in the methods by which quinquennial data may be plotted in the form of pyramids; these are compared in Fig. 166.

The Absolute Method. The data are plotted directly from the absolute quinquennial totals given in the census returns, after a suitable scale has been chosen (example *a*). If a linear scale only is used, all the pyramids will be the same height, the variation will be in their relative widths. If the width of each tier in the pyramid is made proportional to its length, the relative heights of the pyramids will vary as well as their widths (example *b*). The pyramids are then orthomorphic and their shapes may be easily compared.[2] Even so, the pyramids will vary enormously in size, from that of a large town, for example, to that of a small rural community.

[1] *Administrative County of London Development Plan, 1951. Analysis* (London, 1951).
[2] Pyramids for European populations are thus constructed in a map in D. Kirk, op. cit. (1946).

The Comparable Method. Much more practicable to the geographer in his search for regional differences is the comparable method, whereby all age groups are given percentage values. There are two possible ways of carrying out this method. Either each male age group may be expressed as percentages of the total male population in each community, and similarly each female group (example *d*),[1] or, each male

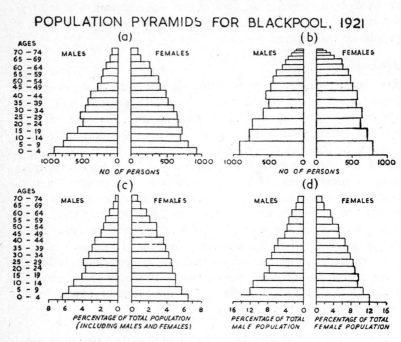

Figure 166. TYPES OF POPULATION PYRAMID

Source of data: *Census of England and Wales, 1921, County of Lancaster* (London, 1923).

and each female group may be expressed as percentages of the total population of each community (example *c*). The latter method has the advantage of giving a realistic total view of sex and age variations in each community considered.

Compound Pyramids

Pyramids may be used to illustrate features other than sex and age structure. For example, the tiers in the pyramid may be made to

[1] For an application, see a series of pyramids in H. Gille, 'Demographic History of the Northern European Countries', *Population Studies*, vol. 3 (Cambridge, 1949–50).

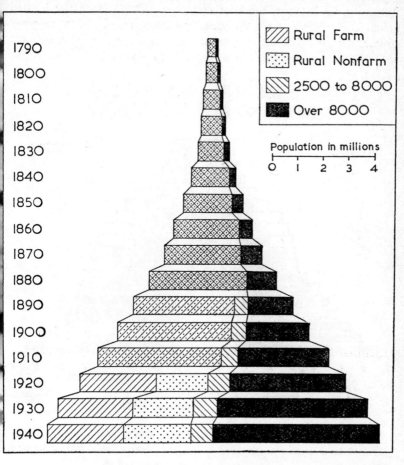

1790
1800
1810
1820
1830
1840
1850
1860
1870
1880
1890
1900
1910
1920
1930
1940

Rural Farm
Rural Nonfarm
2500 to 8000
Over 8000

Population in millions
0 1 2 3 4

Figure 167. COMPOUND PYRAMID

Modified from a diagram in T. L. Smith, *Population Analysis*, p. 318 (New York, 1948).

Each tier has been made proportional in length to the total of population for each census year. The tiers are then subdivided into rural and urban elements in the population according to the categories adopted in the various censuses. From 1790 to 1880 only two categories were recognized and places with a population of over 8,000 were regarded as being urban in character. From 1890 to 1910 there were three divisions when the urban category was extended to include places with a population of over 2,500. From 1920 to 1940 the rural category was subdivided into rural farm and rural non-farm.

represent decennial totals of migrants into a country, with the apex
indicating the earliest, the base the latest returns. Each tier is then
subdivided to show ethnic structure, age structure, etc., of the migrant
stream. Decennial totals of population for a region or for a country
may also be portrayed in this way, the subdivisions in this case being

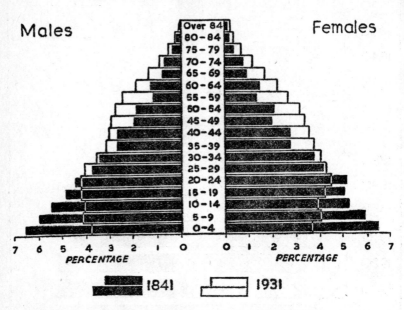

Figure 168. SUPERIMPOSED PYRAMIDS
Source of data: *Census of England and Wales, 1931, General Tables*, p. 151 (London, 1933).
Two comparable pyramids have been superimposed to show changes in the sex
and age structure for two census years.

made to depict rural *versus* urban population. These pyramids are best
constructed with a space left between each tier, across which lines are
drawn to join the component subdivisions (Fig. 167).

Superimposed Pyramids
Age and sex pyramids may be superimposed to show structural
changes in the population of any given region over a period of time.

Similarly, pyramids representing populations of different regions may be superimposed for comparative purposes, or pyramid superimpositions may be made of an actual population upon that of a hypothetical 'stationary' population.[1] The latter procedure has great value in the analysis of the rhythm of population growth. Comparable pyramids must be used, of course, if such comparisons are to be made. Care must be taken, also, that the data utilized applies to the same region, and it is often necessary to consider changes in enumeration areas that may have been made in the interim between the recording of the two sets of data employed.

The 'projecting' of population – a form of forecasting future populations – is frequently illustrated by superimposing present and projected age and sex pyramids for a particular region.[2] Populations are projected for a variety of reasons, principally for planning purposes. Projections may be made on the assumption either that current rates of fertility and mortality are likely to continue into the future, or that they may decline, or that they may behave in accordance with a 'logistic curve', that is, a curve continued according to logical expectations. It is patent that if age–sex specific birth- and death-rates are available, the projection of a population presents no difficulty. But birth- and death-rates do not remain constant for any length of time, and consequently projections made on these assumptions serve only a limited purpose.

DIVIDED RECTANGLES

Rectangles, the lengths of which are usually made to represent 100 per cent, and which are subdivided to show structural aspects of population, are generally employed where comparisons are required. The comparisons may be between one population and another, either for a specific region at different times, or for different regions at the same time.[3]

The method serves well to show shifts in occupations from one decade to another or changes in age structure. Individual rectangles may be constructed to show the data for each decade, or all the data may

[1] For an example, see P. Vincent, 'Retraitres et Immigration', *Population*, vol. 1, p. 234 (Paris, 1946).
[2] See method illustrated in F. W. Notestein, *The Future Population of Europe and the Soviet Union* (Geneva, 1944).
[3] See examples in *Statistical Atlas of the United States*, op. cit. (1935).

be plotted in one rectangle and related plottings joined to form continuous lines (Figs. 169–70). Such a diagram is, of course, a form of
compound graph. The distribution of changes in structure may be
shown by means of rectangles located on a suitable base-map. Thus
the changes in the ethnic structure of Greek Macedonia between 1912

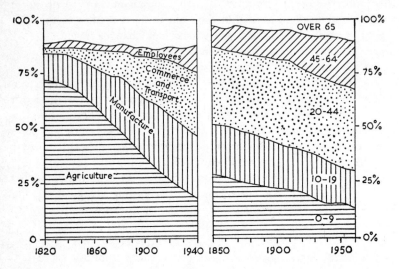

Figures 169–70. DIVIDED RECTANGLES OF OCCUPATION AND AGE STRUC-
TURE

Based on W. S. Thompson, 'La Population des Etats-Unis d'Amérique', *Population*,
vol. 3, pp. 123 and 125 (Paris, 1948).

Figure 169 shows changes in the percentage of paid workers over the age of 10 in
the major occupational groups of the United States according to data derived from
various census returns. The composite category of 'Others' is left unshaded.

Figure 170 shows changes in the age structure of the population of the United States
according both to various census returns and to Thompson's own estimates for
1950–60.

and 1926 have been shown by a series of twin rectangles, one of the
structure in 1912 and the other of the structure in 1926.[1]

Divided Strips

The divided strip method of showing distributional aspects of population structure is a variant of the divided rectangle method. It may

[1] 'Ethnographical Map of Greek Macedonia showing the proportion of the different
ethnographical elements in 1912 and in 1926', *Greek Refugee Settlement* (Geneva, 1926).

be exemplified with reference to the ethnic structure of population. If the distribution of the relative proportions of various nationalities in an ethnically complex population has to be depicted on the basis of administrative units, the proportions of each nationality in each

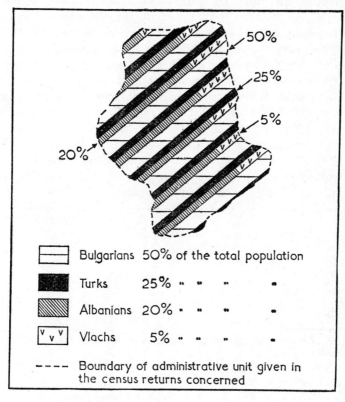

Figure 171. DIVIDED STRIP TECHNIQUE
Based on H. R. Wilkinson, op. cit. (1951).

unit are first of all calculated on a percentage basis. They are then plotted on the base-map, which consists of an outline of the administrative units, but with each individual unit divided into oblique strips of uniform width. Each adjacent strip represents 100 per cent, and these are divided to depict the ethnic structure of an administrative unit (Fig. 171).

STAR-DIAGRAMS

Star-diagrams may be used in population studies to illustrate affinities of race, age structure, etc., between diverse population groups. Affinities are often measured, for convenience in mapping, by means of indices. Racial affinities as measured by a blood-group index may be considered here as an example of these methods. A variety of blood-group indices have been devised. Four major variables have to be considered – the proportions of *A*, *O*, *B* and *AB* elements, respectively, present in each population group. In the population of the world in general, random tests suggest that these groups are present in the average proportions of *A*, 38 per cent; *O*, 37 per cent; *B*, 18 per cent; and *AB*, 7 per cent, although the proportions in local populations differ considerably from these averages.[1] N. Lahovary put forward the idea that one way of establishing the affinity of race between any two populations might be to compare their blood-group ratios and express the results in the form of an index, as follows:

	Population X	*Population Y*	*Differences*
A	53	40	13
O	27	45	18
B	13	10	3
AB	7	5	2
Totals	100	100	36

The 36-point difference acts as an index of racial affinity.

If a number of population groups is compared on this basis with any one population group, the resultant indices may be plotted diagrammatically as in Fig. 173. Alternatively, local deviations from the world average ratio may be calculated.

INFORMATION DIAGRAMS

In recent years geographers have become increasingly interested in the processes of information distribution as a spatial phenomenon. One of the basic works on this theme, written in 1953 by T. Hägerstrand, has recently been translated into English.[2] It contains a number of

[1] N. Lahovary, *Les Peuples Européens* (Paris, 1946).
[2] Torsten Hägerstrand, *Innovation Diffusion as a Spatial Process*, translation and postscript by Allan Pred (Chicago, 1967).

interesting maps and diagrams, illustrative, analytical and predictive. Attention is drawn here to his use of 'profile bands' based on grid maps and to his specialized use of isarithmic maps. In interposing isarithms he used a method which involved a transformation of logarithmic diagrams into map form.

R. L. Morrill and F. R. Pitts have used marriage and migration data to construct mean information fields for different communities in Europe, the United States and Japan.[1] They include an interesting illustration to show the spatial influences underlying marital selection. It consists of a series of lines extending between addresses of the bride and groom. In the diagram a general pattern emerges from the aggregation of individual actions. The aggregation of individual information fields can be grouped into a community or mean information field. These studies are useful in showing how information and ideas spread throughout populated areas and are helpful in the general explanation of social behaviour of large groups of people.

THREE-DIMENSIONAL DIAGRAMS

Isometric Block-diagram of Age Structure

M. R. C. Coulson has devised an age structure index as a single measure of age structure.[2] Using an idealized age structure histogram he establishes the size relationship between age groups as a straight line. When the histogram for a young population is generalized, a steeper slope results; for old population the line is flatter. The angle of slope changes according to the distribution of population among various age groups. Using this device it is possible to show the spatial distribution of age structure in a variety of ways; for example, it can be shown in the form of an isometric block-diagram, the surface being developed in this case from an isoline map of spatial age structure data using the index (Fig. 172).

Three-dimensional diagrams are frequently used in geographical analysis concerned with descriptive models of spatial regularities. They are particularly useful to illustrate the fitting of surfaces. For example,

[1] Richard L. Morrill and Forrest R. Pitts, 'Marriage Migration and the Mean Information Field: A Study in Uniqueness and Generality', *Annals of the Association of American Geographers*, vol. 57, pp. 401–22 (Lancaster, Pa., 1967).

[2] M. R. C. Coulson, 'The Distribution of Population Age Structures in Kansas City', *Annals of the Association of American Geographers*, vol. 58, pp. 155–76 (Lancaster, Pa., 1968).

it is possible to produce a three-dimensional surface to show the distribution of firms in a particular locality, the variations in population density or variations in rates of growth. This work can be carried further by using trend analysis techniques for providing statistical

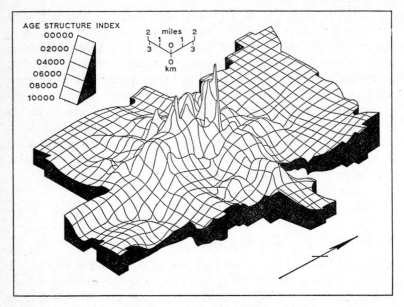

Figure 172. THREE-DIMENSIONAL DIAGRAM TO SHOW THE ISOMETRIC
DISTRIBUTION OF THE AGE STRUCTURE INDEX

Based on M. R. C. Coulson, 'The Distribution of Population Age Structures in Kansas City', *Annals of the Association of American Geographers*, vol. 58, pp. 155–76 (Lancaster, Pa., 1968). The figure refers to Kansas City S.M.S.A. (Tracted Area) and the data were derived from the *U.S. Census of Population and Housing, 1950*.

descriptions of geographical data. These models are, of course, predictive rather than explanatory but they can be used to indicate various growth patterns. For a useful illustration of these procedures attention is drawn in particular to W. R. Tobler's work.[1] Reference is made here in particular to a series of three-dimensional diagrams to show the distribution of population in part of the Detroit region, U.S.A., between 1930–1960.

[1] Waldo R. Tobler, 'Geographical Filters and Their Inverses', *Geographical Analysis*, vol. 1, no. 3, pp. 234–53 (Ohio State University, 1969).

ARROWS

Arrows are used in population studies chiefly as a method of showing the movement or migration of population. Only certain categories of data can be illustrated by this method because, to avoid confusion, only a limited number of arrows may be placed on any one map. They may be used to show: (*a*) the main currents of movement of seasonal

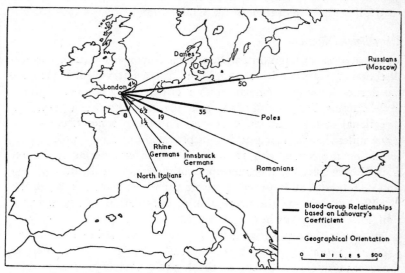

Figure 173. A STAR-DIAGRAM OF RACIAL AFFINITIES

Source of data: N. Lahovary, *Les Peuples Européens* (Paris, 1946).

The map shows racial affinities between a selected group of English (from London) and certain other European groups as measured by blood-group reactions. The lengths of the thickened lines have in each case been made proportional to the co-efficient, e.g., for Poles, the coefficient is 35, therefore the thick line represents 35 units. The shorter the thick line, the closer is the affinity between any two groups.

workers, for example, in the case of the United States, those of fruit and truck workers, berry-crop harvesters, and wheat harvesters;[1] (*b*) seasonal movements of nomadic or semi-nomadic communities;[2] (*c*) daily movement of workers from place of residence to place of work;[3]

[1] Differentiated arrows are used by T. L. Smith, op. cit. (1948), to show such movements in a map entitled, 'Streams of Inter-state Farm Labor Migration [in the United States]'.

[2] See such movements shown in a map, 'Formes de la Propriété Rurale,' in J. Cvijić, *La Péninsule Balkanique* (Paris, 1918).

[3] A good example of such a map, using arrows of proportional thickness, is to be found in F. Longstreth Thompson, *Merseyside Plan, 1944* (London, 1945).

(d) the balance of migration between countries and between states;[1] (e) migration from countryside to town; and (f) generalized patterns of direction of migrations.[2]

Where arrows proportional in thickness to totals of migrating population are used, the map tends to become obscured by a few thick arrows. To overcome this difficulty arrows and proportional symbols may be combined. This method has been used to good effect in a very full study of migration in the United States.[3]

Directional Diagrams

In the analysis of migratory streams some importance attaches to the direction of migration in relationship to distance travelled. J. Wolpert has devised a useful graphical solution for describing flows of migrants in these terms. This consists of a circular diagram with twelve 30-degree directional sectors and ten exponential distance zones (from 25 to 3,500 miles).[4] A circular grid is thus provided based on a polar transformation projection with 120 cells. Thus, in considering migration from Kansas City to other metropolitan migration fields, the distance and direction of all out-migrants were calculated so that it would be possible to determine the proportion of out-migrants whose destination, for example, was in one of the cells bounded by the 0–30-degree azimuths and the distance zone from 158 to 250 miles. The information about migration may thus be economically ordered and the method makes it possible to speculate on the degree of directional bias in relationship to distance travelled. It is possible to use the same grid diagrams to establish median centres and also to show frequencies for comparative purposes.

GRAPHS

In recent years the application of statistical techniques (see Appendix) to population analysis has broadened and deepened the scope of demographic studies. Graphs can obviously be used to establish lineal

---

[1] Illustrated by a map in E. M. Kulischer, *Europe on the Move* (New York, 1948).

[2] J. Cvijić, 'Courants Métanastasiques dans le Peuplement des Pays Serbe', op. cit. (1918).

[3] C. J. Galpin and T. B. Marny, *Inter-State Migrations among the Native White Populations as indicated by Difference between State of Birth and State of Residence. A Series of Maps based on the Census, 1870–1930* (U.S. Dept. of Agriculture, Washington, 1934).

[4] Julian Wolpert, 'Distance and Directional Bias in Inter-Urban Migratory Streams', *Annals of the Association of American Geographers*, vol. 57, pp. 605-6(Lancaster, Pa., 1967).

relationships between different sets of related data as well as illustrating particular trends over a period of time. So great is the variety of graphs employed to illustrate population data that only a selected number can be discussed here. The elemental principles of graphical analysis are summarized, and a few rather special types of graph are discussed in more detail.

Simple Line-graphs

Absolute Growth Graphs. Probably the most familiar graph used in population studies shows changes over a period of time. Totals of population are plotted as ordinates and time values, usually in decennial periods, as abscissae (Fig. 174). After the points have been plotted, they are conventionally joined by short straight lines, although strictly speaking the implication of uniform growth during intercensal periods to be inferred from such a procedure is erroneous.

Percentage Increase Graphs. Instead of plotting absolute totals of population, percentage increase values are plotted as ordinates. The points are joined and the resulting line is conventionally assumed to show rates of growth, which may be either negative or positive (Fig. 174).

Polygraphs

Trends in Birth-rate, Death-rate and Natural Replacement. In all cases, rates in values per thousand are plotted as ordinates and time values as abscissae. Values may be plotted independently and points joined to form simple line-graphs, but more usually birth- and death-rates are plotted on one diagram and natural replacement inferred by inspection (Fig. 175).

Birth, death and replacement rates may be broken down for further analysis into age-sex specific rates. The graphs are constructed in the same way, but the values for different age and sex groups are plotted as ordinates instead of graphical values. Age–sex specific rates may be still further broken down to give racial values, such as coloured and white.[1] It is customary to show data for a complete generation on one graph, so that some care is needed to differentiate between any multiple lines which might cross (see p. 35). Age–sex specific rates are generally plotted for quinquennial rather than for individual yearly values, but an exception is often made in the case of infants under one year of

[1] For example, see graphs in T. L. Smith, op. cit. (1948).

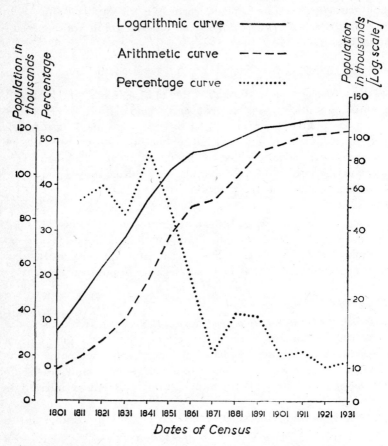

Figure 174. VARIOUS LINE-GRAPHS OF POPULATION GROWTH

Source of data: *Census of England and Wales* (various dates).

The curves show the growth of the town of Preston in Lancashire in terms of the size of the population within its administrative boundary at the dates of the various censuses. Since the administrative boundaries have undergone certain changes, the curves do not necessarily reflect increases in population density.

age, because of the relative frequency of deaths at this time of life compared with other ages.

Distribution of Population and Employment. E. M. Hoover and Raymond Vernon have made a special study of changes in the distribution of population and employment in the New York Metropolitan Area.[1]

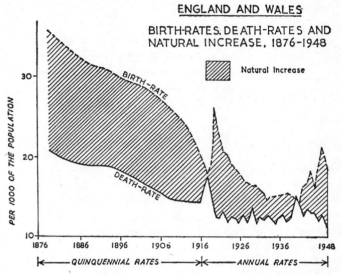

ENGLAND AND WALES

BIRTH-RATES, DEATH-RATES AND NATURAL INCREASE, 1876-1948

Figure 175. A POLYGRAPH OF BIRTH- AND DEATH-RATES

Source of data: *The Registrar General's Statistical Review of England and Wales for the Year 1948* (London, 1950).

Note the smoothing effect on the curve produced by using quinquennial rates as compared with annual rates.

They divided the area into regions for this purpose and the value of the study largely derives from their method of regional comparison. A number of graphs has been used to illustrate particular trends. The most effective are those which show curves of population growth and decline for different parts of the town on the same base, and, similarly, changes in retail distribution, number and kind of jobs in different regions, areal distributions of incomes, and comparative journeys to work.

[1] E. M. Hoover and Raymond Vernon, *Anatomy of a Metropolis: The Changing Distribution of People and Jobs within the New York Metropolitan Area* (Cambridge, Mass., 1959).

Located Graphs

Line-graphs may be located at the relevant points on a map in a manner comparable to located columnar diagrams (see p. 363). An example is a map by A. Geddes, who located a graph of population change from 1900 to 1950 for each state of the United States on a single map.[1]

Smoothed Curves

Where long-term trends in population growth or in fertility are being analysed, running means may be employed to smooth out curves (see p. 229). Curve-fitting is also occasionally used; the inherent problems are too complex to discuss here, and in any case, the number of occasions on which curve-fitting is a justifiable procedure are very few indeed.[2]

Frequency Graphs

These are used more particularly in analyses of changes in mortality and survival rates. For example, years of age from 0 to 100 may be plotted as abscissae, and the number of male or female survivors out of 10,000 as ordinates. Using this framework, data may then be plotted for a particular year, and points joined by a line to give a frequency curve. On the same framework other data may be plotted for other years and the points joined by lines pecked or dotted for clarity.[3]

Any other population data can, of course, be analysed on a frequency basis. Thus A. Stevens (op. cit., 1946) made liberal use of frequency curves in his analysis of the rural population of Great Britain. Using the percentages of parishes and rural districts as ordinates, and mean densities per square mile as abscissae, he was able to establish modal densities of population for England and Wales and for Scotland.

Triangular Graphs

Triangular graphs are used where three variables in the composition of population require analysis. They are useful, for example, in the depiction of the young, the middle-aged and the old in any population. The percentage proportions of all three elements are first cal-

---

[1] A. Geddes, 'Variability in Change of Population in the United States and Canada', *Geographical Review*, vol. 44, p. 91 (New York, 1954).

[2] See examples of curve-fitting in H. Gille, op. cit. (1949–50).

[3] See examples of these graphs in L. Henry and J. Voranger, 'La Situation Démographique', *Population*, vol. 5, pp. 141–54 (Paris, 1950).

culated; the percentage of the young is plotted along one axis of the triangular graph, of the middle-aged along another axis, and of the old along the third axis. Because the component values all add up to 100 per cent, all three values may be plotted on the graph by a single point. The position of this point within the triangle gives an immediate indication of the threefold age structure of the population. A number

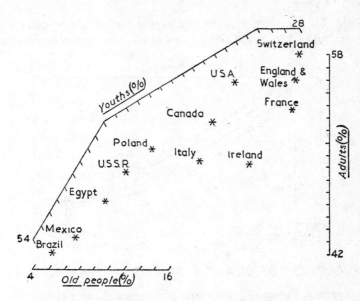

AGE-COMPOSITION OF DIFFERENT COUNTRIES

Figure 176. A TRIANGULAR GRAPH OF AGE STRUCTURE

Based on P. Vincent 'Une interessante Application du Diagramme triangulaire', *Population*, vol. 2 (Paris, 1947).

The critical ages have been taken as 20 and 60 years, and the graph refers to the year 1940.

of points plotted in the same triangle enables comparisons to be made between the age-structures of different populations (Fig. 176).

The percentage of working population in agriculture, manufacturing and services can also be illustrated by a triangular graph. J. O. M. Broek and J. W. Webb[1] illustrate patterns of livelihood – variations from one country to another in these employment categories – by using a triangular graph. The countries can then be grouped according

[1] J. O. M. Broek and J. W. Webb, *A Geography of Mankind*, p. 322 (New York, 1968).

to their location within the triangle. For example, the European countries together with the United Kingdom, the United States, Australia,
New Zealand, Canada and Japan form one distinctive group, whereas
China, India, Ecuador, Mexico and Ghana form another. Such
information can, of course, be mapped on this basis if required.

Projected Curves

A device for adjusting trend curves so that they might be compared
is portrayed in Fig. 177. Comparisons of this kind prove useful when

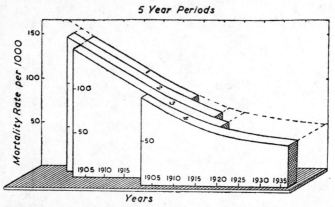

EVOLUTION OF INFANT MORTALITY RATE IN FRANCE

Figure 177. A DEVICE FOR THE PROJECTING OF CURVES OF INFANT MOR-
TALITY RATE

Based on J. Bourgeois, 'La Situation démographique', *Population*, vol. 1 (Paris, 1946).
   The four curves shown are contemporaneous and refer to conditions in four communes. The pecked line in the background represents the sum tendency of all four
curves as arrived at by arranging the four curves in the manner indicated above.

it is required to demonstrate that a particular trend, which has existed
in the past in one population, is true at the present time for another
population. They are of help also in projecting curves into the future.

Scatter-diagrams

Scatter-diagrams are used in population studies to give a graphic
indication of the amount of correlation between two sets of statistical
data. If the exact degree of correlation were required it would, of
course, have to be computed mathematically (see pp. 504–5).
   The first step in the construction of a scatter-diagram is to group

the values under consideration into convenient classes arranged in order of magnitude. One set of data is then plotted as ordinates and the other as abscissae. If, when the values are plotted, they tend to be grouped together along a line, some degree of correlation is manifest. If the correlation is perfect, all points would lie on one line, but this condition is rarely achieved when dealing with human affairs.

To illustrate the construction of a scatter-diagram, fertility ratio

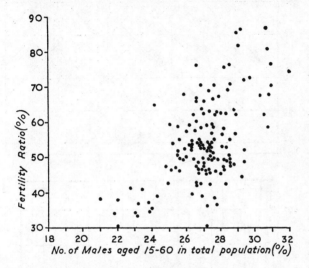

Figure 178. A SCATTER-DIAGRAM

Source of data: *Census of England and Wales, 1881*, vol. 3 (London, 1883).

The fertility ratio in each administrative district in Lancashire in 1881 (as measured by *No. of Children under 5 | No. of Females aged 15–50* × *100*) has been plotted against the relative proportion of males in the population aged 15–60 (as measured by *No. of Males aged 15–60 | Total Population* × *100*).

in Lancashire has been correlated with the proportion of males aged 15–60 in the total population of the county (Fig. 178). Some degree of correlation is to be observed from the diagonal scatter of the dots. The fertility ratio tends to rise as the proportion of males in the population increases.

Many examples of scatter-diagrams appear in demographic texts, but an interesting application of their use in the field of population geography is to be found in A. Stevens' paper on rural population op. cit. (1946).

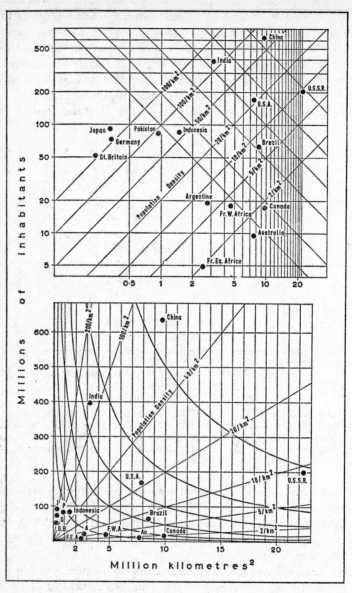

Figure 179. RELATIVE POPULATION TOTALS AND DENSITIES

Based on K. Witthauer, op. cit. (1958).

A comparison of the two graphs shows the advantages of the logarithmic scale in plotting data in the lower values.

Logarithmic and Semi-logarithmic Scales

Graphs based on logarithmic or semi-logarithmic scales may be preferable for certain purposes to those in which arithmetic scales are employed.

Population Growth. If, for example, population growth is to be graphed, the time values are plotted as abscissae on an arithmetic scale, and population values as ordinates on a logarithmic scale. The points are then joined. The resulting line gives not only an indication of total growth, but also a visual impression of rates of growth because the slope of the line is always proportional to the rate of growth (Fig. 174).[1]

Mortality and Survival Trends. Logarithmic scales are preferred to arithmetic scales when details of mortality and survival trends are being plotted, because fluctuations in the rates for the first year of life are very important and they can be clearly plotted on a logarithmic scale, whereas they tend to be compressed on an arithmetic scale.[2]

Comparative Densities. These can be most effectively and clearly shown on logarithmic scales because the scatter is more uniform. The same data plotted on an arithmetic scale results in the graph becoming congested in the area of the lower values (Fig. 179). The theoretical assumptions about population densities may be tested with the aid of simple graphs, for example, Colin Clark's theory of spatial distribution of population densities within cities given by the formula, $d_x = d_o e^{-bx}$, where d_x is population density d at distance x from the city centre, d_o is central density, as extrapolated, and b is the density gradient indicating the rate of diminution of density with distance, a negative exponential decline. The graphs which are used include plotting the

---

[1] (1) H. F. Dickie, 'The Use of Logarithmic Paper for Plotting Geographical Statistics', *Geography*, vol. 24 (London, 1939); and A. Davies, 'Logarithmic Analysis and Population Studies', *Geography*, vol. 33, pp. 53–60 (London, 1948). (2) For applications of this technique with special reference to internal changes of population distribution and location of employment in the New York Metropolitan Area, see E. M. Hoover and R. Vernon, *Anatomy of a Metropolis* (Cambridge, Mass., 1959).

[2] Logarithmic scales are employed in graphs of specific death-rate trends in E. Lessof, 'Mortality in New Zealand and England and Wales', *Population Studies*, vol. 3 (Cambridge, 1949–50).

natural logarithm of density against the distance from the city centre
at different dates.[1]

Regression Diagrams. Theoretical explanations of population arrangement and distribution and movement, sometimes referred to as 'exercises in social physics', depend heavily upon statistical methods, and in

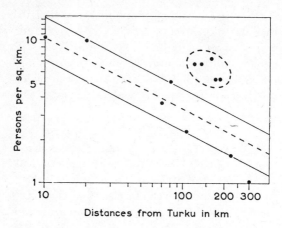

Figure 180. REGRESSION DIAGRAM OF THE DISTRIBUTION OF POPULATION
DENSITY IN FINLAND IN ABOUT 1750

Based on Reino Ajo, op. cit. (1953).

The main centre of population was at Turku, and distances have been measured
from Turku to various settlements on the inland route towards the north-east. The
dotted line encloses exceptional values which refer to settlements in the *kaski* (burnt
clearings), reputedly of very high productivity. Other values fall within the band of
regression with a coefficient of slope of $-\frac{1}{2}$.

particular upon those which help to establish the nature of the relationships between the variables of population density, area, distances from
urban market and so on. J. H. von Thünen provided a theoretical basis
for such speculations many years ago.[2] Reino Ajo has shown that in
the middle of the eighteenth century population distribution in Finland resembled von Thünen's theoretical model. If population densities

[1] Colin Clark, 'Urban Population Densities', *Journal of the Royal Statistical Society*,
ser. A., vol. 114, pp. 490–6 (London, 1951), and Brian J. L. Berry, James W. Simmons
and Robert L. Tennant, 'Urban Population Densities: Structure and Change',
Geographical Review, vol. 53, pp. 389–405 (New York, 1963).

[2] J. H. von Thünen, *Der Isolierte Staat in Beziehung auf Landwirtschaft und National-
ökonomie I-III* (Hamburg and Berlin, 1826–63). See also an enlightening paper, B. J. L.
Berry *et al.*, 'Urban Population Densities: Structure and Change', *Geographical Review*,
vol. 53, pp. 389–405 (New York, 1963), for an interesting extension.

from the coast of Turku to the interior are plotted against distances from the centre (the Turku entrepôt and market) on a logarithmic scale, the scatter of values could be delimited with important exceptions by two parallel straight lines – the so called 'band of regression' (Fig. 180). Thus a linear relationship is deduced which can be expressed mathematically (see p. 505). In this case as the coefficient of slope is $-\frac{1}{2}$, the population density can be said to vary in inverse proportion to the square root of the distance. The exceptional values falling outside the band of regression have to be explained and this is where geography enters. In this particular example, the exceptional values represent settlements in the forest or burnt clearings able to sustain a notably higher proportion than did field culture at that time.[1] B. J. L. Berry et al., op. cit. (1963) include graphs of this type for density–distance relationships and density gradients in western and non-western cities. There is a number of introductory texts which *inter alia* consider the use of simple regression and correlation analysis in population geography[2] (see also pp. 504–10).

Curve Fitting

In addition to linear curves others can be fitted to observed distributions in the analysis of population data. For example, in consideration of journey to work data, it may be necessary to identify the relationships between given sets of variables, e.g. the length of trip, mode of travel, access to bus service, etc. F. R. Wilson makes effective use of diversion curves in his study of transportation in Coventry and London.[3] By plotting the percentage of all work trips by public transport against the length of the trip in miles, it is possible to get a series of diversion curves related to different accessibility indexes, i.e. the accessibility to public transport, the object being to show how trip length operates as a factor in the choice of the mode of travel. The curves tend to rise to an optimum and then to fall away (Fig. 181). The curve that would most closely fit this description is one associated with gamma distribution of the form $y = ce^{-ax} x^b$. Diversion curves of a different shape can be fitted to show the relationship between the

---

[1] Reino Ajo, 'Contribution to Social Physics. A programme sketch with special regard to national planning', *Lund Studies in Geography*, Series B. *Human Geography*, no. 11 (Lund, 1953).

[2] See, for example, Leslie J. King, *Statistical Analysis in Geography*, p. 117–33 (London, 1969).

[3] F. R. Wilson, *Journey to Work – Modal Split* (London, 1967).

percentage of all work trips by public transport to door-to-door travel time as measured by the ratio, Time by bus/Time by car. The data in this case tended to curve upwards to the y axis and indicated that an expression of the form $y = ax^{-b}$ might fit the data closer than a straight line. It is possible to produce diversion curves for the entire city which are of great use in the formulation of predictive models which can then be tested in new situations (Fig. 181).

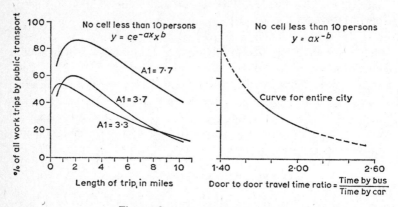

Figure 181. DIVERSION CURVES

Based on F. R. Wilson, *Journey to Work – Modal Split*, pp. 157, 165 (London, 1967).
    On the left, Public Transport Use Diversion Curves. These are shown as a function of trip length and public transport service accessibility index (A.I.) at destination. In this case the curves refer to central business district (C.B.D.) firms only. On the right, a different form of curve has been fitted to show public transport use as a function of travel–time ratio for all firms included in a survey of Coventry work trips.

Deviational Graphs

Graphs which show deviations from average conditions help to elucidate regional variations in sex and age structure and in family structure. For example, percentage deviations from the national norm may be plotted in the form of a graph to give a picture of local variations in age and sex structure (Fig. 182). Local variations are plotted for quinquennial age groups, for males and females respectively, about a straight line which serves the dual function of representing national norms of male and female percentages. The graph then shows not only age variations but also variations in the balance of the sexes from one region to another.[1] The method does successfully show important

[1] This graphical method is used by T. K. Smith, op. cit. (1948), and he has devised the term 'Index Numbers of Population' to describe it.

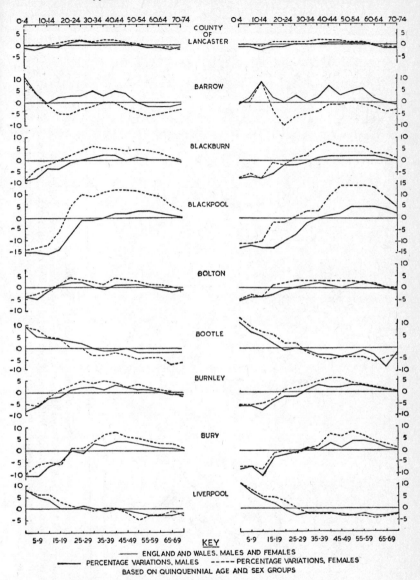

Figure 182. DEVIATIONAL GRAPHS OF SEX AND AGE STRUCTURE
Source of data: *Census of England and Wales* (various dates).

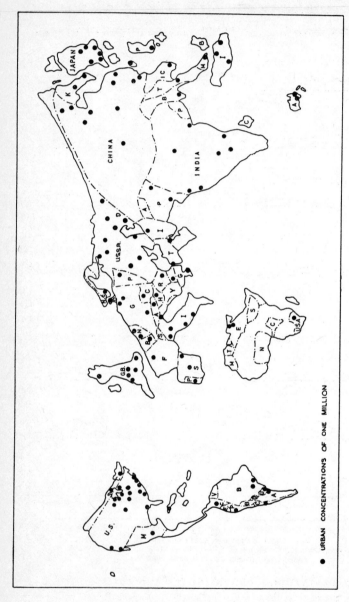

● URBAN CONCENTRATIONS OF ONE MILLION

Figure 183. WORLD POPULATION

Based on maps in W. S. and E. S. Woytinsky, op. cit. (1953), and on miscellaneous statistical sources. The urban concentrations shown include conurbations. Population directly adjacent to large cities has also been included. Countries are depicted in proportion to the size of their populations. This is an excellent example of a topological map (see p. 85).

variations. For example, in Fig. 182 Lancashire as a whole does not differ greatly from the average figures for England and Wales, but in the case of Blackpool both males and females in the younger age-groups are very much under-represented; on the other hand, females in the older age-groups form a much larger element in the population than is normal in the country as a whole. The disproportion between the sexes is also much higher than average. Compare the population of Blackpool and that of Liverpool where the balance of sexes is about normal but where younger age-groups dominate.

Deviation can be evaluated statistically (see pp. 210, 486) in terms of central tendency and variance about the mode, median or mean. The *standard deviation* is frequently used as a measure of variation by geographers but care is needed because in the case of skewed distributions the standard deviation will be weighted by extreme values.

Cumulative Graphs

The application of cumulative graphs to population studies might be exemplified with reference to the problem of population distribution. Practically all the difficulties which face the cartographer in the drawing of population maps arise out of the great concentrations of population which occur in the cities and conurbations of the world (Fig. 183). These concentrations themselves vary in size and statisticians have gone so far as to postulate a law of population concentration, the *Rank-Size Rule for Cities*.[1] Whether or not the law holds good in all cases, its application to the cities of the United States reveals useful evidence about concentrations of population in relation to area. These concentrations of population in any region may be demonstrated by means of cumulative graphs in which area is plotted against population.[2]

The administrative areas within the region, the cities, towns and rural districts are first of all put in order of rank, according to the size of their populations. Areas and populations are then plotted cumulatively. If all the population lives in one place in 5 per cent of

[1] G. K. Zipf, *National Unity and Disunity* (Bloomington, Ind., 1941). See also F. Auerbach, 'Das Gesetz der Bevölkerungskonzentration', *Petermanns Mitteilungen*, vol. 59, pp. 74–6 and Tafel 14 (Gotha, 1913).
[2] The computation of coefficients from such graphs is discussed by J. K. Wright, 'Some Measures of Distributions', *Annals of the Association of American Geographers*, vol. 27, pp. 177–211 (Lancaster, Pa., 1937). He applies his methods in 'Certain Changes in Population Distribution in the United States', *Geographical Review*, vol. 31, p. 488–90 (New York, 1941).

the total area of the region, the resultant curve would be an almost vertical line X (Fig. 184). If the population is spread out in a uniform manner, the resultant curve would be a straight line Y. The curve Q shows a population mostly concentrated in big cities but with a thin spread of rural population over the region as a whole. Cumulative graphs of this type may be used, not only to show the uniformity of the spread, i.e. the concentration of population within specific regions, but also the trends of population concentration over a number of years; they may be adopted also to express in quantitative terms the

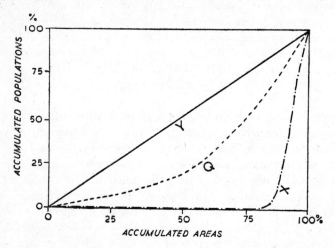

Figure 184. THE PRINCIPLE OF THE CUMULATIVE GRAPH

'degree of areal association of different distributions, i.e. the degree to which two distributions "fit together" or tend to occupy different areas'.[1]

Typology of Population Change

Growth categories of population have already been discussed (see p. 345 *et seq.*). The two components, natural increase of population and net migration, plotted on a Cartesian co-ordinate, may serve as the basis for a typology of population change. By this means J. W. Webb distinguished eight types of population change based on the variables of net gain or loss by natural increase or by migration. He used these in an analysis of the population changes in England and Wales, 1921–31.

[1] J. K. Wright, 'Some Measures of Distributions', op. cit., p. 178 (1937).

The recognition of these types enabled them to be plotted on a regional basis by means of symbols and choropleths.[1]

CENTROGRAMS

Regional trends in the distribution of population can be graphed by means of *centrograms*. A number of points is plotted on a map which coincide as nearly as possible with the successive centres of gravity, or with the median points, of the population of a particular country as enumerated in decennial census returns. These points are then joined by a line which reflects any tendency of the centre of gravity, or of the median point, to shift from one decade to another. The centre of population is defined as a point upon which the country concerned 'would balance if it were a rigid plane without weight and the population were distributed thereon, each individual being assumed to have equal weight and to exert an influence on the central point proportional to his distance from the point'.[2] Where the median-point method is used, distance need not be taken into account. Needless to say, either calculation is laborious, but the results have significance for regional planning.[3]

STEREOGRAMS

W. R. Mead in illustrating population growth in Sweden makes use of a 'stereogram', a three-dimensional diagram to show increase in population and changing age structure over a period of time. This principle can obviously be applied to other aspects of the changing structure of population with time.[4]

AUTOMATIC POPULATION MAPPING

Isopleth Maps

Various techniques have been devised to show population distribution using isopleths, as, for example, those based on NORI and NORIP.

[1] John W. Webb, 'The Natural and Migrational Components of Population Change in England and Wales, 1921–31', *Economic Geography*, vol. 39, pp. 130–48 (Worcester, Mass., 1963).

[2] E. Raisz, op. cit., p. 260 (1948).

[3] For elaborations on the technique, see E. E. Sviatlovsky and W. C. Eells, 'The Centrographical Method and Regional Analysis', *Geographical Review*, vol. 27, pp. 240–54 (New York, 1937).

[4] W. R. Mead, *An Economic Geography of the Scandinavian States and Finland*, p. 73 (London, 1957).

This system employs a triangular grid net since this yields unambiguously localized isopleths or isarithms. It is important to note in this connection that when computers are being used the solutions provided

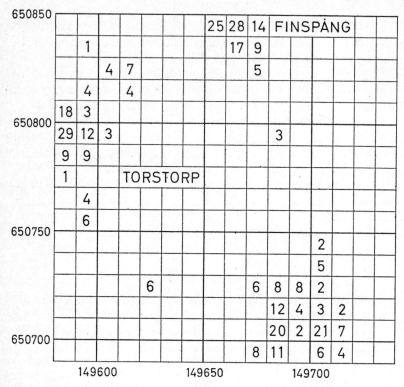

Figure 185. A COMPUTER-DRAWN MAP (NORK) IN THE FORM OF A SQUARE-NET MAP SHOWING POPULATION DISTRIBUTION IN A RURAL AREA IN SWEDEN

Based on Stig Nordbeck, 'Coordinate Mapping Technics', *Plan*, vol. 22, p. 103 (Stockholm, 1968).

This map simply shows total population for each grid square. The system of coordinates used for this purpose is that of the Swedish land-use map on a scale of 1 : 10,000.

by manual cartography are not always suitable. For example in manual cartography the isopleths could be based on the functional values for each grid point. Instead of doing this the computer can be asked to search for grid points which lie within particular influence areas. These values are calculated for grid points arranged in a triangular net which also allows reference areas to overlap, thus reducing the error in localizing the isopleths.[1]

Square Net Maps

Reference has already been made to various systems of automatic cartography (see pp. 81–2) and population maps can now be automatically produced. One of the simplest programmes is called NORK,[1] which produces an ordinary square-net map. Cheap to produce and easy to read (Fig. 185), they are based on co-ordinate mapping techniques. In this case six-figure co-ordinates are used and the total number of persons is shown in each square produced by the co-ordinates. These maps can of course be converted to choropleth or isarithmic maps by using the grid map as a base.

---

[1] These methods are described in Stig Nordbeck, 'Coordinate Mapping Technics', *Plan, Tidskrift för planering av landsbygd och tätorter*, special issue, pp. 101–17 (Stockholm, 1968).

6

Maps and Diagrams of Settlements

Distribution and Forms

There is a vast range of data which relates to the distribution of settlements and to their forms. As problems of distribution and forms of modern settlement are bound up with problems of origins a historical as well as a geographical approach to settlement is imperative, and this immeasurably widens the scope of inquiry. Sources can be grouped into a number of categories.

Maps and Plans. Printed maps on a scale of 1 : 100,000 or larger invariably portray all settlements, either by conventional symbols, or by plan. For details of actual form, however, maps on a scale of about 1 : 25,000 or greater are necessary. In Britain the current 1- and 6-inch (which is being changed to the 1 : 10,000 scale) series serve admirably to show distribution and form of settlements respectively. For field plotting, base-maps which give details of individual buildings must be used and plans on a scale of 1 : 2,500 or even 1 : 1,250 are necessary.[1]

Past editions of topographical maps give some historical perspective. For example, the first editions of the British Ordnance Survey 1- and 6-inch Series are invaluable in the study of distributions during the first and succeeding decades of the nineteenth century. Tithe Award maps are also invaluable where available, as they supplement and sometimes pre-date the data given in Ordnance Survey maps. Often they are on a greater scale and give more detail of building plans, field boundaries and place-names, as well as throwing light upon conditions of tenure and land utilization.[2]

---

[1] The Ordnance Survey publishes scale-plans of built-up areas (1 : 1,250 and 1 : 2,500) specifically for this purpose in connection with town-planning.

[2] Copies of Tithe Award maps are sometimes to be found in the local Parish Church. Copies and originals may be consulted at county record offices. See, for example,

In addition to these, other map sources include old county maps,[1] printed town plans and manuscript maps of various kinds, in particular estate maps. The Land Utilisation Survey maps of Britain must not be overlooked, because they contain information about total land taken up by settlements which cannot be found elsewhere. Geological maps, particularly drift-maps, give invaluable details about siting of settlement. The 1- and 6-inch geological drift-maps of Britain should be consulted wherever available. Unfortunately some of these maps are scarce today, but in cases of necessity the original manuscript survey maps may be viewed at the Geological Survey.

Air Photographs. Air photographs provide the means of bringing settlement patterns up to date, and they are of basic value in the study of settlement where cadastral survey is lacking altogether. The Ordnance Survey use 1 : 2,500-scale aerial photographs for revision. They are of use too for providing detailed supplementary information about settlements which does not appear on maps, for example, relationships of 'ridge and furrow' to settlement form, relationship of settlement to land-use and to landforms,[2] and evidence of former settlements.[3]

Documentary Sources. The origins of different forms of settlement are frequently bound up with obsolete conditions of land tenure, with former agricultural practices, with particular stages in the evolution of the economic structure of a region, with original field systems, with frontier functions and so on. A particular form of settlement may be purely accidental or its site fortuitous. Only its history affords an explanation of the circumstances of its siting and form. The most important single document affording evidence about medieval settlement in Britain is the Domesday Book.[4] Enclosure awards incorporate detailed information for many parts of the British Isles and collectively

---

Catalogue of Maps in the Essex Record Office, 1566–1855, edited by F. G. Emmison (Chelmsford, 1947).

[1] See R. E. Dickinson's use of Saxton's maps in 'The Distribution and Functions of Smaller Urban Settlements of East Anglia', *Geography*, vol. 17, pp. 19–31 (London, 1932).

[2] See the remarkable series of block diagrams illustrating settlement features, drawn directly from air photographs in T. Hagen, 'Wissenschaftliche Luftbild-Interpretation', *Geographica Helvetica*, vol. 5 (Bern, 1950).

[3] See, for example, the air photographs in W. G. Hoskins (ed.), *Studies in Leicestershire History* (Liverpool, 1950).

[4] This evidence has been systematically analysed by H. C. Darby, and is in course of publication as *The Domesday Geography of England* (Cambridge), in six volumes, under his general editorship.

provide the basic source for the study of rural settlements in a particularly important formative period.[1]

Place-names. Place-names are of paramount importance in the study
of the origins of the distribution and of the spread, as well as of the
character, of early settlements. Great care, however, must be exercised
in the use of place-names. The random plotting of any place-names
which happen to figure on ordinary topographical maps is to be depre-
cated, because many of the names plotted may not be genuine place-
names. Fortunately, the English Place-Name Society has published a
number of county volumes on English place-names, and they provide
authentic lists of names.[2]

Archaeological Evidence. The reconstruction of the distribution and
form of prehistoric settlements is dependent upon adequate archaeological evidence. This may be found in reports of excavation of settlements, for example, the excavation of Roman towns and of Iron Age
hill forts. The general distribution of prehistoric settlement may often
be inferred from plottings made of locations of archaeological finds.
Cyril Fox's work still provides the best introduction to the general
cartographical treatment of archaeological data as applied to settlement distributions.[3] As the result of excavations in the Middle East and
elsewhere, it has been found possible to reconstruct the plans of some
of the earliest towns. In this way the origins of certain of the functions
of towns can be traced, as well as morphological and structural features,
some of which are found, albeit in modified forms, in our present day
towns.[4]

Structure of Settlements

The geographer is interested not only in the distribution and arrangement of settlements, but also in the functions which the settlements
perform. Information about the functional structure may be gained
from a number of sources.

[1] A. Harris, *The Rural Landscape of the East Riding of Yorkshire, 1700–1850* (Oxford,
1961).
[2] E. Ekwall's *The Concise Oxford Dictionary of English Place-Names* (2nd edition, Oxford,
1940) is a useful supplementary source. Before working with place-names, A. Mawer
and F. M. Stenton, *Introduction to the Survey of English Place-Names* (Cambridge, 1924),
should be read.
[3] C. Fox, *The Personality of Britain* (4th edition, Cardiff, 1943).
[4] R. M. Adams, 'The Origin of Cities', *Scientific American*, vol. 203, 10 pp. reprint
(September, 1960).

Population Statistics. Details of the occupational and industrial structure of the population of different localities can be analysed to show the functional character of the settlements (see p. 361–4).

Building-use and Urban Land-use Surveys. Detailed information about the distribution of function in any settlement can only be obtained by making a survey of the uses to which the component buildings are put. Under the Town and Country Planning Acts, Local Planning Authorities (i.e. County and County Borough Councils) are obliged to make such surveys for themselves in connection with Development Plans, and information may be obtained from them. Surveys of this kind may be limited to the plotting of the use of the buildings themselves, or they may extend to the detailed plotting of land-use in built-up areas.

Directories. Apart from field observation, a great deal of information about settlement structure can be gleaned from local directories. Current editions of local directories and of the Telephone Trade Directory are useful, but just as important are earlier editions which constitute historical sources and give information about past conditions which often cannot be obtained elsewhere. Since settlement structure is dynamic, an historical approach is essential to the appreciation of changes and trends.[1]

Inventories of Property. Inventories of property made by civic authorities or by estate agents and landlords contain useful supplementary data about building-use. A good example of such source material is the inventory of property in New York made by 'white collar' unemployed during the period of depression.[2]

Relationships of Settlements

Information about the relationships between various settlements is drawn from miscellaneous sources, but mention may be made of a few of the more important.

The network of communications as portrayed on topographical maps is an obvious reflection of relationships between settlements, but telephonic communications are not to be neglected in this connection (see p. 420).

[1] W. K. C. Davies, J. A. Giggs and D. T. Herbert, 'Directories, Rate Books and the Commercial Structure of Towns', *Geography*, vol. 53, pp. 41–54 (Sheffield, 1968).
[2] See J. K. Wright, 'Diversity of New York City', *Geographical Review*, vol. 26, pp. 620–39 (New York, 1936).

Traffic Censuses. The communications network gives no information about traffic density, but some indication is given in traffic censuses which are held from time to time, conducted by government agencies, by the Planning Offices of County and County Borough Councils, or in the case of Britain, by the Automobile Association and the Royal Automobile Club.

Time-tables. Apart from traffic censuses, local bus and train time-tables give information about density of traffic, although such time-tables are difficult to interpret.[1]

Population Statistics. Movement of population between settlements has obviously to be considered in the study of their relationships. Migration statistics have already been referred to (see p. 326). Just as significant is the information given in census returns about the daily movement of population between place of work and place of residence.[2]

Newspaper Circulation. Data about the area served by local newspapers are used in examining the relationships between small and large settlements, especially in the determination of the limits of urban fields.[3]

Details of Special Services. Details about settlements served by special centralized retailing establishments such as the local Co-operative Society, the local hospital and the local Defence Headquarters, are sometimes available to throw light on special problems of urban and rural relationships.

Market Areas. The size of, and the area served by, local markets reflect relationships between farming settlements and urban centres. Local Authorities often keep records of sales, and details concerning them may also be obtained from auctioneers' lists, etc.[4]

---

[1] For details of the use of bus time-tables, see F. H. W. Green, 'Town and Country in Northern Ireland from a Study of Motor-bus Services', *Geography*, vol. 34, pp. 89–96 (London, 1949).

[2] Such data are contained, for example, in the *Workplaces* volume of the Census of England and Wales, 1921 (H.M.S.O., 1925).

[3] J. P. Haughton, 'Irish Local Newspapers: A Geographical Study', *Irish Geography*, vol. 2, pp. 52–7 (Dublin, 1950).

[4] R. E. Dickinson, 'The Markets and Market Areas of East Anglia', *Economic Geography*, vol. 10, pp. 172–82 (Worcester, Mass., 1934).

Questionnaires and Field-work. Details about local shopping habits, about the extent of local hospital services, about local entertainment facilities and so on, are not easily come by, and some considerable field investigation is necessary before sufficient data can be assembled to allow aspects of the relationships between rural and urban settlements to be mapped. Questionnaires which are circulated via local organizations, such as women's clubs and schools, can be successfully utilized in this connection,[1] and there is a strong feeling in some Planning Departments that stopping people in the street is the only positive answer for sphere of influence studies.

FACSIMILES

Since the distribution of buildings is shown on topographical maps of various scales, as well as on plans and manuscript maps, it follows that settlement problems can be directly illustrated as facsimiles taken from such maps.[2] Moreover, facsimile reproductions have the advantage of authenticity, particularly in the case of historical records.[3]

But although facsimile reproductions of maps are often used, it does not follow that these sources are always either satisfactory or suitable. Facsimile maps are expensive because they may necessitate the use of colour or half-tones and attempts made to reproduce coloured maps in black and white are usually unsuccessful. More frequently reduction of scale is attempted with disastrous results as far as the legibility of the map is concerned.

Facsimiles which are intended for reduction should usually be redrawn (Fig. 186). Problems of scale and technique to be employed in the photographic reproduction of settlements from maps deserve

---

[1] Relationships between Worcester and neighbouring villages were investigated by this method. See J. Glaisyer *et al.*, *County Town* (London, 1946). A committee sponsored by the Geographical Association and under the chairmanship of A. E. Smailes investigated 'urban fields' in Britain, and great use was made of printed questionnaires. See also H. E. Bracey, *Social Provision in Rural Wiltshire* (London, 1952).

[2] There are many examples of this. A. Demangeon's maps based on the French 1 : 10,000 series might be cited in 'Types de Peuplement rural en France', *Annales de Géographie*, vol. 48 (Paris, 1939). Facsimiles of the 6-inch series are to be found in H. Thorpe, 'Some Aspects of Rural Settlement in County Durham', *Geography*, vol. 35, pp. 244–55 (London, 1950). Numerous facsimiles to illustrate settlement problems are reproduced in M. A. Lefèvre, *Principes et Problèmes de Géographie Humaine* (Brussels, 1946).

[3] M. A. Lefèvre, op. cit. (1946), for example, makes use of facsimiles from J. de Deventer, *Atlas des Villes de la Belge au XVIᵉ Siècle*, to illustrate aspects of the morphology of towns.

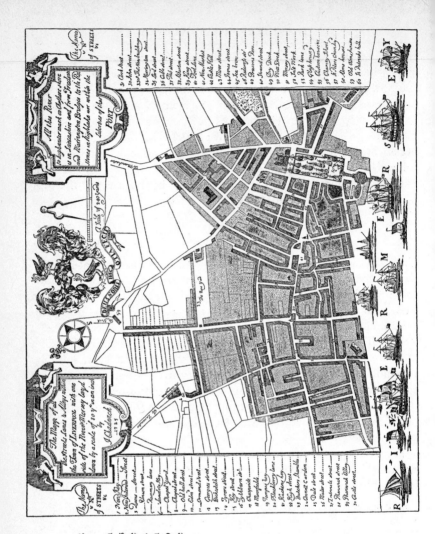

Figure 186. A FACSIMILE MAP OF LIVERPOOL

The map has been redrawn and then reduced to less than one quarter of the scale of the original, i.e. from 21·75 to 5·0 inches in length. Details on the map may still be read with the aid of a magnifying glass.

Figure 187. A PHOTO-FACSIMILE

Based on *Atlas topographique de la Suisse (Atlas Siegfried)*, 1 : 25,000, sheet no. 451 (Berne, 1934).
This half-tone shows part of the town of Geneva, but has been produced in such a way as to emphasize the plan of the town.

careful consideration. A useful technique in the case of town plans is to superimpose a positive and a negative, emulsion to emulsion and slightly out of register. When printed the shadow effect so obtained creates an illusion of relief which adds to the clarity of the facsimile (Fig. 187).

CHOROCHROMATIC MAPS

Chorochromatic maps can be used to show the regional distribution of various forms and types of settlement. They are of necessity generalized; areas with contrasting types of settlement, for example, are tinted to show major variations, or delimiting lines between settlement types are drawn and the areas so delimited labelled accordingly.[1]

Instead of tints or shading, symbols may be employed to show distributions. Maps of this type are sometimes termed *choroschematic*. They have the advantage in the case of settlement of showing admixture of types (Fig. 188).

Maps of this type employing pictorial symbols are used to portray the distribution of house-types and forms of buildings. For example, the distributions of closed and open farm-buildings, of stone, wattle and timber houses, of roof-type, may be suitably depicted by symbols; very often pictorial symbols are most effective to convey regional variations in the plan or elevation of buildings, or to convey variations in the building materials used.

The chorochromatic technique has a special application in the illustration of the differentiation of urban land-use (Fig. 189). These maps are of particular value to the town-planner. They are usually executed in colour and are compiled from field surveys. Land-use maps can be combined with population distribution maps, particularly in the consideration of land-use in urban areas. For example, residential land use can be shown by dots with a given population value (see p. 352).[2]

TRACES

It is often unnecessary to reproduce facsimile all the detail of specific maps when settlement patterns are being considered. Indeed, the very

---

[1] See an example in M. A. Lefèvre, op. cit. (1946).
[2] W. Applebaum, 'A Technique for Constructing a Population and Urban Land-use Map', *Economic Geography*, vol. 28, pp. 240–3 (Worcester, Mass., 1952).

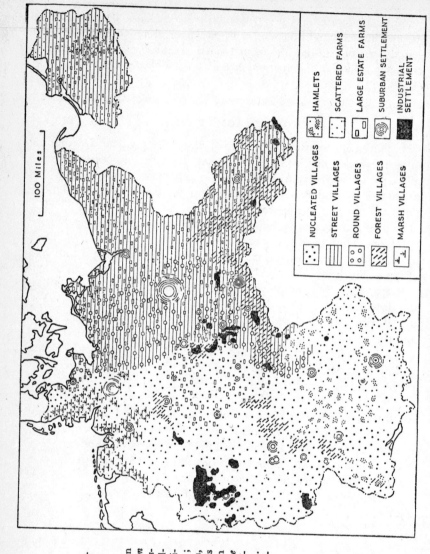

Figure 188.
A CHOROSCHEM-
ATIC MAP OF
SETTLEMENT

Based partly on
maps in (1) E. von
Seydlitzsche Geogra-
phie, vol. 4, 'Mittel-
europa' (second edi-
tion, Breslau, 1930);
(2) R. E. Dickinson,
'Rural Settlements
in the German
Lands', Annals of the
Association of Ameri-
can Geographers, vol.
39, p. 260 (Lan-
caster, Pa., 1949).

100 Miles

NUCLEATED VILLAGES

STREET VILLAGES

ROUND VILLAGES

FOREST VILLAGES

MARSH VILLAGES

HAMLETS

SCATTERED FARMS

LARGE ESTATE FARMS

SUBURBAN SETTLEMENT

INDUSTRIAL
SETTLEMENT

process of selecting only certain items on the map and extracting them for reproduction purposes by means of a trace may aid analysis by shifting the emphasis from the total map to certain distributions.

Building Patterns

The extraction of the building pattern to the exclusion of all other topographical detail serves to clarify the arrangement of the buildings,

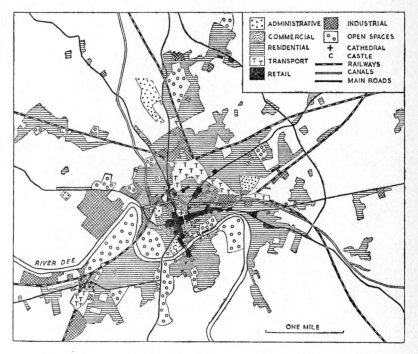

Figure 189. A CHOROCHROMATIC MAP OF URBAN LAND-USE

The map relates to conditions in the county town of Chester in 1949. Land used for agricultural purposes has been left unshaded. Data were derived from field surveys.

and significant features of their arrangement may more easily be noted. The extraction is readily accomplished by tracing off all buildings and marking them black (Fig. 190). Alternatively, it may be useful for certain purposes to trace all the area taken up by buildings, including gardens, industrial yards and installations of various kinds. In the case of British settlement, this can be accomplished by tracing off the Land Utilisation Survey maps all land marked in red and purple, and marking such areas black on the trace.[1] The procedure has special merit in that reduction of the tracings can be effected without great loss of detail, but it does not serve always to show the forms of the settlements.

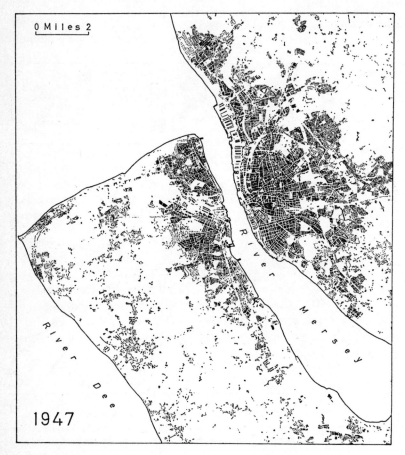

Figure 190. TRACE OF THE SETTLEMENT PATTERN OF MERSEYSIDE IN 1947

Based on Ordnance Survey, One-Inch Series, Sixth Edition, sheet 100.

The subsequent analysis of the causes and nature of regional differences in the settlement pattern may be aided if specific topographical

[1] Plenty of examples of this technique may be seen in the various volumes of the *Land Utilisation Survey of Britain.*

features are added to the trace, in addition to the building pattern. The inclusion of the road pattern may, for example, elucidate ribbon development, or it may help to explain forms of rural settlement such as the street-villages of northern Belgium,[1] the forest-villages of northern Germany, and the round-villages of eastern Germany. Certain types of rural settlement, moreover, may themselves be associated with distinctive road patterns (Fig. 191). In the case of towns, the pattern of roads is often the most conservative feature of urban morphology. A tracing of the road pattern of a town may be sufficient to show details of its origin and form without the added labour of tracing the numerous buildings of the town. It must not be forgotten, either, that railways constitute an important feature in the structure of towns, and the tracing of the railway pattern has value in this connection.[2]

Forms of settlement in certain parts of the world have had their origins in relation to various field systems. Hence the inclusion of field boundaries may sometimes help to illustrate the conditions of origin of particular settlements (Fig. 192). Enclosures have, of course, greatly modified British field systems over a long period of years. Traces from enclosure maps and old estate maps provide useful evidence where enclosure of land has taken place relatively recently.[3]

The addition of parish, commune or township boundaries in the case of rural settlement and of administrative boundaries in the case of towns, may be necessary to clarify certain aspects of the distribution of settlements. Thus the addition of parish boundaries, for example, serves to show the siting of villages in relation to the resources of the parish. The parish is not always a unit of settlement in the British Isles, but where there is reason to believe that the parish boundaries do, or once did, delimit one parish economically from another, then obviously these boundaries become significant features.

Natural vegetation, particularly the prevalence of marsh, heathland and forest, often affects forms of regional settlement. For example, the forest-villages of Silesia and the marsh-villages of the coastlands of Northern Europe, are related to the colonization and exploitation of

---

[1] M. A. Lefèvre, *L'Habitat rural en Belgique. Etude de Géographie humaine* (Liége, 1926).

[2] See maps in S. H. Beaver, 'Railways of Great Cities', *Geography*, vol. 22, pp. 116–21 (London, 1937).

[3] See the fine selection of maps in M. Bloch, *Les Charactères Originaux de l'Histoire Rurale Française* (Oslo, London, etc., 1931); also M. A. Davies, 'Selected Types of Settlements and Field Shapes in Pembrokeshire', *Land Utilisation Survey of Britain*, pt 32 (London, 1939).

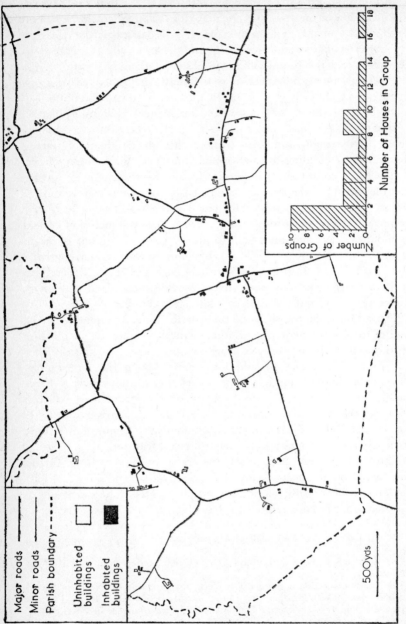

Figure 191. A SELECTIVE TRACE OF HOLYWELL, A DISPERSED VILLAGE IN THE VALE OF BLACKMORE
The inset illustrates a method of analysing the house-groupings in the parish.

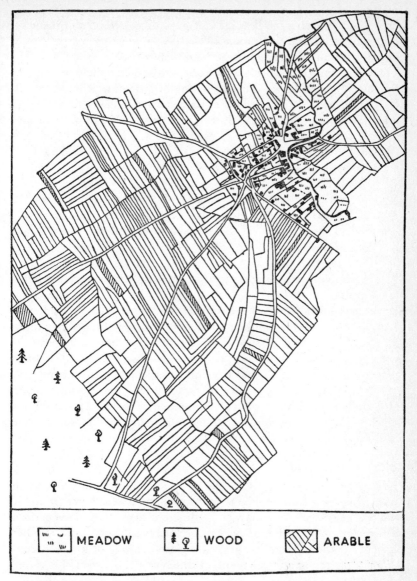

MEADOW   WOOD   ARABLE

Figure 192. A TRACE OF A VILLAGE WITH FIELD BOUNDARIES

Based on S. Ilešič, 'Vasi na Ljubljanskem polju in njegovem obrobu', *Geografski Vestnik*, vol. 10 (Ljubljana, 1934).

    The village is Spodnje Brnik in Slovenia, a typical *Gewanndorf*. Individual field boundaries are shown, and strips belonging to one farm have been shaded to indicate the extent of the fragmentation of holdings.

different types of natural environment. Consequently, the inclusion of details of natural vegetation and of land utilization may add considerably to an understanding of the settlement forms.[1]

Tracings from geological maps, both solid and drift, help to illustrate the location of villages. The distribution of spring-line villages, or villages on drift-free sites, is best explained by references to geological maps. Moreover, geological distributions may exert some influence, directly or indirectly, on the form of villages. The linear agglomerations in the valleys of the chalklands of Dorset differ considerably in form from the dispersed settlements of the neighbouring claylands of the Vale of Blackmore. P. L. Michotte's maps of Belgium which show the distribution of settlements upon geological base-maps might be cited as a good example of this type of illustration.[2] These maps are coloured to show geological distributions, and settlements are shown by enlarged and clarified symbols.

Significant geological distributions may often be shown by outlines and by stipples, providing that such details do not obscure the settlement symbols.

The siting and form of settlements may be related to landforms. Landforms as reflected in the pattern of drainage and relief may therefore be depicted to show specific relationships. Meander-site towns, for example, settlements on river terraces, gap towns, towns defensively placed on isolated hills, and dry-valley settlements, are all instances of such relationships. Usually the tracing of certain contour-lines and drainage channels suffices to summarize particular landforms. Hachures, it might be noted, are frequently more helpful than contours in demonstrating certain aspects of site.

Elements of the Settlement Pattern

So far the settlement pattern has been considered with reference to the various kinds of buildings and installations which make up the pattern. The buildings may be of different ages, they may perform different functions, they may vary architecturally one from another. Except for the information given by place-names, not much attempt is made in topographical maps to differentiate between the variety

[1] For further examples, see E. H. G. Dobby, 'Settlement Patterns in Malaya', *Geographical Review*, vol. 32, pp. 211–32 (New York, 1942); and K. C. Edwards *et al.*, 'The Nowy Targ Basin of the Polish Tatra', *Scottish Geographical Magazine*, vol. 51, pp. 215–28 (Edinburgh, 1935).

[2] P. L. Michotte, *Belgique. Echantillons-types de Régions Géographiques.* Collection Jean Brunhes, 1 : 20,000 (Paris, various dates).

of elements in the building pattern. Symbols are used to show churches, and in the case of the Swedish 1 : 100,000 series, individual farms are differentiated by symbols.[1] Different housing patterns are also shown symbolically in specific editions of the British 1-inch series,[2] and much can be inferred about the residential character of different neighbourhoods by an inspection of the building patterns given on maps (Fig. 190).

For a proper analysis of the elements in the building pattern, however, a certain amount of field-work is necessary. Elements in the building pattern which might come in for special study are age elements, architectural elements (for example, height of buildings[3] or style) and functional elements. The treatment of age elements proffers certain problems in the selection of dates to delimit age categories which are to have significance; inquiry into the architectural elements may prove of use in the demonstration of the changing functions of buildings, and in the reconstruction of historical functions performed by a particular settlement. It is not intended to consider age or architectural analysis here, but something further may be added about functional analysis.

Functional Elements. Mention has already been made of building-use surveys (see p. 399). Such surveys provide the data for functional analyses of settlement pattern. The essence of the cartographical treatment is to produce distribution maps of the functional elements. This may be achieved by tracing the settlement pattern in detail and tinting individual buildings and installations according to the nature of their functions, or as a more practicable method, particular elements may be extracted from the pattern and considered separately (Figs. 193–4).

The fundamental problem is to show functional elements over a wide area, without over-reducing the tracing so that isolated buildings are lost in the process. The following method is applicable in the case of dispersed buildings such as farms. Suppose, for example, it is required to show the distribution of the agricultural element in the building

---

[1] For maps of distribution of farms extracted from the Norwegian topographical maps see H. W. Ahlmann, 'The Study of Settlements', *Geographical Review*, vol. 18, pp. 93–128 (New York, 1928).

[2] See H. Rees, 'The Representation of the Housing Patterns in Fifth and Sixth Editions of the Ordnance Survey One-inch', *Geography*, vol. 31, pp. 110–16 (London, 1946).

[3] See W. W. Atwood, 'A Geographical Study of the Mesa Verde', *Annals of the Association of American Geographers*, vol. 1, pp. 95–100 (Lancaster, Pa., 1910).

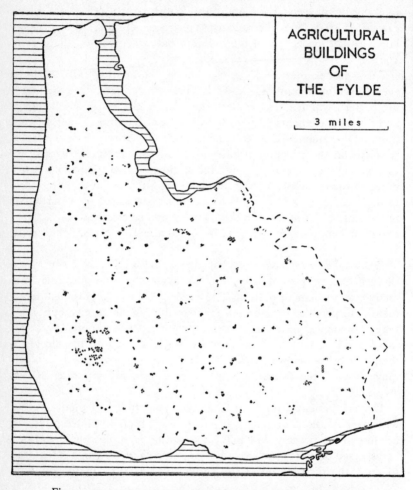

Figure 193. A TRACE OF FARM-BUILDINGS IN THE FYLDE

Based on field reconnaissance with 6-inch series, Ordnance Survey maps. Glasshouses are indicated by open symbols, and farm-buildings by solid black symbols. All symbols have been exaggerated for the sake of clarity.

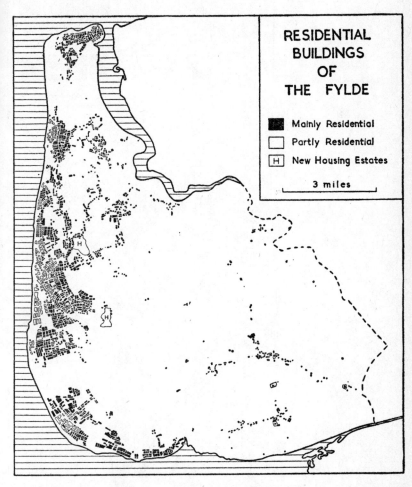

Figure 194. A TRACE OF RESIDENTIAL BUILDINGS IN THE FYLDE

Based on field reconnaissance with 6-inch series, Ordnance Survey maps. For the purpose of this map, farm-buildings have not been included as residences.

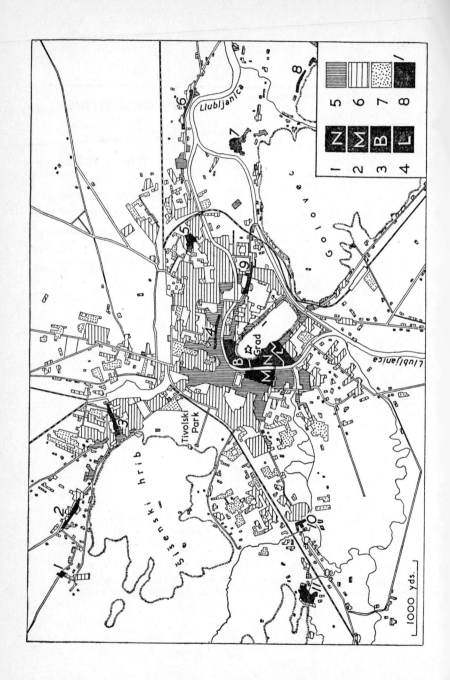

Ljubljanica

Golovec

Ljubljanica

Grad

B M N

Tivolski
Park

Šišenski hrib

5
6
7
8

N M B L
1 2 3 4

1000 yds.

pattern of a region in Britain. The data are compiled by field survey on a 6-inch base-map. The symbols representing farm buildings may then be transferred on to a 1 : 25,000 base-map, or even on to a 1-inch map, providing that care is taken to locate the symbols accurately (Fig. 193). This method cannot be used for agglomerated buildings, but in this case the problems of reduction are not so acute.

A further difficulty in mapping functional elements in the building pattern arises from the necessity of deciding upon the function of a building used for more than one purpose. In Fig. 194 a solution to this problem is offered. The proportion of a specific function, residential in this case, performed by each block of buildings has been estimated, and the building symbols shaded accordingly.

Elemental functional analysis of settlement pattern facilitates consideration of the factors determining both the distribution of settlements and the form of the settlement pattern. It is often a useful exercise to compare the arrangements of a particular element in the pattern, for example, the central business district, in different towns,[1] or to compare the arrangement of agricultural buildings in different regions. The principles of plotting elements in the settlement pattern may be extended by considering the detailed composition of each element. The residential element, for example, is capable of subdivision into various grades of property.

It must not be forgotten, however, that settlement pattern is an organic whole. Elemental analysis should be employed as much to demonstrate the relationships between its various parts, as to analyse the variety of ingredients in its make-up.

Growth Maps

Mention has been made of date-of-building traces. Such traces give a rough guide to the growth of particular settlement units, but they

[1] See, for example, G. W. Hartmann, 'The Central Business District: A Study in Urban Geography', *Economic Geography*, vol. 26 (Worcester, Mass., 1950).

Figure 195. A GROWTH MAP OF LJUBLJANA

Based on A. Melik, 'Razvoj Ljubljane', *Geografski Vestnik*, vols. 5–6 (Ljubljana, 1930).

The references in the key are as follows: **1.** the medieval nucleus; **2.** the new market built in the 13th century; **3.** the town on the edges of Mestni street and around the bishop's palace, built at the beginning of the 14th century and later; **4.** the manorial quarter attached to the town in 1533; **5.** the growth of the town up to 1825; **6.** growth to 1914; **7.** growth to 1929; **8.** original nuclei of surrounding rural settlements, consecutively numbered.

are not always completely satisfactory in this respect. Often much rebuilding has taken place in the centre of a town, the age of which is no guide to the growth of the town from a specific nucleus. With the aid of a series of topographical maps, composite traces may be constructed which provide a truer picture of the expansion of the town and of the rate of its growth (Fig. 195). One-inch maps which give the limits of built-up areas are usually sufficiently detailed to plot growth by means of tracing the building pattern at successive decades.[1]

SYMBOLS

Strictly speaking, even the tracing of settlement patterns from topographical maps, which is referred to above (p. 403), makes use of symbolic illustration, in so much as the settlements on topographical maps are sometimes conventionally depicted because of scale problems. However, the symbols on these maps usually reflect the plan of the original buildings where possible. The symbols dealt with in this section take the form of dots, proportional circles, or differentiated unquantitative symbols such as stars and squares. These symbols are used generally to depict the distribution of settlements over a wide area, where traces would be impracticable, to depict elements within the settlement pattern, to indicate different functions of settlements and of individual buildings, to relate settlements to population, to illustrate problems of siting, form and structure, and to show relationships between settlements.

Individual Buildings

Symbols to show individual buildings have their uses. The distribution of habitations, or of farms, can be shown by dot maps where one dot is equivalent to one unit.[2]

Differential symbols are useful for depicting the distribution of individual buildings according to their function, in order to illustrate the

[1] A remarkable series of growth-maps for Minneapolis-St. Paul was produced by J. R. Borchert, who drew maps of the 'Twin Cities' for 1900, 1940 and 1956, demarcating areas of high-, medium- and low-density settlement. He added a projected pattern for 1980. See J. R. Borchert, 'The Twin Cities Urbanized Area: Past, Present and Future', *Geographical Review*, vol. 51, pp. 47–70 (New York, 1961). This paper abounds in various cartographic and diagrammatic methods of settlement representation.

[2] See Edgar Kant, *Quelques Problèmes Concernant la Représentation de la Densité des Habitations Rurales*, p. 455. *Apophoreta Tartuensia* (Stockholm, 1949).

plan and structure of rural settlements. The dots may represent houses, churches, schools or administrative buildings. G. T. Trewartha used this method to demonstrate varieties of hamlets in the United States.[1]

Urban Structure. Symbols are frequently employed to show the distribution of offices, industrial undertakings, shops and banks in towns. Spheres proportional in size to the number of employees in a firm enable important differences in the size of undertakings to be depicted.[2] Factory locations may be plotted according to the number of hands they employ or according to the amount of capital invested in them.[3]

Shop Rent Index. Typical of the problems which the urban geographer encounters in his mapping of town structure is the difficulty of representing the distribution of the intensity of shopping within a particular town. Proportional symbols have to be devised because dot maps to show the distribution of shops do not reveal shopping intensity. Data on which to base the symbols are hard to come by. Annual turnover is often inaccessible, the size of window is not a reliable guide, and rents vary according to the size of the buildings. Even the number of employees is misleading because of the different values of goods handled by particular shops. In his analysis of Stockholm, Olssen offered one solution of the problem. He devised a shop rent index which took into account the total shop frontage in a particular street, together with the rents paid by individual shops. He suggested the following formula:

$$Shop \; rent \; index = \frac{Total \; shop \; rents \; in \; the \; street \; frontage}{Length \; of \; street \; frontage}$$

The shop rent index is thus the shop rent per unit of frontage. Proportional symbols can be constructed on the basis of these data in the form of rectangles, the depth of which vary according to the index (Fig. 196). The rectangles may then be shaded according to the different activities of the shops. These maps may be simplified to show shopping intensity road by road throughout the town, by making the

[1] G. T. Trewartha, 'The Unincorporated Hamlet', *Annals of the Association of American Geographers*, vol. 33, pp. 32–81 (Lancaster, Pa., 1943).
[2] Aspects of the urban structure are shown by these methods in the analysis of Stockholm, by W. William-Olssen and others, of which a summary is given in 'Stockholm', *Geographical Review*, vol. 30, pp. 420–38 (New York, 1940).
[3] J. E. Orchard, 'Shanghai', *Geographical Review*, vol. 26, pp. 1–31 (New York, 1936). See also J. R. Passonneau and R. S. Wurman, *Urban Atlas: Twenty American Cities* (M.I.T., 1966).

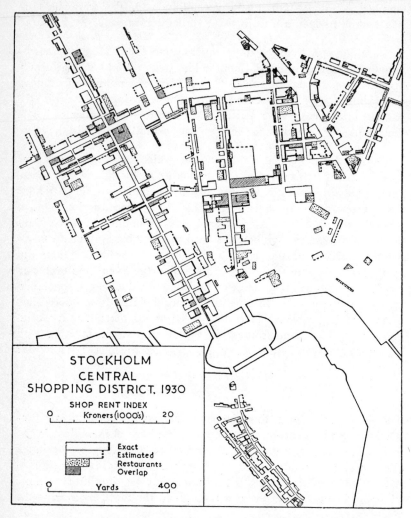

Figure 196. SHOP RENT INDEX SYMBOLS

Based on W. William-Olssen *et al.*, 'Stockholm', *Geographical Review*, vol. 30, p. 246 (New York, 1940).

Intensity of shopping is proportional to size of symbol. Overlap of symbols has been shaded as a further measure of intensity.

width of each road conform to the varying proportions of shop front index. Most U.K. Planning Offices use rateable values in this kind of work (see also Fig. 217).

Town and Village Symbols

Population Totals. The convention of representing the distribution of towns and villages by dots, small circles, small squares, dots in squares and similar symbols has a wide application. These symbols are designed to show the importance of each town as measured by the size of its population, according to the total enumerated within the civic boundaries. These totals may not be fully representative of a town's regional importance, because the built-up limits of towns are not coincident with the civic administrative boundaries. Too often, also, a purely arbitrary classification of population totals is employed, such as towns with a population over one million, over 100,000, over 50,000 or over 10,000 inhabitants.[1]

Certain significant population totals can serve to indicate the function of particular settlements and to point to their place in the urban

Settlement	Average Population
Small Market Towns (*Marktort*)	1,000
Market Towns (*Amtsort*)	2,000
Local Centres (*Kreisstadt*)	4,000
District Centres (*Bezirkstadt*)	10,000
Large District Centres (*Gaustadt*)	30,000
Provincial Centres (*Provincestadt*)	100,000
Regional Centres (*Landstadt*)	500,000

hierarchy. For example, a study of settlements in the United States suggests that purely rural centres (hamlets) usually have populations of less than 100, that villages with essential services usually have a population of over 200 and seldom reach 1,200, that towns offering a range of services have populations of 1,200–6,000, and that 'cities' with a full complement of services have populations usually near or in excess of 6,000. Important 'regional metropoles' have populations near or in

---

[1] Sometimes a useful purpose is served by mapping towns according to a progressive classification of population totals. An outstanding example of the application of such a method is to be found in M. Jefferson, 'Some Considerations of the Geographical Provinces of the United States', *Annals of the Association of American Geographers*, vol. 7, pp. 3–15 (Lancaster, Pa., 1917), which is illustrated by several very detailed maps showing the distribution of towns by symbols to represent populations from 1,000 progressively to 1 million and over.

excess of 100,000.[1] It is interesting to compare this classification with that of W. Christaller[2] for southern Germany on p. 419.

Centrality. The size of its population is, however, only one indication of the regional significance of a town. Some towns with a large population, such as the agricultural villages in parts of the Danubian plain, have relatively under-developed regional functions. Urban geographers have therefore devised many other methods of measuring the centrality of a town, in addition to employing the size of its population. Christaller[2] devised an index of centrality based on telephone services as follows:

$$Z_z = T_z - E_z \left(\frac{Tg}{Eg} \right)$$

where $Z = Centrality$, $Tz = number\ of\ telephone\ connections\ in\ the\ town$ *(central place)*, $Ez = number\ of\ inhabitants\ of\ town$, $Tg = number\ of\ telephone\ connections\ in\ the\ region$, $Eg = number\ of\ inhabitants\ in\ the\ region$.

Centrality is thus the difference between the expression $E_z \left(\dfrac{Tg}{Eg} \right)$, the *expected importance*, and T_z, the *actual importance*. It can be plotted by means of proportional symbols. E. Neef has recently criticized this method on the grounds that telephone ratios indicate only certain regional functions, and that Christaller fails to take into account important geographical differences between regions in his concept.[3]

Christaller's work has been expanded by A. Lösch whose theoretical concepts of market area, networks of market, and systems of network have added a new dimension to the theory of central places.[4] This work has been followed up by American geographers, with particular reference to settlements in America.[5] Some interesting maps of the distribution of settlements in parts of the United States, according to

---

[1] Largely deduced from B. J. L. Berry, H. M. Mayer *et al.*, 'Retail Location and Consumer Behaviour', in a report, 'Comparative Studies of Central Place Systems', for the Geography Branch, U.S. Office of Naval Research.

[2] W. Christaller, *Central Places in Southern Germany*, trans. by W. Baskin (New Jersey, 1966).

[3] E. Neef, 'Das Problem der Zentralen Orte', *Petermanns Geographische Mitteilungen*, vol. 94, pp. 6–17 (Gotha, 1950).

[4] A. Lösch, *The Economics of Location*, translated by W. H. Woglom *et al.*, (New Haven, 1954). See also W. Isard, *Location and Space Economy* (New York, 1956).

[5] A most useful reference to the bibliography is to be found in B. J. L. Berry and A. Pred, *Central Place Studies; A Bibliography of Theory and Applications* (Philadelphia, Regional Science Research Institute, 1961).

place in a hierarchy, have been compiled.[1] They are on a symbolic basis. J. E. Brush also shows area served. A. K. Philbrick's system of seven 'orders' of towns provides a useful basis for functional classification (see below).

Nature of Buildings. Irrespective of the nature and size of its population, a town's buildings may afford an indication of its place in the urban hierarchy. The number of banks in a town, for example, is to some extent a measure of its regional importance relative to other towns. The presence or otherwise of specialized retail stores, or department stores, of a newspaper, of a hospital, or of a university are other features to be taken into consideration (Fig. 200). Major cities, for example, according to A. E. Smailes, should contain certain government departments, a stock exchange, and a main branch of the Bank of England; they should have a daily morning newspaper, a university or university college and a medical school. Main towns should incorporate three or four banks, a grammar school (Fig. 197), a cinema, a hospital and a weekly newspaper. Sub-towns lack one or two of these services and urban villages lack two or three.[2]

Functional Classification of Towns. Symbols may be employed to show the distribution of towns according to their major functions (Fig. 198). These may be determined by a number of methods. Analysis of the occupational and industrial structure of urban populations is the method principally adopted to determine functional type. Functional analysis along these lines has already been mentioned in Chapter 5 (pp. 318–20). It should be stressed here that one of the difficulties in analysing occupational structure is the deficiency of detailed population data based on place of work.

The chief types of towns comprise manufacturing centres, retail centres, wholesale centres, transport centres, mining centres, health resorts, entertainment and residential centres, educational centres, administrative centres and large towns with diversified functions. Of course, all towns do perform a diversity of functions in varying pro-

[1] Two interesting examples are (1) J. E. Brush, 'The Hierarchy of Central Places in South-western Wisconsin', *Geographical Review*, vol. 43, pp. 380–402 (New York, 1953); and (2) A. K. Philbrick, 'Principles of Areal Functional Organization in Regional Human Geography', *Economic Geography*, vol. 33, pp. 299–336 (Worcester, Mass., 1957).
[2] For details of a comprehensive classification, see A. E. Smailes, 'The Urban Hierarchy of England and Wales', *Geography*, vol. 24, pp. 45–51 (London, 1944).

portions. The major function of the town is not necessarily that in which the major part of its population is engaged, but that in which an unusually large proportion of its population is engaged (Fig. 164). In considering the classification of towns along these lines, the general

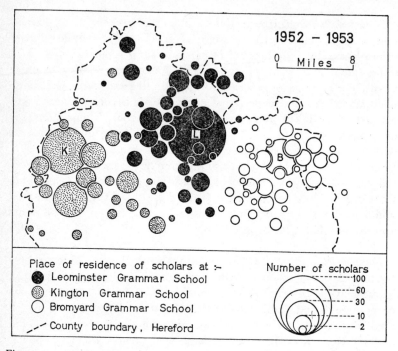

1952 – 1953

0   Miles   8

Place of residence of scholars at :-
● Leominster Grammar School
◉ Kington Grammar School
○ Bromyard Grammar School

╴╴⌒ County boundary, Hereford

Number of scholars
------100
------60
------30
------10
------2

Figure 197. TRIBUTARY AREAS OF GRAMMAR SCHOOLS IN NORTHERN HEREFORDSHIRE BY PROPORTIONAL SYMBOLS

Based on an unpublished map by C. D. Reade in 1954, as part of a series illustrating the urban fields of Kington, Leominster and Bromyard.

averages for England and Wales given above form a useful working basis.

For fuller details of the classification of towns and cities by function with the aid of occupational data, reference may be made to the work of C. D. Harris.[1]

Social and Economic Classification of Towns. For planning purposes it is often necessary to have as much comparative information as possible

[1] C. D. Harris, 'A Functional Classification of Cities in the United States', *Geographical Review*, vol. 33, pp. 86–99 (New York, 1943).

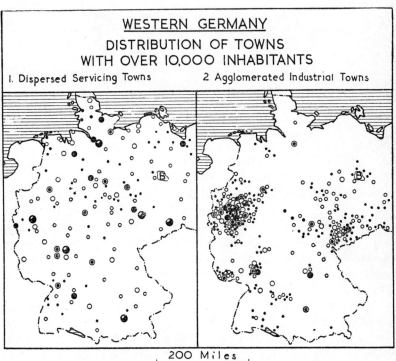

WESTERN GERMANY
DISTRIBUTION OF TOWNS
WITH OVER IO,OOO INHABITANTS
I. Dispersed Servicing Towns   2 Agglomerated Industrial Towns

200 Miles

● Over 500,000   o 50,000 - 99,999
● 200,000-500,000   o 20,000 - 49,999
◉ IOO,OOO-I99,999   • IO,OOO - I9,999

Figure 198. FUNCTIONAL TOWN SYMBOLS

Based on H. Bobek, 'Uber einige functionelle Stadttypen und ihre Beziehungen zum Lande', *Comptes Rendus du Congrès International de Géographie, Amsterdam, 1938*, vol. 2, sect. 3a, p. 102 (Leiden, 1938).

Berlin is indicated on each map by the letter **B**. The maps not only bring out the contrast between the dispersal of servicing towns and the agglomeration of industrial towns but also indicate that the hierarchy of industrial towns is not so well developed as in the case of the servicing towns. There are, for example, few large industrial towns, i.e. with a population of over half a million.

not only on the major functional features of towns, but on their social and demographic characteristics as well. The work of C. A. Moser and Wolf Scott may be cited here as an example.[1] Their classification was

England and Wales: Industrial Structure

Industry	Percentage of Labour Force Employed
Manufacturing	34·5
Servicing of all types	34·5
Financial Services	2
Transport and Communication	7
Administration and Defence	8
Mining	7
Fishing and Agriculture	7

based mainly on data available in census returns. After obtaining correlation coefficients for a variety of data for 157 towns they reduce the data by component analysis to four components: (1) social class difference; (2) inter-censal population change difference; (3) post-1951 development difference; and (4) housing difference. Cross tabulation suggested a final classification of 14 groups with two towns unclassified (see also p. 465).

Variegated symbols may be employed to indicate the different classified types of towns and cities. Because they stand for something rather different from the conventional town symbols, functional symbols might be given a rather different character. For this reason stars and triangles are rather more effective than small dots and squares but proportional symbols are also useful. W. William-Olsson, for example, in his *Economic Map of Europe*, employed 'spherical' symbols to give an indication both of size of towns and their major economic function.[2] W. R. Siddall also makes use of differentiated proportional symbols to give an indication of measure of function as well as classification in an ingenious fashion.[3]

Forms of Settlement. The form of a settlement often, though not always, provides evidence of its origins, together with some indication of the

---

[1] C. A. Moser and Wolf Scott, *British Towns, a Statistical Study of their Social and Economic Differences* (London, 1961).

[2] W. William-Olsson, *Economic Map of Europe*, 1 : 3,250,000 (Stockholm, 1953).

[3] W. R. Siddall, 'Wholesale-Retail Trade Ratios as Indices of Urban Centrality', *Economic Geography*, vol. 37, pp. 124–32 (Worcester, Mass., 1961).

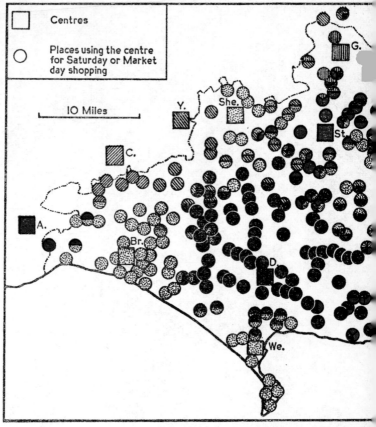

Figure 199. SHOPPING AFFINITIES BY SY

Based on a manuscript map compiled by A. E. Smailes in his capacity as chair graphical Association on urban spheres of influence. The map gives a measure as shopping centres.

Data for the map were obtained from questionnaires circulated to schools in I map are as follows: **A.** Axminster; **Bl.** Blandford; **Bo.** Bournemouth; **Br.** Bridpo lingham; **R.** Ringwood; **Sa.** Salisbury; **Sha.** Shaftesbury; **She.** Sherborne; **St.** S **We.** Weymouth; **Y.** Yeovil.

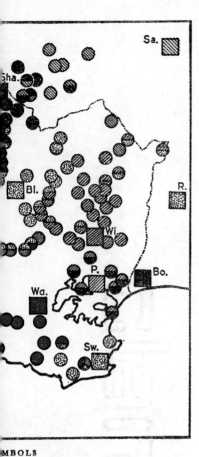

MBOLS

nan of a standing committee of the Geo-
f the function of various towns in Dorset

orset. References to towns centres on the
t; **C.** Crewkerne; **D.** Dorchester; **G.** Gil-
albridge; **Sw.** Swanage; **Wa.** Wareham;

changes which have taken place in its function and character since its foundation. Symbols may be effectively employed to show the forms of settlement when the scale of the map is not sufficient to show the actual street-plan and building patterns. The classification of forms of settlement to be adopted may be based on traditional types such as street-villages, green-villages, *bastides*, 'checker-board' towns and concentric towns. More elaborate classifications may be employed for specific purposes, such as that used by M. R. G. Conzen in his portrayal of the forms of settlement in North-east England.[1]

Site. The mapping of the distribution of different types of settlement sites may be executed most easily and efficiently by symbols. Sites which make use of the defensive possibilities of hill tops, of islands and of meanders, which are at wet or dry points, sites which exploit the commercial possibilities of river crossings or route convergences, are but a few instances of possible types. S. H. Beaver makes use of symbols in maps to show the distribution of hilltop, valley-bottom and spring-line villages in Northamptonshire.[2]

Dates of Foundation. The plotting of the dates of the foundation of towns may afford useful evidence concerning the spread of settlement and of economic organization in frontier lands. A map of this type has been compiled by R. E. Dickinson.[3]

The Sphere of Influence of Towns. Certain aspects of the sphere of influence of towns may be effectively portrayed by symbols. Variegated symbols, for example, may be employed to show the distribution of the sources of daily labour of neighbouring towns. The affinities of villages to various regional cities, as shown by shopping habits, by newspaper circulation, by the provision of special services such as higher educational institutes and hospitals, can be shown by means of symbols (Fig. 199). The symbolic method has certain advantages over the lineal delimitation of urban fields, because it reflects both the over-

---

[1] M. R. G. Conzen, 'Modern Settlement', *Scientific Survey of North-Eastern England*, British Association for the Advancement of Science (Newcastle, 1949).For the use of symbols to show stages in the growth and development of mining settlements, see P. N. Jones, *Colliery Settlements in the South Wales Coalfield, 1850–1966. A study in genesis and form*, Occasional Papers in Geography, no. 14 (University of Hull, 1970).

[2] S. H. Beaver, *The Land of Britain*, pt 58 (London, 1943).

[3] R. E. Dickinson, 'The Development and Distribution of the Medieval German Town', *Geography*, vol. 27, p. 49 (London, 1942).

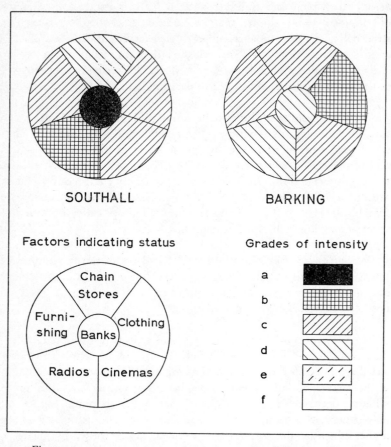

Figure 200. SYMBOLS TO SHOW STATUS OF SERVICE CENTRES

Based on W. J. Carruthers, 'Service Centres in Greater London', *Town Planning Review*, vol. 33 (Liverpool, 1962).

The symbols have been differentiated to show the intensity of the varied services in each centre. For example, more than eight banks would be classified as **a,** 1–3 banks would be **f,** with four intermediate grades between. Similarly, 38 clothing stores or over would be classified as **a,** whereas six or less would be **f,** and so on.

lap and specialization of services provided by different regional centres.[1] Proportional symbols also have a place in the analysis of urban fields, and are particularly useful in providing a quantitative basis for the comparison of the regional functions of different centres (Figs. 197, 200).

Place-name Elements

The plotting of place-name elements by means of symbols is of considerable help in the evolutionary study of settlement. Mention has already been made (see p. 398) of the English Place-Name Society and its series of county volumes, which may be used to plot the distribution of certain early place-name elements.[2] For example, Celtic, Anglo-Saxon and Scandinavian elements have been plotted in Fig. 201. One-inch or 1 : 25,000 topographical maps must be consulted in order that certain place-names can be located. It should be borne in mind that it is not always necessary to plot every variety of place-name to reconstruct the distribution of early settlement. One or two types are often sufficient to give the broad distribution, and detailed plotting serves only to emphasize this distribution. Interpretation of the patterns of distributions in the light of geographical circumstances can be facilitated by mapping distribution of relief, drift geology, soils and drainage, on a scale similar to that of the distribution map.

Superimposed Symbols

It may be useful occasionally to so design symbols that they can be superimposed one over the other without loss of clarity. This may be helpful in showing characteristics of particular locations where more than one characteristic needs to be shown. For example, these symbols can be used to portray historical features in place-name studies, the range of central place functions at given places, or the variety of morphological features associated with particular settlements.[3] J. R. Borchert

---

[1] For a good example of the effective use of symbols combined with lineal demarcation, see H. E. Bracey, op. cit. From a series of symbolic maps he is able to derive 'median' boundaries between the spheres of influence of neighbouring towns. See also A. E. Smailes, 'The Analysis and Delimitation of Urban Fields', *Geography*, vol. 32, pp. 151–61 (London, 1947).

[2] See an application in H. C. Darby (ed.), *An Historical Geography of England before 1800*, p. 112 (Cambridge, 1936).

[3] Superimposed symbols are used to show particular features of the port history of the Adriatic coast in a paper by H. R. Wilkinson, 'Regional features in Yugoslavia', *Liverpool Essays in Geography, A Jubilee Collection*, ed. by R. W. Steel and R. Lawton, p. 569 (London, 1967).

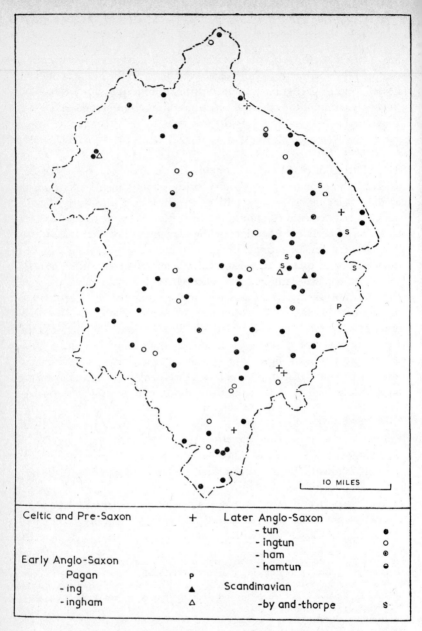

Celtic and Pre-Saxon	+	Later Anglo-Saxon	
		- tun	●
		- ingtun	○
Early Anglo-Saxon		- ham	◉
Pagan	P	- hamtun	⊖
- ing	▲	Scandinavian	
- ingham	△	-by and -thorpe	s

Figure 201. PLACE-NAME SYMBOLS

Source of data: J. E. B. Gover *et al.*, 'The Place-Names of Warwickshire', *English Place-Name Society*, vol. 13 (Cambridge, 1936).

The map shows the distribution of certain place-names in Warwickshire, with the aim of indicating the pattern of early settlement in the county.

uses superimposed symbols to show rates of differential growth of towns in the United States.[1] The total population of each town is shown epoch by epoch by means of a proportional circle, the outer ring representing the situation in 1960. The compression or expansion of the concentric circles provides immediate visual appreciation of phases of growth.

CHOROPLETH MAPS

Dispersion and Concentration of Settlements

In the study of settlement over wide areas, building traces become cumbersome to use, and it is often impossible to analyse the scatter of settlements by inspection alone. It becomes imperative to devise some means of quantitative analysis of the settlement pattern, and students of settlement have long concerned themselves with the problem of evolving formulae for this purpose. In considering dispersion and concentration of settlements, for example, a number of variables have to be considered. These include: (a) the number of settlements; (b) the number of houses in each settlement; (c) the size of the population in each settlement; (d) the area of the region served by each settlement; and (e) the distance apart of settlements. No one formula has ever been satisfactorily devised which would take all these variables into consideration, but a number of indices, in which two or three variables are involved, may be used to map certain distributional aspects. A selection of these may be briefly considered.

Demangeon's Coefficient of Dispersion. This is obtained from the formula:

$$C = \frac{E \times N}{T}$$

where E = *population of écarts (population of commune minus that of its chief place),* N = *number of écarts* − 1, T = *total population.*

This calculation is only made possible because in French censuses returns of population for each commune give two totals – one for the chief place in the commune and one for the population of the *écarts* or isolated settlements. This formula has been applied, with excellent results, in a map of the dispersion of settlements in France in the French National Atlas (op. cit.).

[1] J. R. Borchert, 'American Metropolitan Evolution', *Geographical Review*, vol. 57, p. 326 (New York, 1967).

Simple Index of Dispersion. Unfortunately, British census returns do not differentiate between chief places and isolated settlements in any one parish, but some idea of settlement dispersion may be deduced from the formula:

$$I = \frac{S}{H}$$

where I is the Index of dispersion, S the number of settlements in any parish, and H the number of isolated houses.

A house which is situated a quarter of a mile from any other house is conveniently taken to be isolated.

Bernhard's Index of Concentration. Rather more elaborate and not equally applicable to all regions is J. Bernhard's formula, which takes into account the number of houses (H), the area under consideration (A) and the number of settlements (S). Bernhard argues that concentration of settlement is a function of the mean number of houses in each settlement H/S and of the number of settlements in a given area A/S.[1] Considering both functions, he obtains the formula:

$$C = \left(\frac{H}{S}\right)\left(\frac{A}{S}\right) = \frac{HA}{S^2}$$

Pawlowski's Indices of Concentration. S. Pawlowski in a study of settlement concentration in Poland[2] used the following formulae derived from settlement traces which had been gridded with a network of 25-km squares:

$$\text{Mean value of Concentration} = \frac{As}{S} \qquad (1)$$

where As = area occupied by the settlements in km^2 and S = number of settlements.

This value will vary between 0 and 1.

$$\text{Coefficient of Concentration} = \frac{A}{A_s} \qquad (2)$$

where A = the area of one square (25 km^2) and A_s = area occupied by settlements.

---

[1] Jean Bernhard, 'Une Formule pour la Cartographie de l'Habitat Rural avec Application au Département de l'Yonne', *Comptes Rendus du Congrès International de Géographie, Paris, 1931*, vol. 3, pp. 108–17 (Paris, 1934).

[2] S. Pawlowski, 'Encore une Méthode de Représentation Cartographique Générale de l'Habitat Rural', *Comptes Rendus du Congrès International de Géographie, Amsterdam, 1938*, vol. 2, sect. A, pp. 129–30 (Leiden, 1938).

The area occupied by settlements includes the area of gardens, etc., is comparable with areas marked red and purple on the British Land Utilisation Survey maps.

Kant's Index of Concentration. A most ingenious formula to show density of rural settlements was devised by E. Kant (op. cit.) to illustrate settlement in Esthonia. It was designed to be used to reduce a map showing distribution of habitations by means of non-quantitative dot-symbols to one in which dispersion and concentration of settlement was more precisely reflected in terms of distance between the habitations. The formula is:

$$X = \frac{1}{M}\sqrt{\frac{A}{D}}$$

where X is the relative interval between two settlements, $1/M$ is the scale of the map, A is equal to the area under consideration, and D the density of habitations.

Near Neighbourhood Statistic. This is a value arrived at statistically and is based on near neighbourhood analysis. It indicates the degree to which the settlements of any area are agglomerated or dispersed either uniformly or at random.[1]

Settlement Groupings

The size of settlements and the grouping of houses in settlements may be indicated by adopting the following coefficient of grouping (C):

$$C = \frac{\textit{Number of inhabitants in the parish}}{\textit{Number of settlements in the parish}}$$

Population totals in this case are taken from census returns, and the number of settlements in a parish have to be counted from contemporary topographical maps. Another possible coefficient (C_1) is as follows:

$$C_1 = \frac{\textit{Number of houses in the parish}}{\textit{Number of settlements in the parish}}$$

Information about numbers of houses is also yielded by census returns.

[1] L. J. King, 'A Quantitative Expression of the Pattern of Urban Settlements in Selected Areas of the United States', *Tijdschrift voor Economische en Sociale Geografie*, vol. 53, pp. 1–7 (Rotterdam, 1962).

Density of Housing

The density of housing in any district is conventionally expressed as the ratio of the number of houses, or alternatively the number of habitable rooms (i.e. all but halls, kitchens and bathrooms) per unit of area (Fig. 202). This ratio, however, is not always reliable, because it

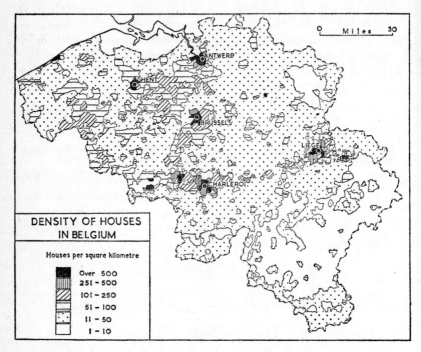

Figure 202. A CHOROPLETH MAP OF HOUSING DENSITY
Based on M. A. Lefèvre, 'Carte de Densités des Maisons en Belgique', *L'Habitat rural en Belgique* (Liége, 1926).

is very often invalidated by the inclusion of extensive areas of parkland, land taken up by industrial sites, cemeteries and so on. For some purposes a *comparative density of housing* is useful, which is expressed by the ratio of the number of houses per unit of area of land used for residential purposes only.

Residential and Social Structure (Social Area Analysis)

The census returns for the wards, blocks of buildings, or 'tracts' of big cities provide data whereby the social and residential structure of the city may be analysed geographically, largely by means of choropleth maps based on certain indices. J. Moscheles, the Czech geographer, was a pioneer worker in this field.[1] Significant indices comprise the percentage of males over 25, to give an immediate indication of the distribution of sex-ratio in the town, and the number of domestic servants per 100 houses, to give an indication of grades of residential property. Other indices include the percentage of the residential population gainfully employed, the number of families per house and so on. Reference should be made here to E. Jones' work on Belfast which contains a number of maps devised to illustrate the social structure of the city (see p. 339).

Much urban analysis, whether concerned with the elucidation of the physical, the economic, or the social structure of the city, aims to distinguish sub-regions of various degrees of homogeneity. For this purpose there is a large amount of data available on a ward, enumeration area, social tract or block basis. The development of statistical techniques, some borrowed from other disciplines such as botany and psychology, together with the present-day facilities for computation, has meant that detailed studies of a wide variety of spatial conditions within towns can be attempted. The problem involves the analysis of a large number of variables, both dependent and independent, to establish spatial affinities including contiguity. Among the various methods used, that which aims to identify a diagnostic variable such as gross rateable value provides a useful preliminary approach. This method has been applied by D. T. Herbert in a study of Newcastle under Lyme.[2] Ultimately choropleth maps are produced showing different categories of social rank within the town.

E. Shevky and W. Bell, in their studies of American towns, categorize census tract populations by using indices to measure (a) social rank (occupation, schooling, rent), (b) urbanization (fertility, women at work, single family-dwelling units), and (c) segregation (racial and national groups in relative isolation).[3] Grouping analysis involving taxonomic methods can also be used for distinguishing sub-areas within cities. P. J. Taylor, for example, in an analysis of Liverpool in 1851,

---

[1] J. Moscheles, 'Prague', *Geographical Review*, vol. 27, pp. 414–29 (New York, 1937).

[2] D. T. Herbert, 'Approaches to Social Area Analysis', *I.B.G., Study Group in Urban Geography, The Social Structure of Cities*, pp. 15–25, (Liverpool, 1966).

[3] E. Shevky and W. Bell, *Social Area Analysis, Theory, Illustrative Application and Computational Procedures*, Stanford Sociological Series, no. 1 (Stanford, 1955).

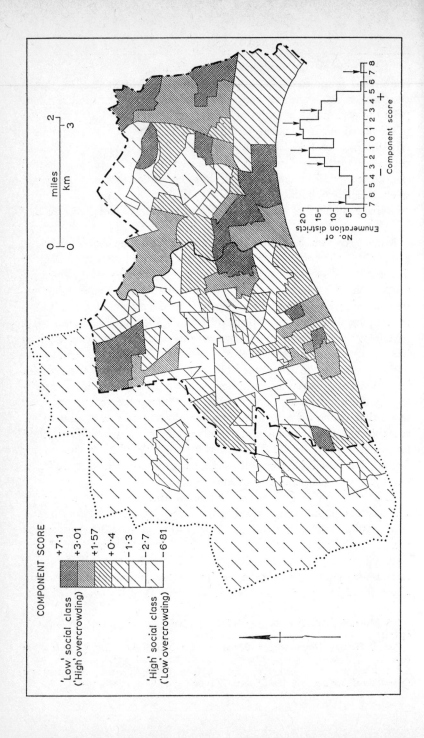

COMPONENT SCORE

+7·1
+3·01
'Low' social class
('High' overcrowding)
+1·57
+0·4
−1·3
−2·7
'High' social class
('Low' overcrowding)
−6·81

No. of Enumeration districts

Component score

divides up 302 enumeration districts into 8 compact social regions upon this basis.[1]

Multivariate Analysis of Urban Areas

There are other approaches to the study of a typology of social areas than social area analysis. For example, if matrices of correlation co-efficients between each pair of all the variables selected for study are constructed it becomes apparent that causal relations between any single independent and a number of dependent variables is not indi-cated. The matrices in fact exhibit different degrees of correlation between the variables selected. In such circumstances component analysis provides a particularly useful approach to the study of a large universe of variables where the object is to collapse them into fewer and more manageable groups. The components are so ordered that they have a high degree of intra-correlation but a low degree of inter-correlation. Moreover, it is possible to measure precisely the amount of variance accounted for by each component. As the components are linear combinations of the original data, we end up with as many components as there are original variables but in practice the first

[1] Peter J. Taylor, 'The Location Variable in Taxonomy', *Geographical Analysis*, vol. 1, pp. 181–95 (Ohio State University, 1969).

Figure 203. THE DISTRIBUTION VALUES FOR COMPONENT I, HULL AND
HALTEMPRICE, 1966

The data for the above map were drawn from the 1966 Census, Enumeration District tabulations; some thirty-seven variables were considered including indices of popula-tion structure, household structure and tenure, and socio-economic structure. The data were analysed by principal components analysis. This particular map shows the distri-bution of component 1, accounting for 23·3 per cent of the total variance. The scores ranged from +7·117799 to −6·807521. The scores were divided into six categories on the basis of an equal number of enumeration districts within each category. The distribution in relation to frequency is shown on the chart inset, the arrows indicating the limits of each division. It would, of course, have been possible to map the compo-nent score on a frequency basis or alternatively to use a variety of methods of grouping different component scores to optimize regional divisions. This particular distribution shown does, however, tend to emphasize certain features of the social structure of Hull and Haltemprice. The high scores denote the areas of low social class and high over-crowding and these tend to be concentrated in the inner ring of nineteenth-century by-law housing and also to extend into the areas associated with Hull's fish docks along Hessle Road. A second element in the pattern is the high scores associated with corpora-tion housing estates in the north of the city, in the east of the city and, to some extent, isolated patches in west Hull. The low scores indicate high social class with little over-crowding. They tend to be concentrated in Haltemprice and in the area of private housing in west Hull and in Sutton in east Hull.

three or four components usually explain the greater part of the variation in the data under consideration. When components have been identified it is possible to plot them on a map either individually, using choropleth maps, or by grouping components according to their degree of association with particular enumeration areas. For examples of applications see Brian T. Robson,[1] D. T. Herbert[2] and B. J. L. Berry.[3]

B. T. Robson, in a useful discussion of the application of principal component analysis to Sunderland, proposes a graphical cross tabulation of the first four components to produce sub-areas.[4] He includes a map to illustrate how a shading system can be adopted to show the distribution of the quartiles of more than one component on one and the same map.

The method has been applied to the delimitation of social areas in Hull (Fig. 203). In the preparation of the data for this map some thirty-seven variables were used, drawn from the 1966 census enumeration data. In certain enumeration districts in the centre of the township populations were excluded from the study. Only the distribution of the first component is shown here. This is a measure of social class and in the case of Hull shows a strong tendency to a concentric pattern from lower class outwards to higher class, but modified on the outskirts by the distorting effect of corporation housing estates.

The Use of Grid Maps in Urban Analysis

Instead of using wards and enumeration districts as bases for showing distribution of different features of urban geography, the town can be divided on the basis of grids of various dimensions. As an example in his work on the distribution of by-law housing in Hull, C. A. Forster used a grid, each square of which formed a quarter of the Ordnance Survey 25-inch plan, that is, an area of about 61·7 acres.[5] Using this as a base, he was able to produce a series of maps showing different

[1] Brian T. Robson, 'Multivariate Analysis of Urban Areas', *I.B.G. Study Group in Urban Geography, Social Structure of Cities*, pp. 1–14 (Liverpool, 1966).

[2] D. T. Herbert, 'The Application of Principal Components Analysis of Studies of Urban-social Structure', *I.B.G. Study Group in Urban Geography, Techniques in Urban Geography*, pp. 106–14 (Salford, 1968).

[3] B. J. L. Berry and P. Rees, 'The Factorial Ecology of Calcutta', *American Journal of Sociology*, vol. 74, pp. 445–91 (Chicago, 1969).

[4] B. T. Robson, *Urban Analysis*, p. 108 (Cambridge, 1969).

[5] C. A. Forster, 'The history, development and present-day significance of by-law-housing morphology with particular reference to Hull, York and Middlesbrough', unpublished Ph.D. thesis (University of Hull, 1969).

features of residential development in the centre of the town. One of these (Fig. 204), for example, shows the area of cul-de-sac terraces in Hull as a percentage of total residential building in 1965. Grid maps of this kind make it possible both to analyse locational features in

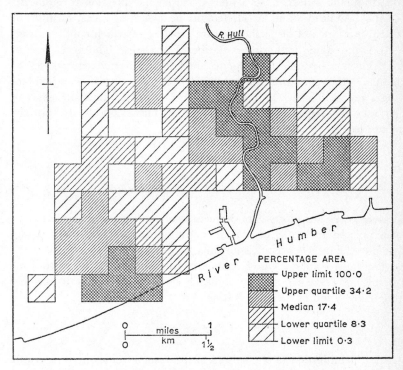

Figure 204. AN URBAN GRID MAP

Based on C. A. Forster, 'The Historical Development and Present-Day Significance of By-law-housing Morphology, with particular reference to Hull, York and Middlesbrough', unpublished Ph.D. thesis (University of Hull, 1969). This map shows features of the urban morphology of Hull on the basis of the Ordnance Survey 25-inch plan. The map shows cul-de-sac terraces as a percentage of total residential building in 1965.

detail and to compare development from one period to another and from one town to another.

D. Keeble has some interesting comments on the use of grid maps to show various features of the geography of towns, including the possibility of using these as the basis of simulation models.[1]

[1] D. Keeble, 'School Teaching and Urban Geography: Some New Approaches', *Geography*, vol. 54, pp. 18–33 (London, 1969).

ISOPLETH MAPS

Isopleths do not have a very wide application in the cartographical study of settlements. They are used largely as a means of simplifying distributional aspects of settlements where generalization is required.[1] Thus J. A. Barnes and A. H. Robinson devised a special formula to obtain an index of dispersion of rural settlement which could be used as a point-value for the interpolation of isopleths.[2] The best measure of dispersion between farm houses, or for that matter between settlements, is their distance apart, but to measure such distances involves great labour. Barnes and Robinson demonstrated that the average distance (D) between farms in any administrative unit closely approximates to:

$$ 1 \cdot 11 \sqrt{\frac{A}{n}} $$

where A is the area of the unit and n the number of farms.

Values of D are plotted in the centre of the unit and isopleths are interpolated in tenths of a mile, or at any distance suitable to the scale of the map (cf. Kant's formula, p. 431).

Polish geographers, who have tended rather to specialize in problems of rural settlements, often combine isopleths with choropleths to show the distribution of rural habitations.[3]

Isochrones

In the delimitation of the sphere of influence of cities, it is desirable to know the time taken in travelling between the city and the countryside. Times can be ascertained for a number of points and isopleths then interpolated for intervals of an hour or so (Fig. 205). Isopleths which join places having the same travelling time to the centre of the

[1] See the maps of the distribution of hamlets in the United States by G. T. Trewartha, op. cit. (1943). M. Jésman used isopleths (interpolated from points fixed in each commune according to the mostly densely settled part) to show the general distribution of habitations in Valkynia in 'Gestość Zabudowania Województwa Wotynskiego', (Density of Buildings in Valkynia), *Prace Wykonane w Zakladzie Geograficznym Uniwersytetu Wilénskiego*, no. 1 (Vilno, 1936).

[2] J. A. Barnes and A. H. Robinson, 'A New Method for the Representation of Dispersed Rural Settlement', *Geographical Review*, vol. 30, pp. 134-7 (New York, 1940).

[3] M. Zdobnicka, 'Metoda izarytmiczna w grafice statystycznej', *Pokfosie Geograficzne* (Warsaw, 1925).

city are sometimes called *isochrones* (see also p. 280).[1] This term is obviously capable of extension to include various isopleths based on time values.

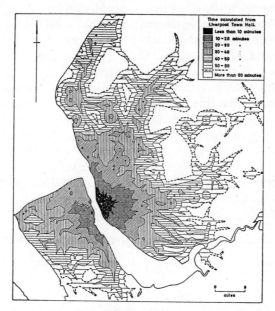

Figure 205. ISOCHRONIC MAP OF THE JOURNEY
TO WORK IN CENTRAL LIVERPOOL, 1953

Based on K. Rowe, 'The Journey to work on Merseyside',
Advancement of Science, vol. 10 (London, 1953–4).
    Data for the map were drawn from all public transport except express train services and due allowance was made for intervals of walking.

Isostades

The isopleth method may also be used to show the spread of organized settlement in frontier districts. Dates of foundation of particular towns are plotted and isopleths interpolated for significant dates. These have been called *isostades*.[2]

---

[1] For an example, see E. Kant, a map entitled 'Urban Hinterlands in Estonia with Policentric Isochrones', in *Environment and Population Problems in Estonia* (Tartu, 1934).

[2] G. Conzen, 'East Prussia', *Geography*, vol. 30, pp. 1–10 (London, 1945). See J. R. Borchert, 'The Twin Cities Urbanized Area: Past, Present and Future', *Geographical Review*, vol. 51, folder-map facing p. 62 (New York, 1961). Three maps show isostades of 'the advancing frontier' of high-, medium- and low-density settlement for 1900, 1940, 1956 and (projected) 1980.

Mental Maps

It may be of some importance in attempting to measure various preferences to consider what has been termed environmental perception.
Everyone carries round in their mind mental maps of particular areas
and images of places which may, or may not, stand up to realistic
appraisal, but which are nevertheless important in decision-making
on their part. Such attitudes for example may determine where people
decide to live and where they decide to emigrate, and are also important
as measures of what they feel should be conserved in the environment.
It is possible to produce such mental maps by structuring the information from questionnaires concerning individual ideas about the
environment. Thus P. Gould has been able to produce, for example, a
contoured perception surface map of the United States of America to
show residential preferences of a group of professional students, using
both ordinal and interval scales to measure preference.[1]

COLUMNAR DIAGRAMS AND DIVIDED RECTANGLES

Starting from the axiom that 'a qualitative analysis of rural settlement
has scientific value only if supported by a quantitative statistical
foundation',[2] Miss B. M. Swainson analysed the dispersion and
agglomeration of rural settlement in Somerset by means of columnar
diagrams, and it is principally in connection with such analyses that
these diagrams have so far been employed. But they are also of use
in the comparative study of the population structure of different settlements and in the study of differential growth of urban and rural
population.

Population and House-groupings

Miss Swainson devised a system of eight columns to represent a series
of house-groupings for contrasting blocks of rural parishes in different
parts of Somerset. The heights of the columns were made proportional
to the percentage of the total population in each block of parishes.

---

[1] See, for example, P. Gould (1) 'On Mental Maps', *Michigan Inter-University
Community of Mathematical Geographers*, vol. 9, pp. 1–54 (Michigan State University,
1966); (2) 'Problems of Space Preference Measures and Relationships', *Geographical
Analysis*, vol. 1, pp. 31–44 (Ohio State University, 1969).

[2] B. M. Swainson, 'Dispersion and Agglomeration of Rural Settlement in Somerset',
Geography, vol. 29, pp. 1–8 (London, 1944).

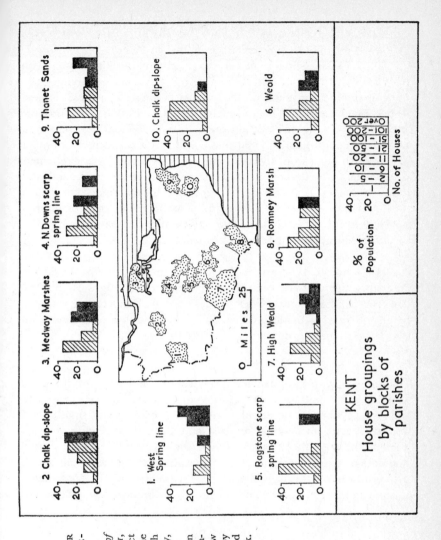

Figure 206. COLUMNAR DIAGRAMS OF HOUSE-GROUPINGS

Source of data: (1) *Census of England and Wales, 1841,* Enumeration Abstract (London, 1843); (2) the relevant sheets of the 1-inch series, Ordnance Survey, first edition.

Columns have been shaded to facilitate comparison: groupings below 21 houses are shown by diagonal shading, and above that figure in black.

The method is demonstrated in Fig. 206, which has been constructed to show features of the settlement-groupings in Kent.

The data about population and the number of dwellings in a parish may be obtained from census returns, but the house-groupings have to be ascertained by counting the number of houses in each settlement from large-scale maps. There is often difficulty in doing this because individual dwellings are not always distinguished separately on maps, and it frequently happens that anomalies occur between the number of houses counted from the map and the number of houses returned on the census. Some field-work, therefore, is necessary to smooth out these anomalies. The censuses of 1951 and 1961 provide data which make possible analyses of house-groupings in rural areas of interest not only in the study of the evolution of settlement patterns, but also in the classification of modern settlements.

House-groupings. Where it is not possible to obtain maps which are contemporaneous with census returns, columnar diagrams may be constructed along similar lines to those outlined above, but with the height of columns made proportional to the number of occasions on which a particular grouping occurs (Fig. 191).

Compound Columnar Diagrams

Diagrams may be constructed proportional in height to the total population of units of area at different periods, and subdivided to show proportions of the population living in large towns, small towns and villages. Some census returns lend themselves readily to such analyses because they contain tables showing totals of enumerated population living in communities of specific sizes, i.e. less than 2,500, 2,500–10,000 and so on.[1] Alternatively, columns may be constructed to show number of houses, and divided to distinguish types of houses.[2]

Divided Rectangles

Divided rectangles may be employed in much the same way as compound columnar diagrams to indicate changes in settlement structure as reflected in the urban and rural status of the population. In Fig. 207, for example, comparable rectangles have been constructed to show

[1] See an application in Admiralty Geographical Handbook, *Germany*, vol. 3, p. 80 (London, 1944).
[2] See a series of such diagrams arranged in the form of a profile in R. Finley and E. M. Scott, 'A Great Lakes-to-Gulf Profile of Dispersed Dwelling Types', *Geographical Review*, vol. 30, pp. 412–19 (New York, 1940).

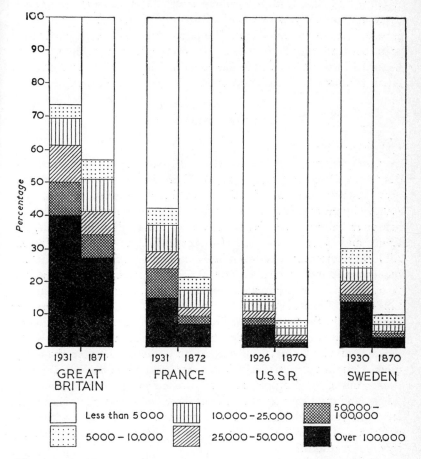

Figure 207. DIVIDED RECTANGLES OF RURAL AND URBAN POPULATION

Source of data: J. Haliczer, 'Agglomeration und Verteilung der Bevölkerung in Europa auf Grund einer neuen Karte in Masse 1 : 4,000,000', *Comptes Rendus du Congrès International de Géographie, Amsterdam, 1938,* vol. 2, sect. 3a, pp. 167–80 (Leiden, 1938).

The diagrams show the relative proportions of the population of various European countries living in places of less than 5,000 inhabitants or in larger places, as the case might be. Changes occurring within the space of about sixty years are also shown, to enable the progress of urbanization to be visually assessed.

the state and progress of urbanization in selected European countries
between 1870 and 1930.

These diagrams used as an index of urbanization have been effectively employed by C. M. Law in a study of the urban population
growth in England and Wales. The population of the different census
years between 1801 and 1911 is shown by columns subdivided into
non-urban population on the one hand and into various categories of
urban population according to population size on the other.[1]

SPECIAL DIAGRAMS

Flow-line Maps

Traffic-flow maps and the principles of their construction are fully
dealt with elsewhere (see pp. 307–11). They deserve a separate mention
here, however, because of their use in the delimitation of town and
country by means of motor-bus services.[2] The delimitation has a reconnaissance value in the determination of the median hinterlands of
towns, and in the evaluation of shopping facilities of urban centres.
All local services must be taken into consideration. First of all, 'motor-
bus centres', from which radiate at least some regular bus services
which serve no town larger than themselves, are ascertained from the
time-tables. The frequencies of buses employed on the various services
are plotted in the form of flow-lines (Fig. 208). After an inspection of
the diagrams, it is possible to decide where boundaries may be drawn to
separate areas within which one centre is more accessible than any
other. Certain difficulties arise in the execution of the method. Double-
decker buses, for example, may be used on some routes and not on
others, and duplicate services are not shown on time-tables. Nevertheless, the compilation of these diagrams constitutes a constructive
exercise and paves the way for further analyses of urban hinterlands
by other methods.[3]

[1] C. M. Law, 'Urban Population Growth, England and Wales', *Institute of British
Geographers, Transactions* no. 41, p. 134 (London, 1967).
[2] F. H. W. Green, 'Town and Country in Northern Ireland', *Geography*, vol. 34,
pp. 89–96 (London, 1949); and 'Motor-bus Centres in South-west England considered in relation to Population and Shopping Facilities', *Institute of British Geographers, Transactions and Papers, 1948*, Publication no. 14, pp. 59–68 (London, 1949).
[3] Note an interesting map depicting mid-nineteenth century urban hinterlands in
Leicestershire, using flow-lines (based on numbers of carriers' carts derived from a
county directory), radiating from the centres, by P. R. Odell, 'Urban Spheres of
Influence in Leicestershire in the Mid-nineteenth Century', *Geographical Studies*, vol.
4, no. 1, p. 38 (London, 1957).

Ray-diagrams

Ray-diagrams are used to illustrate aspects of the sphere of influence of towns. The affinities between town centres and villages in the surrounding countryside may be indicated by a line joining the village to the town centre. Thus if a town functions as the shopping centre for a number of neighbouring villages, its function is reflected by a

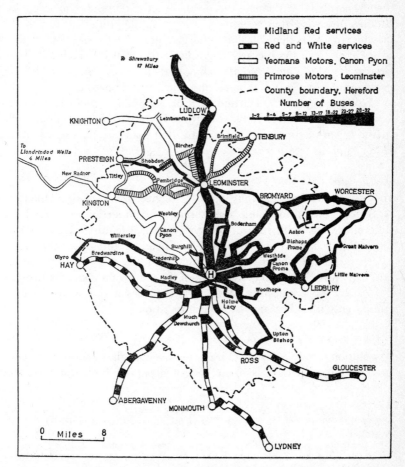

Figure 208. FREQUENCY OF PRINCIPAL WEEKDAY BUS SERVICES FROM HEREFORD, LEOMINSTER, KINGTON AND BROMYARD

Based on an unpublished map compiled by C. D. Reade in 1954. The data exclude additional and special market and weekend services. **H.** refers to Hereford. The number of buses is per day.

number of rays radiating from the town to the villages.[1] It is an advantage of this method in the delimitation of urban fields that it allows overlap to be shown. It can also be adapted to show a great variety of affinities in a simple but effective manner, for example, the circulation of newspapers (Fig. 210).[2]

Ranking Diagrams

The ranking of settlements according to their population size, the number of services they contain, etc., is a helpful preliminary step in the classification of settlements and in the grouping of settlements according to their place in the hierarchy. A simple diagram in which shopping centres have been arranged according to their score as measured by the presence of selected stores, banks and other criteria has been devised by A. E. Smailes and G. Hartley to rank shopping centres in Greater London.[3] This method enables centres to be graded by inspection. Such methods can be supplemented statistically by such techniques as direct factor analysis (see p. 510).

Morphographical Stars

In the field of biogeography a useful device has been developed by A. S. Kostrowicki to enable a simultaneous comparison to be made of different faunas with regard to their structural features. He calls this a morphological star and it is capable of being modified for other data, for example, characteristics and structural features of different populations. On each ray the percentage of the value of a given feature has been marked. Connecting these parts provides a star which can be compared with other stars by superimposition.[4]

Association Analysis Charts

Association analysis can be used in various classifications of geographical material. It is a method of grouping individuals on the

[1] See, for example, M. W. Mikesell, 'Market Centers of North-eastern Spain', *Geographical Review*, vol. 50, p. 248 (New York, 1960). (Map prepared by G. L. Augustine.)

[2] For a series of such maps, see J. R. Tarrant, *Retail Distribution in Eastern Yorkshire in Relation to Central Place Theory: A Methodological Study*, Occasional Papers in Geography, no. 8 (University of Hull, 1967).

[3] A. E. Smailes and G. Hartley, 'Shopping centres in the Greater London Area', *Transactions and Papers, 1961: Institute of British Geographers*, no. 29 (London, 1961).

[4] Andrzej Samuel Kostrowicki, 'Biogeographical Complexes', *Geographia Polonica*, pp. 185–94 (Warsaw, 1968).

basis of presence or absence of possible attributes or characteristics
originally developed by ecologists to classify biotic communities by
the presence or absence of plant species. In geography it can be used,
for example, to classify shops or houses. In association analysis, the
population (i.e. the total number of individuals to be classified) is
divided on the basis of that attribute, the presence or absence of which
gives rise to the greater dichotomy in the population, calculated by the
formula, $\Sigma \sqrt{\dfrac{x^2}{n}}$, whereby Chi^2 values are calculated in the case of each
attribute for its correlation in occurrence with each other attribute. The
values are then summed, regardless of sign, and the population is split
into two groups, one of which possesses an attribute of the highest
accumulated Chi^2 score, and one which does not. The process is then
repeated for each of these groups to generate four groups and the
subdivision can be repeated indefinitely but is usually contained at a
desired level of significance. The level of significance of each division
is calculated by the highest Chi^2 score between attributes in the groups
just divided, and division usually ceases when the maximum Chi^2
value falls below 6·63, which, with one degree of freedom, represents
1 per cent probability of further grouping being on the basis of chance
associations rather than significant groups. An example is given in
Fig. 209 which is an attempt to group terraces occurring in the built-up
area of central Hull using the method of association analysis. In this
particular case the first dichotomous variable is terrace width. Terraces over 20 feet wide and terraces over 100 feet long are also distinguished. Various other external features of the houses within the
terraces have also been analysed, giving rise ultimately to twenty-two
morphological groups. This kind of analysis makes it possible to trace
evolution of particular types of building both in time and space and is
useful also in tracing origins of certain types of housing, for example
by-law housing in English cities.[1]

Scalograms

The use of various devices for measuring responses is quite common in
perception geography and these methods can be applied not only to the
identification and measurement of attitudes but also to the measurement of non-personal attributes. For example, D. N. Parkes uses the

[1] C. A. Forster, 'The Historical Development and Present-day Significance of
By-law-housing Morphology, with particular reference to Hull, York and Middlesbrough', unpublished Ph.D. thesis (University of Hull, 1969).

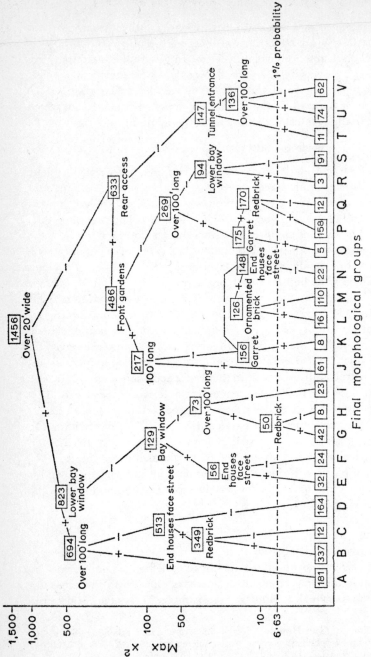

Figure 209. ASSOCIATION ANALYSIS CHART

Based on C. A. Forster, 'The Development of By-law Housing in Kingston upon Hull, an Example of Multivariate Morphological Analysis', the I.B.G. Study Group in Urban Geography, *Techniques in Urban Geography*, p. 122 (Salford, 1968). In this diagram, twenty-two morphological groups are finally distinguished on the basis of various morphological features associated with terraced housing in central Hull.

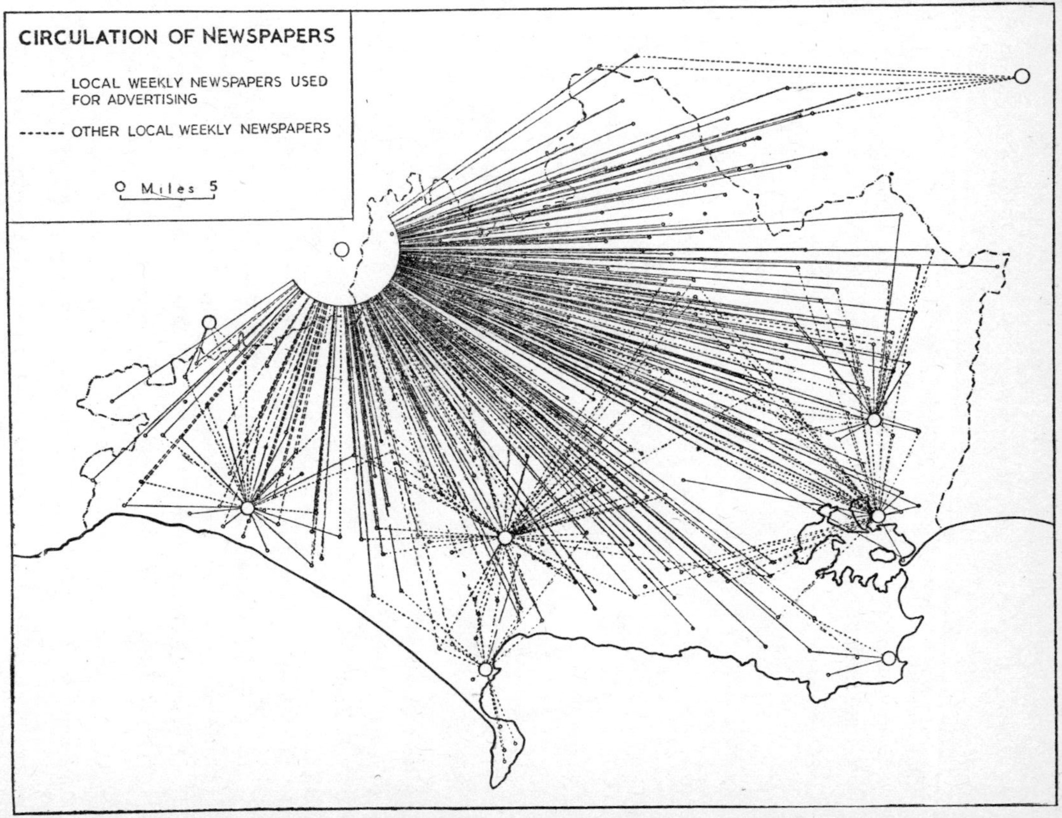

CIRCULATION OF NEWSPAPERS

LOCAL WEEKLY NEWSPAPERS USED
FOR ADVERTISING

------ OTHER LOCAL WEEKLY NEWSPAPERS

0 Miles 5

Figure 210. A RAY-
DIAGRAM OF NEWS-
PAPER CIRCULATION

Based on a manuscript
map compiled by A. E.
Smailes. The map refers to
Dorset, and a key to centres
is provided in Fig. 199.

Guttman scalogram as a means of measuring the quality of residential areas.[1]

Urban Profiles

As a contribution to the study of the morphology of cities, E. van Cleef has made use of 'urban profiles'.[2] They take the form of slightly

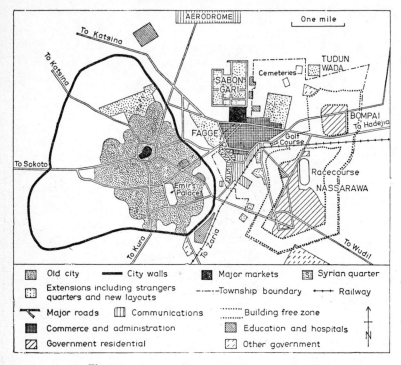

Figure 211. A SKETCH-MAP OF KANO CITY
Based on K. M. Buchanan and J. C. Pugh, *Land and People in Nigeria* (London, 1955).

exaggerated silhouettes of the profiles of cities, with the object of depicting variations in their geographical character in a pictorial manner.

Sketch-maps

The sketch-map has long been an indispensable tool of the geographer in his illustration of the position, siting and character of settlements. It

[1] D. N. Parkes, 'The Guttman Scalogram: an empirical appraisal in urban geography', *Australian Geographical Studies*, vol. 7, pp. 109–36 (Melbourne, 1969).
[2] E. van Cleef, 'The Urban Profile', *Annals of the Association of American Geographers*, vol. 22, pp. 237–41 (Lancaster, Pa., 1932).

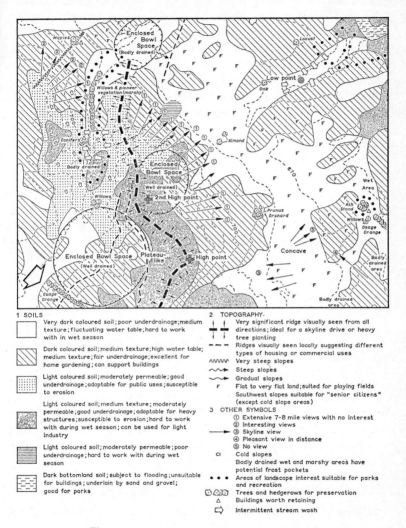

1 SOILS

	Very dark coloured soil; poor underdrainage; medium texture; fluctuating water table; hard to work with in wet season
	Dark coloured soil; medium texture; high water table; medium texture; fair underdrainage; excellent for home gardening; can support buildings
	Light coloured soil; moderately permeable; good underdrainage; adaptable for public uses; susceptible to erosion
	Light coloured soil; medium texture; moderately permeable; good underdrainage; adaptable for heavy structures; susceptible to erosion; hard to work with during wet season; can be used for light industry
	Light coloured soil; moderately permeable; poor underdrainage; hard to work with during wet season
	Dark bottomland soil; subject to flooding; unsuitable for buildings; underlain by sand and gravel; good for parks

2 TOPOGRAPHY

- Very significant ridge visually seen from all directions; ideal for a skyline drive or heavy tree planting
- Ridges visually seen locally suggesting different types of housing or commercial uses
- Very steep slopes
- Steep slopes
- Gradual slopes
- F Flat to very flat land; suited for playing fields
  Southwest slopes suitable for "senior citizens" (except cold slope areas)

3 OTHER SYMBOLS

- ① Extensive 7-8 mile views with no interest
- ② Interesting views
- ③ Skyline view
- ④ Pleasant view in distance
- ⑤ No view
- Cl Cold slopes
  Badly drained wet and marshy areas have potential frost pockets
- ••• Areas of landscape interest suitable for parks and recreation
- Trees and hedgerows for preservation
- △ Buildings worth retaining
- ⇨ Intermittent stream wash

Figure 212. SKETCH OF A NEW TOWN SITE

Based on B. Hackett, 'The Landscape Analysis of New Town Sites', *Journal of Town Planning Institute*, vol. 48, p. 39 (London, 1962). The contour interval is 10 feet.

also serves to clarify analysis of the form and function of settlements.[1]
Distinctive morphological elements, major land uses based on function,
major road patterns, and railway lines may be depicted on a selective
basis to provide a sketch-map which summarizes the principal features
of a town at a glance (Fig. 211). Basic data for sketch-maps may be
derived in the first instance from topographical maps, for example, key
contours, limits of built-up areas, main lines of communication and so
on. Further details are added from other sources, such as historical and
geological maps, and field observation, to illustrate specific points.
Griffith Taylor makes great use of the sketch-map. His use of hachures,
his selection of key phenomena, and his method of annotation might
be noted.[2]

The selection of data for sketch-maps is of course determined by the
purpose for which the map is designed. As an illustration, in the planning of a new town or an estate the natural possibilities of the site must
be exploited to the full. In this connection a sketch-map which emphasizes pertinent features and obvious natural advantages of the site is
helpful (Fig. 212).

Diagrammatic Sketch-maps. Sketch-maps in which the scale is distorted
in order to demonstrate relationships between siting and surface
features, and in which at the same time space is conserved and an
orderly classification effected, have been termed diagrammatic. The
word *cartogram* is sometimes though not exclusively used to describe
such illustrations.[3]

Generalized Sketch-maps. A useful exercise in the analysis of the forms
of towns entails the construction of generalized diagrams, which incorporate distinctive features of the plan and structure of regional types
of towns (Fig. 213). The idea of generalization has been extended
by A. E. Smailes to include the morphology of modern towns (Fig.
214).

The Ecological Model of Residential Areas in a British Town

Generalized diagrams can also be used to sum up conclusions about

---

[1] There are some excellent examples in O. H. K. Spate and E. Ahmad, 'Five Cities
of the Gangetic Plain', *Geographical Review*, vol. 40, pp. 260–78 (New York, 1950); and
R. E. Dickinson, 'Morphology of Medieval German Towns', ibid., vol. 35, pp. 74–97
(1945).

[2] T. G. Taylor, *Urban Geography* (London, 1949).

[3] A good example of its application to a classification of settlements in Sussex is to
be found in *The Land of Britain*, pts 83–4, *Sussex (East and West)* (London, 1942).

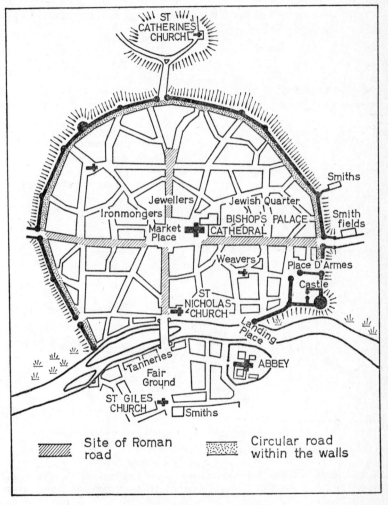

Labels within the figure:

ST CATHERINES CHURCH

Jewellers
Ironmongers
Market Place
Jewish Quarter
BISHOP'S PALACE
CATHEDRAL
Smiths
Smith fields
Weavers
Place D'Armes
Castle
ST NICHOLAS CHURCH
Landing Place
Tanneries
Fair Ground
ABBEY
ST GILES CHURCH
Smiths

Site of Roman road   Circular road within the walls

Figure 213. GENERALIZED DIAGRAM OF A TOWN IN THE PARIS BASIN
Based on a map by H. J. Fleure, *Geography*, vol. 25 (London, 1940). As this is a theoretical concept, no line-scale is attached.

the geography of residential areas and towns, for example, the classic model employed by E. W. Burgess, showing urban areas of Chicago in an article on the growth of the city which first appeared in 1925.[1] B. T. Robson, in a study of residential areas in Sunderland, sums up his conclusions with a general model of ecological areas related to the private housing in Sunderland.[2]

Sketch-blocks

Block-diagrams (see p. 175) can be very effectively employed in the portrayal of the siting of settlements, and in the analysis of geographical

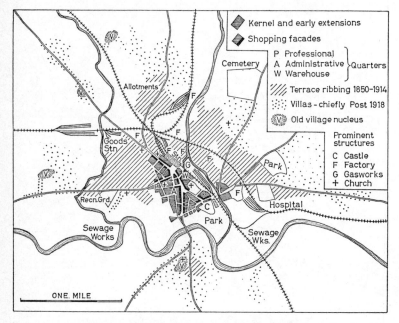

Figure 214. GENERALIZED MORPHOLOGICAL DIAGRAM OF A TYPICAL ENGLISH COUNTY TOWN

Based on A. E. Smailes, 'Some Reflections on the Geographical Description and Analysis of Towns', *Institute of British Geographers: Transactions and Papers*, no. 21 (London, 1955).

[1] Robert E. Park and E. W. Burgess, *The City*, p. 55 (Chicago, 1925).
[2] B. T. Robson, 'Ecological Analysis of the Evolution of Residential Areas in Sunderland', *Urban Studies*, vol. 3, p. 136 (Glasgow, 1966). See also ibid., *Urban Analysis* (Cambridge, 1969).

factors influencing their growth and structure.[1] The introduction of the third dimensional aspect often throws light on the operation of geographical factors which two-dimensional plans tend to conceal altogether.

Stage-Diagrams

Sketch-blocks and sketch-maps may be arranged in tiers to demonstrate the evolution of a settlement at different stages during its period of growth. These have been called *stage-diagrams* by Griffith Taylor. In the construction of such diagrams it is of great help to obtain as many historical maps of the settlement as possible and to reduce these photostatically to the same scale in order to facilitate the process of chronological analysis. The mere mechanical process of reduction and arrangement reveals the operation of geographical factors affecting the growth of settlement which are otherwise easily overlooked.

Growth Charts

The growth of the so-called 'million' cities of the world is effectively shown by A. B. Mountjoy with a diagram which shows the distribution of cities of 1–3 millions, 3–5 millions and over 5 millions by ten-year periods plotted against latitude.[2]

Triangular Graphs

D. Thorpe, in an article discussing shopping centres in Britain, has a number of useful diagrams including correlation and frequency diagrams. He makes use of a triangular graph to show organizational composition of main centre sales as between multiple, co-operative and independent stores.[3] The success of these different types of stores in different groups of towns is admirably shown by this method.

Dendrograms

The quantitative approach to regional geography includes the use of statistical methods to establish what have been called 'multi-factor uniform regions' (see pp. 510–11).[4] Multivariate analysis is frequently

---

[1] For numerous instances of their application, see T. G. Taylor, op. cit. (1949).

[2] A. B. Mountjoy, 'Million Cities: Urbanization and the Developing Countries', *Geography*, vol. 53, p. 369 (London, 1969).

[3] David Thorpe, 'The Main Shopping Centres of Great Britain in 1961: Their Locational and Structural Characteristics', *Urban Studies*, vol. 5, p. 61 (Glasgow, 1968).

[4] D. C. D. Pocock and D. Wishart, 'Methods of Deriving Multi-Factor Uniform Regions', *Institute of British Geographers: Transactions*, no. 47, pp. 73–98 (London, 1969).

made use of in handling the vast amount of regional data involved. Cluster analysis can be used to classify sets of samples. Linkage and agglomerative methods can be employed to fuse progressively popula-tions of samples into a diminishing number of groups, so that relative similarity between individual samples is indicated by the order of their fusion or linkage. The order of fusion can be represented graphically

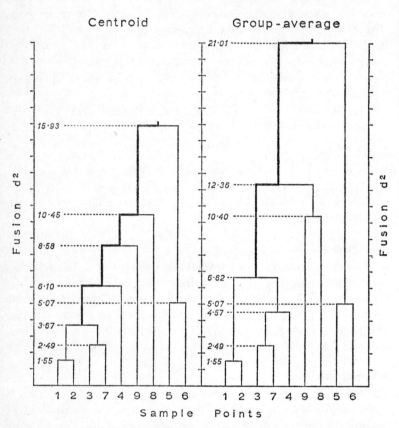

Figure 215. A DENDROGRAM OR LINKAGE TREE (after B. J. L. Berry)
Based on D. C. D. Pocock and D. Wishart, 'Methods of Deriving Multi-Factor Uni-form Regions', *Institute of British Geographers, Transactions* no. 47, pp. 73–98 (London, 1969).

In this diagram centroid and group average methods have been applied to sample data from the 1954 U.S. Census for nine census divisions. Six variables have been con-sidered, each measuring some aspect of 'Services' per 1,000 population (see also Fig. 209).

by a 'dendrogram', 'linkage tree' or dendrite, whereby the fusion of two groups at a particular stage is shown by a joint or node, connecting the two sub-branches which represent the groups.[1] Figure 215 shows a dendrogram of U.S. Census Data for centroid and group-average methods. In this case the least distance measured at each fusion can be used as an indication of the homogeneity of the resulting groups. This is shown on the dendrogram by setting a vertical distance scale and marking every node on the diagram at the distance level corresponding to the squared distances between the two points whose fusion the node represents. (See also Association Analysis Charts, p. 446.)

GRAPHS

Altitude Graphs

Before the Second World War graphs were rarely used in the geographical study of settlements but there were interesting exceptions. H. W. Ahlmann, for example, used an ingenious polygraph to relate settlement and altitude in south-eastern Sicily.[2] He plotted height in metres as ordinates, and area, number of inhabitants and density of population as abscissae. He thus produced three graphs on the same chart to portray the concentrations of settlement at certain altitudes. The method may have fruitful results in the analysis of settlements in relation to plains of erosion, terraces, northward and southward facing slopes, and so on. Over forty years ago the relation between landforms and siting of settlements was the subject of a paper by J. L. Rich, in which he used graphical methods to demonstrate his ideas.[3] He considered the percentage of sites of settlements to be found on the valley bottoms, the slopes and the interfluves respectively of rivers in the period of their youth, their maturity and their old age, and plotted the results graphically (Fig. 216).

Altitude diagrams are used both to enhance description and also to analyse ethnic stratification in a paper by H. Uhlig, who uses diagrams which give landscape profiles showing natural vegetation types,

[1] For a remarkable example of a linkage tree or dendrite of Polish cities, see B. J. L. Berry and Andrzej Wróbel, *Economic Regionalisation and Numerical Methods* (Warsaw, 1968). This is based on the use of taxonomic distances to link some 250 Polish towns on the basis of twelve variables.

[2] H. W. Ahlmann, 'The Geographical Study of Settlement', *Geographical Review*, vol. 18, pp. 93–128 (New York, 1928).

[3] J. L. Rich, 'Cultural Features and the Physiographical Cycle', *Geographical Review*, vol. 4, pp. 297–308 (New York, 1917).

hydrology, the location of settlements, land utilization and the migrations of nomads and herdsmen.[1]

Analysis of the Settlement Net. Another experiment with frequency graphs is worth mentioning as a pointer to their further possibilities.[2] To demonstrate variations in the size and character of settlements in distinctive regions of New South Wales, J. Andrews plotted for each

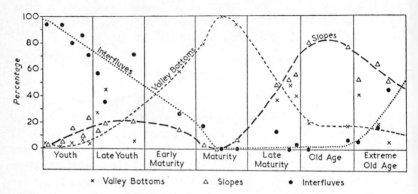

Figure 216. CURVES OF SETTLEMENT LOCATION IN RELATION TO LAND-FORMS

Based on J. L. Rich, 'Cultural Features and the Physiographic Cycle', *Geographical Review*, vol. 4, p. 300 (New York, 1917).
    The frequency with which houses are sited on valley-bottoms, slopes and interfluves has been plotted on a percentage basis, and theoretical curves have been drawn to indicate the relationships which emerge.

region (*a*) the percentage of the total number of agglomerations as ordinates against the population of each agglomeration as abscissae; and (*b*) the percentage of total population as ordinates against the population of each agglomeration as abscissae. He thus obtained two supplementary sets of polygraphs, which revealed distinctive regional variations in the characteristics of the settlements analysed.

Time–Space Diagrams

D. J. Janelle makes use of a logarithmic scale to plot the time taken to travel between any two places in minutes, against the time-scale in

[1] H. Uhlig, 'Hill Tribes and Rice Farmers in the Himalayas and South East Asia: problems of the social and ecological differentiation of agricultural landscape types', *Institute of British Geographers: Transactions*, no. 47, pp. 1–23 (London, 1969).
    [2] J. Andrews, 'The Settlement Net and the Regional Factor', *Australian Geographer*, vol. 3 (Sydney, 1934).

years for stage coach, rail, road and air travel. For example, the scale runs from 20,000 minutes in the case of travel from London to Edinburgh in the early stage coach period to less than 200 minutes by air at the present time.[1]

Three-dimensional Graphs and Diagrams

Reference has already been made to the use of three-dimensional diagrams in connection with population mapping (see pp. 373–4). Three-

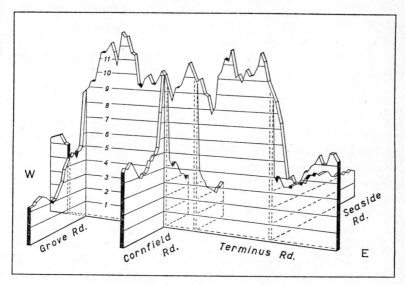

Figure 217. A THREE-DIMENSIONAL DIAGRAM TO SHOW RATEABLE VALUES

Based on an unpublished dissertation by D. W. Hayes.

The diagram shows distribution of rateable value of properties in the central area of Eastbourne. The scale is shown in £'s per square yard., G.R.V.

dimensional graphs and diagrams can also be used in the study of settlements to show specific distributions and to provide surfaces in predictive models; for example, B. J. L. Berry and others made use of generalized land-value surfaces within a city in their analysis of urban commercial structure.[2] Land values reach a grand peak in the centre

[1] Donald G. Janelle, 'Central place development in a time–space framework', *Professional Geographer*, vol. 20, pp. 5–10 (Lawrence, Kansas, 1968).

[2] B. J. L. Berry, R. J. Tennant, B. J. Garner and J. W. Simmons, *Commercial Structure and Commercial Blight*, University of Chicago, Department of Geography, Research Paper no. 85 (Chicago, 1963).

of the city and decrease by varying amounts towards the periphery, but the surface is far from smooth; there are minor peaks and ridges associated with intersections and main routes. Rateable values can be used in an analysis of commercial structure in British towns. This can be done, either by providing generalized surfaces or by producing three-dimensional models with the vertical scale functioning as an index of rateable value. Three-dimensional graphs can also be used (Fig. 217). In this case the main street in Eastbourne and offshoot streets are represented diagrammatically. Values for only one side of the street in running means of 5 have been used in Fig. 217. The high rateable values in the central shopping street reach over £10 per square yard gross rateable value. From Terminus Road values fall away, rising again towards the sea-front.

Statistical Approach to Town Studies

In recent years urban geography has been characterized by an increased emphasis on the statistical approach to the study of the function, size, distribution and hierarchy of towns. The data are largely derived from elaborate field surveys supplemented by census returns. The aim has been to establish the nature of the relationship between population size, functional diversity and spacing of towns. Such analyses have called for more precise definitions of terms, particularly of 'function', 'functional unit', 'establishment', 'central place', etc. The preliminary procedure consists of the definition, recognition and ranking of service functions. Subsequently the number of service functions and units can be plotted against population sizes by means of simple line graphs. H. A. Stafford, for example, made effective use of simple graphs to examine the functional basis of small towns in part of Illinois.[1] He plotted numbers of 'establishments' against town population sizes, numbers of 'functional units' against town population size, and, to obtain some indication of retail specialization with increase in town size, the ratio of number of functional units/number of establishments against number of establishments.

Regression Diagrams

Reference has already been made to this form of graphical analysis in the last chapter (see pp. 386–7). It has assumed particular importance in settlement studies in recent years. G. K. Zipf, whose 'Rank-

[1] Howard A. Stafford, Jr., 'The Functional Basis of Small Towns', *Economic Geography*, vol. 39, pp. 165–75 (Worcester, Mass., 1963).

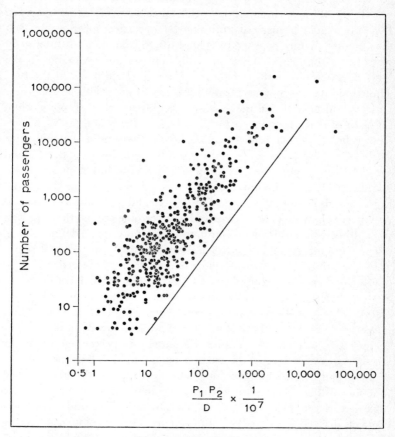

Figure 218. REGRESSION DIAGRAM OF BUS PASSENGER JOURNEYS IN THE
UNITED STATES

Based on G. K. Zipf, *Human Behavior and the Principle of Least Effort* (Cambridge,
Mass., 1949).

The data refers to main road bus passengers between 29 arbitrary cities in the
United States in 1933 and 1934. P_1 is the population of the first, P_2 the population of
the second city in each case, and D is the distance between. It can be seen that the
number of bus passengers declines in a linear fashion as the distance between any pair
of cities increases. This is shown by the line, which has a gradient of 1 : 0·71.

size' rule for cities stimulated interest in the statistical analyses of the size of cities (see p. 419), himself associated size of settlements with economies of distance and the size of markets. He therefore tried to establish rectilinear interactions of distance to explain his concepts. With the use of log linear diagrams he demonstrated, for example, that the number of bus passengers (Fig. 218), the amount of freight by weight, the number of telephone calls between any two cities in the United States (Fig. 218) declines with the distance between the two. This technique has obvious possibilities for testing possible relationships between distance from markets, market areas, size of population, numbers of central services and so on. B. J. L. Berry and H. M. Mayer have in fact made extensive use of log linear diagrams in a comparative study of central place systems in south-western Iowa in the United States.[1] In this study it proved possible not only to link population size quantitatively with the number of 'functional units' and 'central functions', but also to establish sizes of trade area and populations associated with particular ranges of services and so determine critical thresholds in population size. In this way the authors found it possible to define village, town, city and 'regional metropoles' quantitatively in terms of population, location in a system, services offered and area served.

For another good example of the use of regression lines in urban analysis, reference is made to the work of E. L. Ullman and M. F. Dacey. They use the 'minimum requirements method' to classify and measure the economic base of a selected group of towns in the United States.[2] They attempt to establish the minimum percentage of labour required in various sectors of urban economy to maintain the viability of an urban area. This percentage they identify with the servicing function of the city. Any excess on the minimum figure can then be regarded as the 'basic' or 'export' employment. The minimum percentages derived from analysis of the census data are plotted as regression lines using semi-logarithmic scales (Fig. 219). Regression lines can be produced for individual, industrial and service categories. These regression lines can then serve as a model for comparison with specific town situations.

[1] B. J. L. Berry and H. M. Mayer, 'Retail Location and Consumer Behaviour', op. cit. (February, 1962). See also B. J. L. Berry, *Geography of Market Centers and Retail Distribution* (New Jersey, 1967) for many examples related to the analyses of systems of central places.
[2] E. L. Ullman and M. F. Dacey, 'The Minimum Requirements Approach to the Urban and Economic Base', *I.G.U. Symposium in Urban Geography*, pp. 121–43 (Lund, 1960).

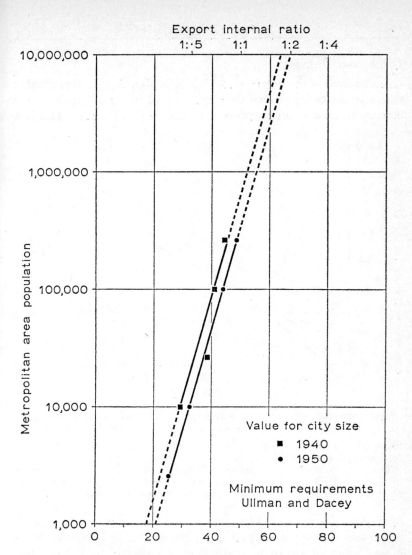

Figure 219. AN EXAMPLE OF THE USE OF REGRESSION LINES IN THE
ANALYSIS OF URBAN EMPLOYMENT

Based on E. L. Ullman and M. F. Dacey, 'The Minimum Requirements Approach to
the Urban and Economic Base', *I.G.U. Symposium in Urban Geography*, pp. 121–43
(Lund, 1960).

  The *x*-axis indicates the sum of minimum employment of forty industries as a
percentage of the total employment. The slope of the curve is roughly the same for
1940–50, but there is a slight shift in the minimum requirements. The export internal
ratio is also shown and it can be seen that this varies with the size of town.

Log Normal Distribution

In an analysis of the distribution of cities by size, B. J. L. Berry makes extensive use of log normal distributions to show variations from one country to another in distribution of city size[1] (Fig. 220). This is a

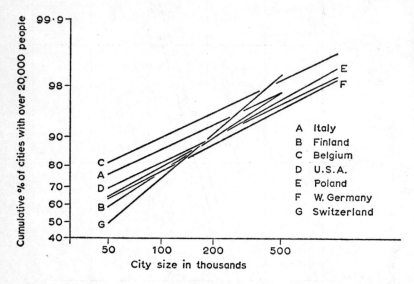

Figure 220. THE USE OF LOG NORMAL CURVES TO SHOW CITY-SIZE DISTRIBUTION

Based on B. J. L. Berry, 'City Size Distribution and Economic Development', *Economic Development and Cultural Change*, vol. 9, pp. 573–88 (Chicago, 1961).

    The countries shown all have log normally distributed city sizes (i.e. the curve is a straight line). The higher the curve on the graph, the greater is the percentage of small cities; the steeper the slope of the curve, the smaller is the largest city.

form of best-fitting curve. The data are plotted in the form of cumulative frequencies on log normal probability paper so that, if city-size distribution is log normal, it takes the form of a straight line. Berry then goes on to produce a developmental model for city-size distributions based on a comparison of log normal curves. The attainment of log normal distribution as distinct from departures is seen as a condition of entropy, defined as circumstances in which forces affecting the distribution are many and act randomly.

    [1] B. J. L. Berry, 'City Size Distributions and Economic Development', *Economic Development and Cultural Change*, vol. 9, pp. 573–88 (Chicago, 1961).

Scatter Diagrams

In the classification of towns by component analysis, such as that done by C. A. Moser and Wolf Scott (op. cit., 1961, p. 257), towns can be grouped according to their similarities on the basis of their score for different components. A visual appreciation of this situation can be achieved by plotting the scores, using two of the components as ordinates and abscissae respectively. The resultant scatter of values gives a measure on a proximity basis of the affinities of the towns plotted.[1]

THE USE OF COMPUTER GRAPHICS

Computer mapping has already been referred to (see p. 81) and the technique has been extended to include a variety of maps dealing with housing and settlement data. Two examples are considered here. The first is a map produced by the Ministry of Housing and Local Government, which utilizes the computer mapping system LINMAP (LINe printer MAPping) to illustrate the use and facility of a co-ordinating reference system for locating statistical information in map form.[2] LINMAP makes use of the high-speed printers associated with modern computers. It can print a variety of characters on any position along the line and can also overprint. The line printer used for LINMAP can print up to 64 lines and up to 136 characters per line on any one map. The system therefore can be adapted to show the location of characters spatially and even to indicate densities by overprinting. There are, however, a number of problems associated with this method of mapping. LINMAP was designed to operate on a geographical data bank, combined with a versatile data processing system, and therefore depends on a rather elaborate organisation. The work of producing

---

[1] For a useful review and diagram, see P. B. Brenikov, '157 Varieties of Towns', *Journal of the Town Planning Institute*, vol. 48, pp. 242–6 (London, 1962). See also B. J. L. Berry, op cit., for a number of scatter diagrams relating area and population served in respect of villages, towns, cities and 'regional capitals' in the United States. B. J. L. Berry and P. Reece, 'The Factorial Ecology of Calcutta', *American Journal of Sociology*, vol. 74, pp. 445–91 (Chicago, 1969), and E. Gittus, 'An Experiment in the Definition of Urban Sub-areas', *Transactions of the Bartlett Society*, vol. 2, pp. 109–35 (Liverpool, 1965), deal with component analysis in the delimitation of urban sub-areas and in particular problems of combining the principal components to optimize regional variations. D. C. D. Pocock and D. Wishart, 'Methods of Deriving Multi-Factor Uniform Regions', *Institute of British Geographers: Transactions*, no. 47, pp. 73–98 (London, 1969), also consider various methods, including 'dense space'.

[2] G. M. Gaits, 'Thematic Mapping by Computer', *Cartographic Journal*, vol. 6, no. 1, pp. 50–68 (London, 1969).

maps includes: (1) the preparation of base maps involving the identification of the correct location of enumeration units, rural parishes,
wards, etc.; (2) digitising; and (3) computer processing, i.e. attributing
values for various items of data to particular regional units.

A variety of maps can be produced by LINMAP; for example, class
intervals of values can be shown by using overprinting to produce a
variety of graphic densities, or a number of a class in which a given
value has fallen can be printed quite simply, those parts of the maps
without values being represented by a dot or a blank space. Either
of the above maps can, of course, be used as data maps for the
construction of finished maps using conventional cartographic
methods.

LINMAP 2 can be used to produce printed thematic maps in colour
(COLMAP), a complicated process which depends on the use of photo-
electric computer type-setting equipment. An interesting colour map
showing the distribution of owner-occupied households as a percentage
of total households has been produced for south–east England by the
Ministry of Housing and Local Government on a scale of 1 : 1,350,000.
This map is much easier to read than the black-and-white line printed
diagrams and approximates much more closely to accepted canons
of good taste in cartographic production.

The second example is based on SYMAP (see p. 82) and shows
residential preferences of Sixth-form students in Hull in the form of a
'surface'. Data in this case are drawn from seven schools, the summed
preferences being expressed as percentages, integrated as isopleths,
and the intervals shaded. Computer-drawn maps of this type can of
course be re-drawn for publication[1] (Fig. 221).

OTHER ASPECTS

Ordering of Centrality Values

In discussion of centrality, W. K. D. Davies employs a number of
interesting charts to illustrate the order of central places within a
hierarchy.[2] In particular, attention is drawn to the use of frequency
distributions of centrality values to show a five-fold order of centres.

[1] See, for example maps of 'Social Class' in Goole in J. D. Porteous, *The Company Town of Goole: An Essay in Urban Genesis*, Occasional Papers in Geography, no. 12 (University of Hull, 1969).
[2] W. K. D. Davies, 'Centrality and the Central Place Hierarchy', *Urban Studies*, vol. 4, pp. 61–79 (Glasgow, 1967).

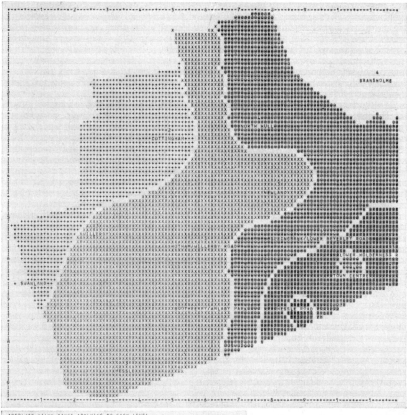

Figure 221. COMPUTER-
DRAWN MAP

This map is based on SYMAP V as modified by P. Adman, Centre for Computer Studies, University of Hull. It shows residential preference of sixth-formers in Haltem-price and West Hull, drawn from a sample of 25 sixth-formers in each of 7 Hull schools. Each sixth-former was asked to rank each of 16 residential areas in order of residential desirability. The values represent the percentage of the total possible preferences (i.e. total possible = 25 × 7 × 16 = 2800 and the summed preferences of each area are expressed as a percentage of this). The lower the percentage the higher the preference.

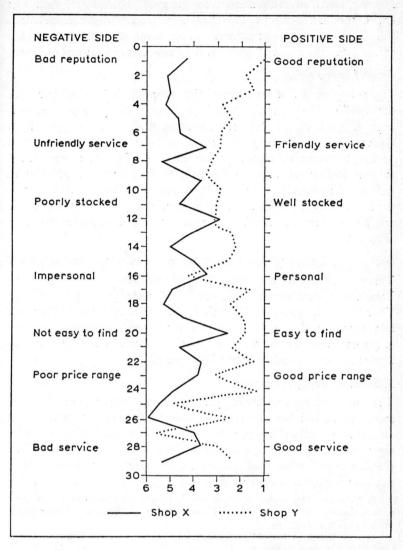

Figure 222. A PROFILE DIAGRAM OF SHOPPING IMAGE

Based on B. J. Garner, 'The Analysis of Qualitative Data in Urban Geography; an example of Shop Quality', *I.B.G. Study Group in Urban Geography, Techniques in Urban Geography*, pp. 16–28 (London, 1968).

For this chart, establishments were grouped according to their first appearance in a hierarchy. This chart helps to establish general relationships between centres of different order, some of the high centrality values being the result of appearance of new functions at certain levels.

Group Image Profiles

The public images of shops may affect shopping habits and influence decisions on where people shop. The study of shopping preference is important within the field of perception studies (see p. 440). B. J. Garner in his study of group images of sample shops makes use of profiles based on a six-point preference scale.[1] Customers were asked to score individual shops on the basis of their service, ranging from unfriendly to friendly, on the basis of their stock ranging from poorly stocked to well stocked, on the basis of price range from poor to good, etc. These scores can then be generalized by means of profile diagrams (Fig. 222).

Grouping of Ranges of Goods to Demonstrate a Hierarchy of Shopping

This method based on B. J. L. Berry's[2] work has been used by J. R. Tarrant[3] to demonstrate a hierarchy of shopping range in eastern Yorkshire. This allows individuals based on samples from villages to be grouped together, with a measure of the loss of accuracy or the degree of generalization involved in the groupings. Fig. 223 shows the groupings of ranges of goods for the village sample.

A Summary Preference Scale

The study of environmental perception has received emphasis in recent years and has many applications in the field of environment and regional analysis (see p. 440). P. R. Gould in considering the use of eigenvalues in geography shows how these values can be used in the plotting of preference scales indicating the desirability of residence in different European countries (Fig. 224).[4]

[1] B. J. Garner, 'The Analysis of Qualitative Data in Urban Geography; an example of Shop Quality', *I.B.G. Study Group in Urban Geography, Techniques in Urban Geography*, pp. 16–28 (London, 1968).

[2] B. J. L. Berry, 'A Method of Defining Multi-factor Uniform Regions', *Przegla Geograficzny* (Warsaw, 1961).

[3] J. R. Tarrant, *Retail Distribution in Eastern Yorkshire in Relation to Central Place Theory: a Methodological Study*, Occasional Papers in Geography, no. 8 (University of Hull, 1967).

[4] P. R. Gould, 'The Geographical Interpretation of Eigenvalues', *Institute of British Geographers, Transactions*, no. 42, p. 81 (London, 1967). See discussion in Appendix pp. 513–15.

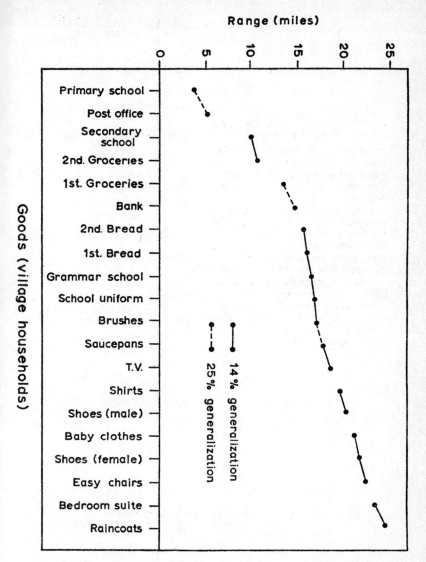

Figure 223. A HIERARCHY OF SHOPPING RANGE

Based on J. R. Tarrant, *Retail Distribution in Eastern Yorkshire in Relation to Central Place Theory: A Methodological Study*, Occasional Papers in Geography, no. 8 (University of Hull, 1967).

The scheme relates to village samples. Goods can be arranged in a hierarchy depending on the maximum probable distance travelled for their purchase.

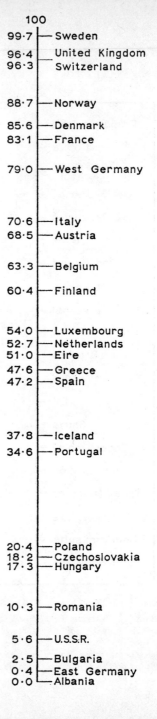

Figure 224. A SUMMARY PRE-
FERENCE SCALE

Based on P. R. Gould, 'The Geographical Interpretations of Eigenvalues', *Institute of British Geographers:
Transactions*, no. 42, p. 81 (London,
1967).

This is a scale measuring residential
desirability in Europe, based on
opinions of Swedish, British, French,
West German and Italian students. In
this case the principal eigenvectors are
extracted from each of the matrices
summarizing preferences of a national
group. The cosines of the angles between each of these make up in turn a
final or higher order matrix whose
principal eigenvector constitutes the
overall summary scale.

100
99·7 ── Sweden
96·4 ── United Kingdom
96·3 ── Switzerland

88·7 ── Norway

85·6 ── Denmark
83·1 ── France

79·0 ── West Germany

70·6 ── Italy
68·5 ── Austria

63·3 ── Belgium

60·4 ── Finland

54·0 ── Luxembourg
52·7 ── Netherlands
51·0 ── Eire
47·6 ── Greece
47·2 ── Spain

37·8 ── Iceland
34·6 ── Portugal

20·4 ── Poland
18·2 ── Czechoslovakia
17·3 ── Hungary

10·3 ── Romania

5·6 ── U.S.S.R.

2·5 ── Bulgaria
0·4 ── East Germany
0·0 ── Albania

Spatial Association Diagrams

These diagrams can be used to establish and demonstrate the degree of association between, for example, retail stores in central city areas or industrial linkages in areas of industrial concentration. A. and J. M. Getis[1] attempted to establish spatial affinities between types of stores in the Philadelphia central business district, using the technique of sequence analysis. The results are set out in the form of a diagram emphasizing significant associations.

[1] A. and J. M. Getis, 'Retail Store Spatial Affinities', *Urban Studies*, vol. 5, p. 320 (Edinburgh, 1968).

An Introduction to Numerical and Mechanical Techniques[1]

R. G. BARRY

In many investigations the geographer is faced either with the problem of handling an almost embarrassing wealth of statistical data, or with the difficulty of deciding how to make selective and yet representative observations. Not infrequently these problems are complementary, for it may be practicable to use only a small proportion of the available statistics. The approach sometimes adopted in geographical work can be termed the 'case-study' method, where a few detailed type-studies are made of the problem in hand. This method has the disadvantage that reliability of generalization is uncertain, since the student has no precise knowledge of the representativeness of his material for the general area within which the case studies are carried out. On the other hand, selective studies which do permit certain generalizations may be undertaken if a statistical sampling basis is adopted for the work. In fact, this is in many cases preferable to a full survey of a par-

[1] An enormous literature is developing on these techniques, reference to much of which is made in footnotes throughout this book. Particular attention is drawn to S. Gregory, *Statistical Methods and the Geographer* (London, first published 1963); and the comprehensive J. P. Cole and C. A. M. King, *Quantitative Geography* (London, 1968); both contain very detailed bibliographies. See also J. P. Cole, 'Mathematics and Geography', *Geography*, vol. 54, pp. 152–64 (London, 1968). A useful compendium is P. W. Porter, *A Bibliography of Statistical Mapping* (Minneapolis, 1964). Special topics are dealt with by W. L. Garrison and D. F. Marble, *Quantitative Geography*. Part I. *Economic and Cultural Topics;* Part II. *Physical and Cartographic Topics* (Evanston, 1967); M. H. Yeates, *An Introduction to Quantitative Analysis in Economic Geography* (New York, 1968); B. J. L. Berry and D. F. Marble, *Spatial Analysis; A Reader in Statistical Geography* (Englewood Cliffs, N.J., 1968); and L. J. King, *Statistical Analysis in Geography* (Englewood Cliffs, N.J., 1969).

ticular problem, for reasons of available time, manpower and finance.

Here we are concerned with some basic aspects of statistical inference as they apply to geography. Principally, these are the use of samples to make inferences about the population from which they were taken or to test hypotheses. A *population* is any set of phenomena defined in terms of certain unique characteristics. It may be finite (e.g., the number of cars manufactured in a given year) or infinite (e.g., the number of waves on the oceans of the world). For practical purposes many populations are assumed to be infinite. A *sample* is a subset of the population selected (in ways described below) so as to be representive of it. Special problems of interpretation occur when all of the available data have to be used because they are very few in number and replication is not possible.[1]

The methods discussed below relate primarily to so-called *parametric* statistics which involve certain assumptions about the parameters of the population, particularly that of a normal distribution. Moreover their application is limited to measurement on an interval or ratio scale. Interval measurements such as degrees Celsius, height above sea level and calendar dates, where the intervals are equal, have an arbitrary zero whereas ratio measurements such as degrees Kelvin, length and time intervals have equality of ratios.[2] *Non-parametric* statistics (where the assumptions about the population are minimal) are, however, appropriate or indeed essential for many types of geographical data since they can be applied to ordinal data (ranked with arbitrary intervals) and even nominal data where there is an arbitrary assignment of classes. Non-parametric methods have been widely used in the social sciences and their value is now more generally appreciated in geography.[3]

[1] See, for example, D. R. Meyer, 'Geographical population data: statistical description not statistical inference', *Professional Geographer*, vol. 24, pp. 26–28 (Washington, D.C., 1972).

[2] Scales of measurement are discussed by H. M. Nelson, 'Measuring Systems: Conception and Design', in E. F. Bradley and O. T. Denmead (eds.), *The Collection and Processing of Field Data*, pp. 311–327 (New York, 1967).

[3] S. Siegel, *Non-parametric Statistics: for the Behavioural Sciences*, 312 pp. (New York, 1956); E. S. Keeping, 'Distribution-free Methods in Statistics', in *Statistical Methods in Hydrology*, pp. 211–247. Proceedings, Hydrology Symposium No. 5, Department of Energy, Mines and Resources (Ottawa, 1967); J. V. Bradley, *Distribution-free Statistical Tests*, 388 pp. (New York 1968); C. H. Kraft and C. Van Eeden, *A Non-Parametric Introduction to Statistics*, 342 pp. (New York, 1968); H. M. French, 'Quantitative Methods and Non-parametric Statistics', in *Quantitative and Qualitative Geography*, pp. 119–128 (Ottawa, 1971).

SAMPLING

There are essentially three ways in which selective or sample surveys may be carried out[1]. The most straightforward technique is the simple *random sample*. The objects of study, which for example may be fields, farms or villages, are listed and each is assigned a number. A random sample of these is then obtained by reference to a table of random numbers.[2] Listing of items is possible when the objects of study form a finite group of discrete units, but the geographer frequently wishes to obtain a sample of a spatially continuous variable, such as height of land, slope angle or soil characteristics. In this case a grid is superimposed on a map of the particular phenomenon and a random series of grid intersections is selected from the table of random numbers. Where the items are discrete, but where for all practical purposes the group is infinite (for example, pebbles in a till or plants in a large community), some form of area sampling must be used.[3]

The random sample method may be inappropriate for the investigation of spatial variations of some phenomena, since it can lead to a biased sample if there is marked spatial clustering of particular categories of the features which are being studied. Unfortunately, landscape elements are rarely scattered at random and consequently caution must be exercised in the use of this method.

The difficulty may be overcome if a *stratified random sample* is taken. The objects to be surveyed are grouped into classes (*strata*), on the basis of one or more characteristics relevant to the problem.[4] For a study of occupation structure, the settlements might be grouped initially according to the size of their population from data in census returns, and random samples are then taken within those groups. Alternatively, the groups may be based on unit areas. Clearly the

---

[1] For a discussion with reference to agricultural geography, see J. M. Blaut, 'Microgeographic Sampling – a Quantitative Approach to Regional Agricultural Geography', *Economic Geography*, vol. 35, pp. 79–88 (Worcester, Mass., 1959).

[2] D. V. Lindley and J. C. P. Miller, *Cambridge Elementary Statistical Tables*, 35 pp. (Cambridge, 1953), Table 8. For an example, see G. W. Hartman and J. C. Hook, 'Substandard Urban Housing in United States: a Quantitative Analysis', *Economic Geography*, vol. 32, pp. 95–114 (Worcester, Mass., 1956).

[3] The botanist employs the method of 'quadrat' study (sample square plots).

[4] W. F. Wood, 'Use of Stratified Random Samples in a Land Use Study', *Annals of the Association of American Geographers*, vol. 45, pp. 350–67 (Lancaster, Pa., 1955).

method of stratified samples requires considerable prior information about the objects of study.[1]

The third sample method, by contrast, necessitates no previous information and is therefore particularly suitable for reconnaissance surveys. The region to be examined is divided into equal units on an areal basis, and a *systematic sample* is taken at regular intervals. Birch[2] uses this method for a 24 per cent sample of farms over 25 acres in the Isle of Man. The sample is spaced at regular intervals, as far as possible, over the farming landscape by using the National Grid. A farm is selected if the grid intersection occurs on its land, ensuring that large holdings are represented by a large proportion of the sample. Alter-

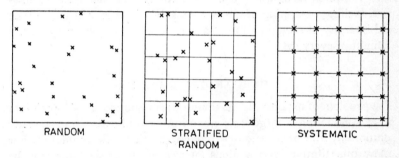

RANDOM   STRATIFIED   SYSTEMATIC
                RANDOM

Figure 225. DISTRIBUTION OF POINTS OBTAINED BY TAKING RANDOM, STRATIFIED RANDOM AND SYSTEMATIC SAMPLES ON AN AREAL BASIS

natively, a systematic sample may be taken by sampling at regular intervals from a list. Birch makes use of this method also for a comparative sample of farms from agricultural returns. The method is less satisfactory than systematic sampling from a map, since the areal distribution of the samples is not controlled to the same extent.

Examples of the distribution of 25 points obtained by taking random, stratified random and systematic samples on an areal basis are shown in Fig. 225. The first point of the systematic grid is obtained from random numbers.

[1] A discussion of the applicability of several methods of stratification for agricultural returns is given by W. B. Taylor and D. V. P. Clement, 'The New Zealand Agricultural Sample Survey', *Journal of the Royal Statistical Society*, A, vol. 119, pp. 409–24 (London, 1956).

[2] J. W. Birch, 'Observations on the Delimitation of Farming-type Regions, with special reference to the Isle of Man', *Transactions and Papers, 1954: Institute of British Geographers*, no. 20, pp. 141–58 (London, 1954).

The stratified random sample generally increases the accuracy of sample surveys, compared with the simple random sample.[1] The former may, in many cases, be further improved by the use of a varying sample fraction for different strata. The sample size can be made proportional to the known or suspected variability within the strata or to the areal extent if spatial phenomena are involved. These methods sometimes give even better results with stratified systematic samples than with stratified random samples.[2]

For particular problems, the requisite sample size can be calculated and an estimate made of the margin of error, at least for random samples. The procedure for determining these is more complex for the other sampling techniques. The systematic sample method does however, facilitate mapping of the sample data.[3] Decisions as to the type and intensity of sampling – the 'sample design' – require very careful thought in every case.[4]

MECHANICAL METHODS

The problem of making the maximum use of abundant statistical data or of conducting a large-scale quantitative study may find a ready solution by the application of modern mechanical techniques. For many quantitative investigations an electric desk-calculator may be sufficient. There is a great range of models, but they are essentially similar in operation and are now widely available.[5]

The simplest machine in the range is the adding machine, which prints on paper the figures added or subtracted, and will give sub-totals as well as final totals. The desk-calculators perform multiplication, division and cumulative multiplication,[6] in addition to the simpler

---

[1] See, for example, P. F. Bourdeau, 'A Test of Random versus Systematic Ecological Sampling', *Ecology*, vol. 34, pp. 499–512 (Durham, N.C., 1953).

[2] These problems are discussed fully by W. G. Cochran, *Sampling Techniques* (New York, 1953).

[3] For example, J. W. Birch, op. cit. (1954).

[4] F. Yates, *Sampling Methods for Censuses and Surveys* (London, 1949). Geomorphological examples are discussed by A. N. Strahler, 'Statistical Analysis in Geomorphic Research', *Journal of Geology*, vol. 62, pp. 1–25 (Chicago, 1954); and W. C. Krumbein, 'The "Geological Population" as a Framework for Analysing Numerical Data in Geology', *Liverpool and Manchester Geological Journal*, vol. 2, pp. 341–68 (Liverpool, 1960).

[5] M. O. Harley, 'The Application of some Commercial Calculating Machines to certain Statistical Calculations , *Supplement to Journal of the Royal Statistical Society*, vol. 8, pp. 154–73 (London, 1946).

[6] Operations of the type $(a \times p) + (b \times q) + (c \times r)$.

operations. The facilities which are provided depend upon the individual model. Some machines carry only the current stage of the calculations on visual registers, whereas other models also print the separate stages and the final answer on paper, thereby providing a useful record for checking purposes.

The use of punched cards allows greater scope for full utilization of the data. The standard Hollerith card has eighty columns and twelve rows[1] (Fig. 226). An individual item of data may be punched in each

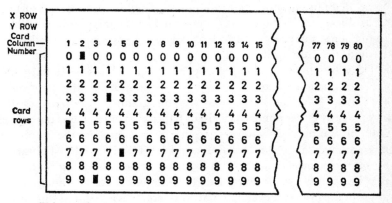

Figure 226. FORMAT OF THE HOLLERITH PUNCHED-CARD

column with an electrical Card Punch machine, or with a manual Key Punch, both of which have keys similar to a typewriter. Holes are punched to represent each digit. In Fig. 226 the number 50937 is punched in columns one to five. Usually one or two columns are sufficient to contain items of data for a specified category. Negative numbers are indicated by a punch in the X row, either above the number (an overpunch), or in the column preceding the number. There is also a combination of an overpunch in the Y, X and O rows with a number in the 1–9 rows to designate letters.[2]

The principal advantage of having data on cards is that the material may be sorted into a variety of classifications with an electrically operated Card Sorter. This machine detects the position (row) of the

[1] There are also 38-column Hollerith cards and 65 column Powers-Samas cards. The latter were used in the 1951 Census of Population in Great Britain.

[2] R. G. Barry, 'The Punched Card and its Application in Geographical Research', *Erdkunde, Band* 15, pp. 140–2 (Bonn, 1961).

hole punched in a column which is specified for each sorting by the operator and directs the card into the appropriate pocket for that number. There is a pocket for each of the twelve rows and a 'reject' pocket for columns with no punch. Negative and alphabetic over-punches require double sorting on each column. Rapid sorting may be especially useful to organize data into categories for plotting on maps. Furthermore. the joint occurrences or association of two or more factors are quickly calculable, since sorting speeds are approximately 500 cards per minute and the cards are counted automatically. Results may then be printed out on paper and totals determined by putting the sorted cards through a Tabulator. Consequently, inter-relationships within large bodies of data are readily determined if the appropriate statistical methods are employed.

The ultimate tool, not only for speeding up calculations but also for the introduction of new techniques, is undoubtedly the electronic computer.[1] Data and instructions (the 'programme') are put into the computer by either punched cards, or paper tape or magnetic tape. The modern computer invariably accepts programmes in the standard international 'languages' ALGOL and FORTRAN[2], and vast libraries of statistical routines are now in existence at all computing centres so that it is only necessary to provide the data under consideration in the appropriate format. For simple statistical analyses a desk-top computer, which is usually programmed with a magnetic card, will often be sufficient. The use of computers to compile data and, by means of a line plotter or similar device, to draw isopleth maps is now a common research aid even though a final version of the map may need to be redrawn for publication[3]. The potentialities of these techniques have given rise to the establishment of research groups concerned solely with the development of computer graphics (for further discussion see pp. 81–2 and 465–6).

---

[1] A valuable summary, with examples of application and bibliography, is T. Hägerstrand, 'The Computer and the Geographer', *Transactions of the Institute of British Geographers*, no. 42, pp. 1–19 (London, 1967). An interesting summary of what might be termed the 'computer-barrier' in the field of geography is given by W. G. V. Balchin and A. F. Coleman, *The Cartographer*, vol. 4, pp. 120–7 (Toronto, 1967).

[2] Numerous publications describing these languages exist, some dealing with their application to particular models of computer.

[3] For a climatological example, see L. D. Williams, R. G. Barry and J. T. Andrews, 'Application of Computed Global Radiation for Areas of High Relief', *Journal of Applied Meteorology*, vol. 11, pp. 526–533 (Boston, Massachusetts, 1972).

THE STATISTICAL TREATMENT OF DATA

Isaiah Bowman[1] pointed out many years ago that 'Geographical thought involves measurement', and statistical analysis of quantitative data greatly assists the interpretation of such information. Earlier chapters in this book outline certain statistical formulae and methods for specific types of data. For convenient reference, a review is here presented of the various means of compiling and treating geographical information, both prior to its cartographic expression and as an adjunct to it.

Frequency Distributions

The number of occurrences of different values of temperature, summit-level heights or population density may be usefully shown by a frequency distribution graph. The horizontal axis (abscissa) shows the range in size of the variable and the vertical axis (ordinate) shows either the actual or the percentage frequency for each value of the variable. The abscissa may refer to a discrete variable, the population of towns, the number of rain-days in each month, for example, or to a continuous variable such as area of land or temperature. It is important to ensure that the data constitute a true statistical population.

It is usual to group the values of the variable into classes. The number of classes one selects is dependent on the number of available observations; the number of classes should not exceed five times the logarithm of the number of observations. Thus the maximum number of classes for an hundred observations is ten and for five hundred it is thirteen. It is generally preferable to make the classes of equal size, although this may not always be possible. In fact, for many types of information the sequence of class intervals is geometric (i.e. 1, 2, 4, 8, 16 ...), as a result of the nature of the basic data. Theoretical values for the number of class intervals and their relative size are discussed for arithmetic (i.e. 2, 4, 6, 8, 10 ...) and geometric sequences by Mackay,[2] with reference to class intervals of atlas maps. A procedure which may be of value for a geometric sequence of class intervals in the

---

[1] I. Bowman, *'Geography in Relation to the Social Sciences'*, Part 5, Report of the Commission on the Social Studies to the American Historical Association, p. 1 (New York, 1934).

[2] J. R. Mackay, 'An Analysis of Isopleth and Choropleth Class Intervals', *Economic Geography*, vol. 31, pp. 71–81 (Worcester, Mass., 1955).

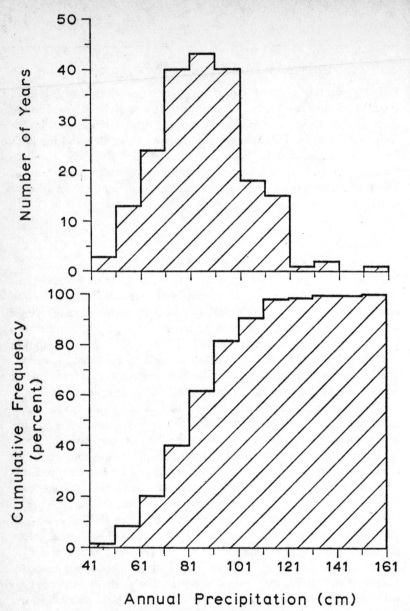

Figure 227. (above) HISTOGRAM OF ANNUAL PRECIPITATION TOTALS AT
PADUA, ITALY, 1725–1924.

Figure 228. (below) A CUMULATIVE PERCENTAGE FREQUENCY GRAPH
DERIVED FROM FIGURE 227.

Based on data in V. Conrad and L. W. Pollak, *Methods in Climatology*
(Harvard, 1950).

basic data is the conversion of the values of the variable into logarithms.[1]

A frequency distribution for totals of annual precipitation at Padua, Italy, for the years 1725–1924 is illustrated by the histogram in Fig. 227. Alternatively, the points can be joined by straight line segments to form a frequency polygon. If the data are in the form of discrete units, number of farms or occurrences of frost, for example, or if the classes are large, it is advisable only to draw columns. On the other hand, when the number of observations is large and the class intervals are small the polygon tends to form a smooth curve. This approximates to the *frequency curve* for the statistical *population* – the complete data series which may be finite or infinite.

Cumulative Frequency

The frequency values may be added to one another successively and then converted into percentages, if they are not initially in this form, to construct a cumulative percentage frequency graph. The S-shaped curve is termed an *ogive*. An example using the data of Fig. 227 is presented in Fig. 228. P. R. Crowe[2] demonstrates the usefulness of this type of analysis in a study of the strength of the monsoon in the Indian Ocean (see also p. 483).

Averages

The most commonly used measure of average or 'central tendency' is the *arithmetic mean*. It is the quotient obtained by totalling the individual values and dividing by the number of occurrences concerned. That is:

$$\text{Arithmetic mean, } \bar{x} = \frac{x_1 + x_2 + x_3 + \dots x_n}{n}$$

$$\text{or } \frac{\sum_{i=1}^{n} x_i}{n}$$

The mean is readily calculated if the numbers are not too large, otherwise it may be convenient to subtract a constant figure from each

---

[1] For an example with cumulative frequencies of particle size, see R. W. Waters and R. H. Johnson, 'The Terraces of the Derbyshire Derwent', *East Midland Geographer*, no. 9, pp. 3–15 (Nottingham, 1958).

[2] P. R. Crowe, 'The Seasonal Variation of the Strength of the Monsoons', *Indian Geographical Society, Silver Jubilee Souvenir*, pp. 186–8 (Madras, 1952); also R. S. Waters and R. H. Johnson, op. cit. (1958).

number and add this to the mean value afterwards. The mean may also be calculated, if the frequency of occurrence of each value is known, by

$$\bar{x} = \frac{\sum\limits_{i=1}^{n} f_i x_i}{\sum\limits_{i=1}^{n} f_i}$$

where f_i is the frequency of each value for $i = 1, 2 \ldots n$.

When the frequency is only known for grouped values, the mean may still be calculated if the classes are of equal size. This method is demonstrated with reference to some hypothetical data in Table 1. An

TABLE I

Class boundaries	Class mid-marks	Frequency (f)	Deviation of mid-marks from assumed mean, x_0, in units of cell intervals (d)	Total deviation of class (f × d)
19·5–22·5	21	2	−3	−6
22·5–25·5	24	5	−2	−10
25·5–28·5	27	10	−1	−10
28·5–31·5	30	10	0	0
31·5–34·5	33	9	1	9
34·5–37·5	36	7	2	14
37·5–40·5	39	4	3	12
40·5–43·5	42	1	4	4
		48		13

assumed mean, x_0, is taken about the middle of the distribution ($x_0 = 30\cdot0$) and the deviation of each class from this assumed mean is calculated in units of the cell interval, c (i.e. size of classes).

The arithmetic mean, $\bar{x} = x_0 + c \cdot \dfrac{\Sigma f \cdot d}{\Sigma f}$

$$= 30\cdot0 + \frac{3 \times 13}{48}$$

$$= 30\cdot0 + 0\cdot81$$

$$= 30\cdot81$$

The mean does not always give the best representation of 'average', especially if there are extreme values in the data which may unduly influence the calculated mean. It is sometimes preferable to employ either the median or the modal value.

The *median* is that value which has half the number of occurrences below it and half above in a scale of values. This central value is consequently a good measure of central tendency in general. It may be readily determined from a cumulative frequency graph (Fig. 229) if

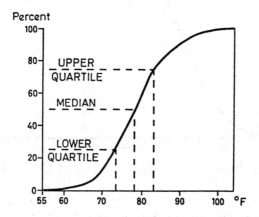

Figure 229. OGIVE SHOWING MEDIAN AND QUARTILE VALUES

the classes are not too large, otherwise it is rather laborious to calculate. The ogive may also be used to obtain the *upper* and *lower quartiles*, which are the central points between the median and the upper and lower extremes. These values are frequently used in conjunction with monthly dispersion graphs to indicate the seasonal march of precipitation values (see p. 246).

The *modal value* or *mode* is that value, or class, which occurs most frequently. It is useful for indicating the most typical figure, when actual numbers of items are involved (i.e. discrete series). The average number of occurrences of thunderstorms, or fog, or of days with snow falling, is most realistically represented by the modal value and it is also a convenient value for expressing the most frequent (prevailing) wind direction. However, the distribution may have more than one mode and the utility of this average is diminished since, unlike the mean and median, it cannot be used for any further mathematical calculations.

The mean, median and mode coincide with one another in the *'Normal'* or *Gaussian Distribution*. This is a mathematically derived frequency distribution which possesses perfect symmetry about the central value (Fig. 230).

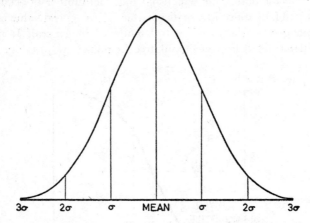

Figure 230. A 'NORMAL' OR 'GAUSSIAN' FREQUENCY DISTRIBUTION

The value of this theoretical curve is developed below. Many frequency curves depart from the symmetry of the normal distribution and are described as possessing *skewness*. A curve has *positive* skewness if the mean exceeds the mode and *negative* skewness if the mode exceeds the mean. These two cases are illustrated schematically in Fig. 231.

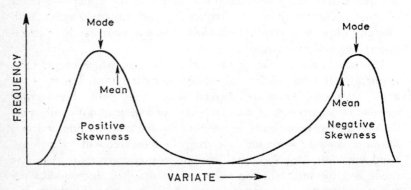

Figure 231. SCHEMATIC ILLUSTRATION OF POSITIVELY AND NEGATIVELY SKEWED FREQUENCY DISTRIBUTIONS

If the skewness is not excessive, the following relationship is known to exist:

$$Mode = Mean - 3(Mean - Median), \text{ } approximately$$

There are several measures of skewness.[1] One is:

$$Skewness = \frac{3(Mean - Median)}{Standard\ deviation}, \text{ } which\ ranges\ from\ -3\ to\ +3$$

and another $= \dfrac{(Q_1 + Q_2 - 2 \times Median)}{(Q_1 - Q_2)}$, which ranges from -1 to $+1$

where Q_1 and Q_2 refer to the upper and lower quartiles respectively.

The concept of central tendency is also applicable for areal distributions. Both mean and median points of the distributions of population, and of agricultural and manufactured products are shown for several countries in a study by Sviatlovsky and Eells.[2] The mean point is very suitable for examining changes in distribution over a period of time; for instance, the changing regional emphasis of mining from 1869–1939 within the Appalachian coal-field is illustrated by plotting decadal positions of the mean point by Murphy and Spittal.[3] The median point is used, however, to show the centres of the cotton area in the south-eastern United States in 1924 and 1944, since it is considered to be the most stable central value if extremes are present in the scale of figures.[4] The median is probably the most satisfactory point for comparing simultaneous distributions of different features within one area. A valuable discussion of the calculation of the different central points is given by Hart,[5] with a cartographic presentation for the population distribution in Georgia, U.S.A. An areal mode cannot be calculated, since areal distributions are commonly bimodal or multi-modal, but Hart determines a point of '*minimum aggregate travel*' for the population of Georgia. This is the point which is reached by all items of a distribution with least total travel in a straight line for all items. Theoretically, the point represents an optimum position for a regional centre.

[1] See, for example, G. U. Yule and M. G. Kendall, *An Introduction to the Theory of Statistics* (fourteenth edition, London, 1950).
[2] E. E. Sviatlovsky and W. C. Eells, 'The Centrographic Method and Regional Analysis', *Geographical Review*, vol. 27, pp. 240–54 (New York, 1937).
[3] R. E. Murphy and H. E. Spittal, 'Movements of the Center of Coalmining in the Appalachian Plateaus', *Geographical Review*, vol. 35, pp. 624–33 (New York, 1945).
[4] M. Prunty, Jr, 'Recent Quantitative Changes in the Cotton Regions of the South-eastern States', *Economic Geography*, vol. 27, pp. 189–208, Fig. 9 (Worcester, Mass., 1951).
[5] J. F. Hart, 'Central Tendency in Areal Distributions', *Economic Geography*, vol. 30, pp. 48–59 (Worcester, Mass., 1954).

Variability

In addition to knowledge of the average it is desirable to be able to indicate the scatter of values about the average. Dispersion graphs with median, quartiles and decile points show departures from average (or *deviation*) in graphical form, but numerical values are also required.

The most simple figure is that expressing the *Range* between the lowest and highest values, but this gives insufficient information for most purposes.

The *Inter-Quartile Deviation* is another rather crude measure of scatter which is used in conjunction with the median. The mean expectation of deviation from the median is expressed by

$$\frac{Q_1 - Q_2}{2}$$

where Q_1 and Q_2 are the upper and lower quartiles respectively. The range between the upper and lower deciles (highest and lowest 10 per cent) may also be of interest.

Mean Deviation

This is the average departure of values from the mean, irrespective of sign.

$$\text{Mean deviation} = \frac{\Sigma|x - \bar{x}|}{n}$$

where the symbol $|\quad|$, which is termed the *modulus*, denotes the absolute value of the deviation. The mean deviation is easily calculated and is indeed commonly used in economic and other statistics. Nevertheless, there is the disadvantage that the mean deviation cannot be used in further statistical formulae, which is not true of the rather more laboriously calculated standard deviation.

Standard Deviation

This is the square root of the sum of the squares of the deviations, divided by the number of items.

$$\text{Standard deviation, } \sigma = \sqrt{\left(\frac{\Sigma(x - \bar{x})^2}{n}\right)}$$

An alternative form is

$$\sigma = \sqrt{\left(\frac{\Sigma x^2}{n} - \bar{x}^2\right)}$$

The expression without the square root sign is called the *variance*.[1]

[1] An example of the use of the variance expression to determine the relative rank of percentage figures for county areas under various crops is given by J. C. Weaver, 'Crop Combination Regions of the Middle West', *Geographical Review*, vol. 44, pp. 175–200 (New York, 1954).

σ refers to the standard deviation of the total population, whereas s is usually employed for the standard deviation of a sample. Invariably one has only a sample of the total population, in which case a 'best estimate', $\hat{\sigma}$, of the true standard deviation is given by

$$\hat{\sigma} = s\sqrt{\frac{n}{n-1}}$$

Thus,

$$\hat{\sigma} = \sqrt{\left(\frac{\Sigma(x - \bar{x}_s)^2}{(n-1)}\right)}$$

where x_s refers to the sample mean.

An approximate expression for standard deviation, which is more readily solved and gives the result with sufficient precision for most practical purposes, is

$$\sigma = c\sqrt{\left\{\frac{\Sigma f . d^2}{\Sigma f} - \left(\frac{\Sigma f . d}{\Sigma f}\right)^2\right\}}$$

The method of using this formula is demonstrated by the example in Table 2.

TABLE 2
Short Method of Calculating the Standard Deviation

Class boundaries	Class mid-marks	Frequency (f)	Deviation of mid-marks from assumed mean, x_0, in units of cell intervals (d)	Total deviation of class (f × d)	(f × d²)
19·5–22·5	21	2	−3	−6	18
22·5–25·5	24	5	−2	−10	20
25·5–28·5	27	10	−1	−10	10
28·5–31·5	30	10	0	0	0
31·5–34·5	33	9	1	9	9
34·5–37·5	36	7	2	14	28
37·5–40·5	39	4	3	12	36
40·5–43·5	42	1	4	4	16
		48		13	137

The value of the standard deviation which is obtained by this short method generally exceeds the true value. Strictly, therefore, a small correction should be made to the result.[1] The amount of correction depends whether the data represent a continuous or a discrete variable.

[1] This arises from the distribution of the actual values within the classes. Fuller details may be found in standard statistical texts.

For the former, $-0.083c^2$ is used, where c is the cell interval (i.e. size of class). For a discrete variable the correction is $-1/12\,(c^2 - 1)$. Table 3 illustrates the size of correction calculated from the latter formula.

TABLE 3

Correction to Standard Deviation of a Discrete Variable calculated by the Grouped Frequency Method

Cell interval	=	1	2	3	4	5
Correction	=	0	−0.25	−0.67	−1.25	−2.00

A cartographical illustration of the use of the standard deviation is given for mean monthly temperatures in North America by Sumner.[1]

$$\sigma = 3\sqrt{\left\{ \left(\frac{137}{48}\right) - \left(\frac{13^2}{48}\right) \right\}}$$
$$= 3\sqrt{2.77}$$
$$= 5.0$$

It should be noted, however, that one cannot compare values of standard deviation unless the means of the two samples are approximately equal. Comparisons are only possible if the deviation is converted into a percentage of the mean. When the standard deviation is used this is termed the *Coefficient of Variation (CV)* and is calculated from

$$CV(\%) = \frac{\sigma}{\bar{x}} \times 100$$

The coefficient is used in climatological analyses, especially studies of rainfall. In this regard its validity is questionable, since the frequency distribution of rainfall is bounded by zero on one side and hence the coefficient tends to infinity as the mean rainfall approaches zero.[2] A recent paper[3] exemplifies the use of the coefficient in a study of the variation of the quality of housing in cities in the United States. Expressions similar to the above may also be based on the inter-quartile and the mean deviations (see also p. 210).[4]

[1] A. R. Sumner, 'Standard Deviation of Mean Monthly Temperatures in Anglo-America', *Geographical Review*, vol. 43, pp. 50–9 (New York, 1953).

[2] An empirical attempt to overcome this problem is made by V. Conrad, 'The Variability of Precipitation', *Monthly Weather Review*, vol. 69, pp. 5–11 (Washington, D.C., 1941).

[3] R. J. Fuchs, 'Intra-urban Variation of Residential Quality', *Economic Geography*, vol. 36, pp. 313–25 (Worcester, Mass., 1960).

[4] See, for example, S. Gregory, 'Some Aspects of the Variability of Annual Rainfall over the British Isles, for the Standard Period 1901–30', *Quarterly Journal of the Royal Meteorological Society*, vol. 81, pp. 257–62 (London, 1955).

In certain cases it may be necessary to devise special measures of variability. A. Geddes[1] (see also p. 349) develops a technique appropriate for use with decadal census data from India for the period 1881–1931. By drawing a smoothed exponential (power series) curve passing through the first and last dates and comparing the theoretical values with the reported figures for the censuses of 1891, 1901, 1911 and 1921, a variability index is obtained. The deviations are summed and divided by five to give the mean deviation for the period. The divisor is five since, although there are four decadal values, the terminal deviations are constructed as zero and as a result probably increase the adjacent deviations.

The concept of dispersion about a mean point is applicable to areal distributions and extends the centrographic methods outlined on p. 485. R. Bachi[2] shows that frequency classes of latitude and longitude for the distribution under investigations may be used to define a mean centre, $\bar{x}\,\bar{y}$ with coordinates

$$\bar{x} = \frac{\Sigma f_a x_a}{\Sigma f_a} \quad \text{and} \quad \bar{y} = \frac{\Sigma f_a x_a}{\Sigma f_a}$$

and with a 'standard distance' d, which is the root of the average of squares of the distance between individual points and the mean centre.

$$d = \sqrt{\left\{ \frac{\Sigma f_a (x_a - \bar{x})^2}{n} + \frac{\Sigma f_a (y_a - \bar{y})^2}{n} \right\}}$$

The technique is valuable for the comparison of different distributions.

Probability

The general concept of probability is a familiar one. It is well known that there is an 'even chance' (50 per cent or 0·5 probability) that a tossed coin will come down 'heads'. The extremes of the probability scale are absolute impossibility (0 per cent or 0·0) and absolute certainty (100 per cent or 1·0).

In the more precise terms, the probability of an event occurring is assessed by dividing the number of occurrences of the event by the total

[1] A. Geddes, 'Half a Century of Population Trends in India: a Regional Study of Net Change and Variability', 1881–1931', Geographical Journal, vol., 98, pp. 228–53 (London, 1941); and 'Variability in Change of Population in the United States and Canada, 1900–1951, Geographical Review, vol. 44, pp. 88–100 (New York, 1954).
[2] R. Bachi, 'Statistical Analysis of Geographical Series', Bulletin de l'Institut International de Statistique', 2nd series, vol. 36, pp. 229–40 (Stockholm, 1958). A summary of methods of areal analysis is given in O. D. Duncan, R. P. Cuzzort and B. Duncan, Statistical Geography: Problems in Analysing Areal Data (Illinois, 1961).

number of cases or trials. Probabilities are summed if the occurrence of one event excludes the occurrence of another. The simplest case of this is the mutual exclusiveness of a tossed coin falling heads or tails up. The total probability, which is 1·0, comprises 0·5 for both eventualities. In other words we are concerned with yes/no problems or cases of occurrence and non-occurrence. However, when simultaneous or associated events are being considered, probabilities are multiplied. For example, the probability of two tossed coins falling heads up is 0·5 × 0·5 = 0·25 and of three tossed coins all showing heads is 0·125. Note that the multiplication of probabilities assumes that the events are independent of one another.

A valuable property of the Normal Distribution is the fact that it specifies the probability of occurrence of any value. Thus, 68·26 per cent of the distribution lies within ±σ of the mean, 95·45 per cent lies within ±2σ and 99·73 per cent within ±3σ of the mean. The probability that the frequency of a value will be greater or less than a stated number of standard deviations *either* above *or* below the mean may also be readily determined. The relationship is shown in Fig. 232. Exact

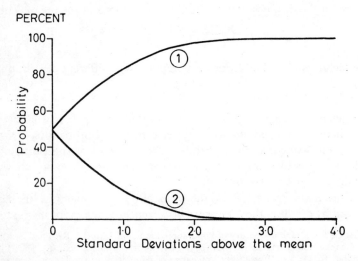

Figure 232. PROBABILITY VALUES FOR DEVIATIONS IN ONE DIRECTION
FROM THE MEAN FOR A NORMAL DISTRIBUTION

Line 1 shows the probability that a value will be *less* than the corresponding number of standard deviations above the mean. Line 2 shows the probability that a value will be *greater* than the corresponding number of standard deviations above the mean.

values may be found in standard statistical tables. It must be stressed
that these probabilities only apply to a near-Normal Distribution,
although many skewed distributions may be 'normalized' (converted
into an approximately Normal Distribution) if the scale is made a
logarithmic one or some other function of the original.[1]

Time Series

Many data of geographical interest exist in the form of monthly or
annual values for a period of years. This is the case with demographic
and climatological records and with figures of agricultural or industrial
production. Such information may be plotted graphically as a function
of time in order to inspect the data series for indications of any trends
or cycles which might be present. A common practice is to calculate
moving averages, or *running means*, to smooth out irregularities in the
graph caused by individual values.[2] Five-year moving averages are
determined successively as follows:

$$\bar{x}_{1-5} = \frac{x_1 + x_2 + x_3 + x_4 + x_5}{5}, \text{ which is the value for year 3}$$

$$\bar{x}_{2-3} = \frac{x_2 + x_3 + x_4 + x_5 + x_6}{5}, \text{ which is the value for year 4}$$

Moving averages may also be calculated for an even number of years,
although this entails an additional step in the working to avoid displacement of the average values between successive observations. The
necessary expressions are:

For a centred two-year moving average

$$\text{for year 2} = \frac{x_1 + 2x_2 + x_3}{4}$$

$$\text{and for year 3} = \frac{x_2 + 2x_3 + x_b}{4}$$

For a centred four-year moving average

$$\text{for year 3} = \frac{x_1 + 2x_2 + 2x_3 + 2x_4 + x_5}{8}$$

$$\text{and for year 4} = \frac{x_2 + 2x_3 + 2x_4 + 2x_5 + x_6}{8}$$

[1] See, for example, E. N. Thomas, 'Areal Associations between Population Growth
and Selected Factors in the Chicago Urbanized Area', *Economic Geography*, vol. 36,
pp. 158–70 (Worcester, Mass., 1960). Examples are given in S. Gregory, *Statistical
Methods and the Geographer, 2nd edition*, pp. 50–52 (London, 1968).

[2] See also p. 229.

Centering is unnecessary if it is only the trend and not the duration of a fluctuation which is of interest.

The smoothing effect of five-year moving averages is illustrated for figures of wine production in France between 1920 and 1945 (Fig. 233).

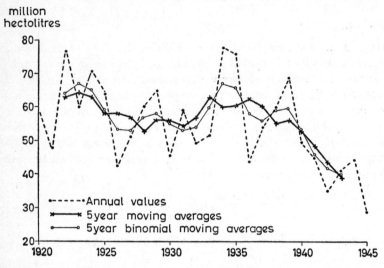

Figure 233. GRAPHS OF WINE PRODUCTION IN FRANCE, 1920–45
Source: *Annuaire Statistique de la France, 1961. Institut National de la Statistique et des Etudes Economiques* (Paris, 1961).

The use of this technique is widespread in searching for climatic fluctuations and trends in temperature and precipitation records, although there are possible dangers in the method. In general, a moving average of 'm' years will largely remove periodicities of length 'm', although Lewis[1] points out that moving averages which are used to smooth sets of hypothetical random data tend to produce irregular periodicities. Such unwanted irregularities can be suppressed or eliminated by the use of a smoothing function with a large central value. For example, the fourth power binomial smoothing function

$$\bar{x}_{1-5} = \frac{x_1 + 4x_2 + 6x_3 + 4x_4 + x_5}{16}$$

[1] P. Lewis, 'The Use of Moving Averages in the Analysis of Time-series', *Weather*, vol. 15, pp. 121–6 (London, 1960). These irregular periodicities are referred to as Slutsky–Yule effects.

may prove more satisfactory than the simple, unweighted five-year average. Figure 233 demonstrates that for the particular data which are being used, there is displacement of maxima and minima with the unweighted average, but not with the weighted one.

An alternative means of studying time series is provided by the calculation of *cumulative residuals*, or deviations from the mean value over the period. The method is widely employed in hydrology and climatology.[1] The deviations, which may be actual or percentage values, are totalled with respect to sign. An example using the previous records of wine production is shown in Fig. 234. It must be emphasized

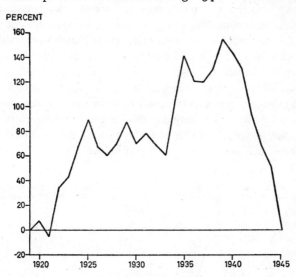

Figure 234. PERCENTAGE CUMULATIVE RESIDUALS OF WINE PRODUCTION
IN FRANCE, 1920–45
Source: *Annuaire Statistique de la France, 1961.*

that it is the inflections of the curve, and not the peaks and troughs, which demarcate changes. Thus, the graph is concave upwards when

A. A. Barnes, 'Rainfall in England: the True Long-Average as Deduced from Symmetry', *Quarterly Journal of the Royal Meteorological Society*, vol. 45, pp. 209–27 (London, 1919); and E. B. Kraus, 'Graphs of Cumulative Residuals', *Quarterly Journal of the Royal Meteorological Society*, vol. 82 pp. 96–8 (London, 1956). For a comparison of moving averages and cumulative residuals, see F. G. Hannell, 'Climatic Fluctuations in Bristol', *British Association for the Advancement of Science*, vol. 12, pp. 373–86 (London, 1956); and E. C. Barrett, 'Some Problems concerned with the Graphical Presentation of Climatic Data', *The Northern Universities Geographical Journal*, no. 3, pp. 16–22 (Nottingham, 1962).

there is an increasing trend and convex upwards for a decreasing
trend.

Both methods of graphical presentation are open to criticism and the
complexities of many data series may not permit confident interpretation of the graphs. More refined techniques are beyond the scope
of this book, although it may be noted that the Student's '*t*' test and
the '*F*' test, which are discussed below, provide methods of checking
data series.[1]

Significance Tests

A difference between two mean values, for example, annual average
precipitation at two neighbouring stations, may arise by chance (i.e.
a random difference). On the other hand, there may be a real difference
and it is extremely useful to be able to assess the likelihood of this
eventuality by statistical tests. The method commonly adopted is to
assume that the two mean values do not differ significantly from each
other and belong to a common population. In other words, any difference between the means is only due to chance. The procedure thereby
sets up a *null hypothesis* which may be tested; means of testing the
significance of differences are developed below.

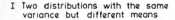

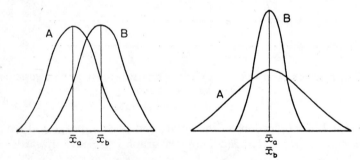

Figure 235. SAMPLE DISTRIBUTIONS

[1] For an illustration of the use of the *F* test (see p. 408) with moving averages, see
J. M. Craddock, 'A Simple Statistical Test for Use in the Study of Climatic Change',
Weather, vol. 12, pp. 252–8 (London, 1957). Climatological times series are discussed
comprehensively in C. E. P. Brooks and N. Carruthers, *Handbook of Statistical Methods in
Meteorology* (H.M.S.O., London 1953); V. Conrad and L. W. Pollock, *Methods in
Climatology* (second edition, Cambridge, Mass., 1950); and H. A. Panofsky and G. W.
Brier, *Some Applications of Statistics to Meteorology* (Pennsylvania State University, 1958).

Figure 235 may help to clarify the type of hypothesis which one may adopt for testing. The sample distributions may possess different mean values and have the same variance (Fig. 235(i)), or have different variances but the same mean (Fig. 235(ii)). The null hypothesis stated above would assume that the difference between the means shown in Fig. 235(i) is not significant. However, it is not always assumed that the variance as well as the mean is common to both samples.

The Concept of Standard Error

In many geographical analyses, data are available for samples and not for the 'population'. If mean values are calculated for each of a number of samples taken from a larger group of data, it is observed that the distribution of the sample means about the mean for the group resembles a normal distribution. The true mean of the whole series is more closely approximated by the centre of the distribution of sample means as the number of samples is increased. The theoretically expected difference between a sample mean and the population mean may be determined by dividing the standard deviation of the population by the square root of the number of observations:

$$\sigma_{\bar{x}} = \frac{\sigma}{\sqrt{n}}$$

This expression is termed the *Standard Error* of the sample mean. Usually the population standard deviation has to be estimated (see p. 486) and the best estimate of the standard error[1] of the mean is given by

$$\hat{\sigma}_{\bar{x}} = \frac{s}{\sqrt{n-1}}$$

Standard errors may also be derived for the variance, standard deviation and coefficient of variation, but the concept is especially valuable when applied to the comparison of two mean values. Two sample means, x_a and x_b, may be compared by calculation of their respective standard errors. Assuming that x_a exceeds x_b, if $x_a - 3\sigma_{\bar{x}a}$ is greater than $\bar{x}_b + 3\hat{\sigma}_{\bar{x}b}$ one can be certain that the sample means are significantly different if the frequency distributions are approximately normal. On the other hand, if the two values differ by less than $(3\hat{\sigma}_{\bar{x}a} + 3\hat{\sigma}_{\bar{x}b})$ it is uncertain whether or not the means differ significantly.

[1] These expressions assume that the observations are not correlated with one another, i.e. that they are 'independent'.

A more useful method is to compare the *standard error of the difference* between the means with the actual difference between the means. It can be shown that the variance of the difference between two sample means is equal to the sum of their variances,[1] or

$$Var. \ (\bar{x}_a - \bar{x}_b) = \frac{\sigma_a^2}{n_a} + \frac{\sigma_b^2}{n_b}$$

Thus,

$$\sigma(\bar{x}_a - \bar{x}_b) = \sqrt{\left(\frac{\sigma_a^2}{n_a} + \frac{\sigma_b^2}{n_b}\right)}$$

The expression is modified when the sample size is small, although it is preferable to use the '*t*' test, which is discussed below. For the purpose of illustration, the data of Table 4 are examined by both methods.

Example: The problem is to assess whether the difference between the average live birth rates of Esch and Remich for 1891–1940 (Table 4) is significant.

TABLE 4

*Live Births per 1,000 Inhabitants, 1891–1940,
in two Luxembourg Cantons*

	Esch	Remich
1891–1895	35·1	22·4
1896–1900	35·8	23·2
1901–1905	36·9	24·6
1906–1910	35·3	24·3
1911–1915	32·0	21·6
1916–1920	20·1	17·0
1921–1925	23·2	16·4
1926–1930	24·5	14·8
1931–1935	18·9	12·4
1936–1940	15·7	12·7
Mean ($\bar{x}$)	27·7	18·9
Variance $\left(\frac{\Sigma(x-\bar{x})^2}{n}\right)$	61·0	20·7

Source: *Annuaire Statistique, 1955, L'Office
de la Statistique Générale* (Luxembourg).

[1] It is assumed that the observations are not correlated with one another. If the observations are not independent, or if the two series are mutually correlated, the expression has to be modified, Further details may be found, for example, in C. C. Peters and W. R. Van Voorhis, *Statistical Procedures and their Mathematical Bases* (New York, 1940), p. 162.

It is assumed that any difference is due to chance for the null hypothesis. The modified expression for standard error of the difference with a small sample is

$$\sigma(\bar{x}_a - \bar{x}_b) = \sqrt{\left(\frac{\sigma_a{}^2}{n_a - 1} + \frac{\sigma_b{}^2}{n_b - 1}\right)}$$

$$= \sqrt{\left(\frac{61 \cdot 0}{9} + \frac{20 \cdot 7}{9}\right)}$$

$$= \quad 3 \cdot 0$$

The difference between the mean values is 8·8, which is almost three times the standard error of the difference. When the difference between the means is less than twice the standard error of the difference the null hypothesis is not in doubt, but the difference is definitely significant if it exceeds three times the standard error of the difference. The difference is probably significant in the present example.[1]

Student's 't' Test

This test is especially valuable in assessing the significance of the difference between means for very small samples.[2] It is essential when employing the t test to ensure, for the purposes of the null hypothesis, that the sample means have approximately the same variance by using Snedecor's F test.[3] The ratio of the variances is compared by

$$F = \frac{\text{Greater estimate of the variance}}{\text{Lesser estimate of the variance}}$$

Since the population variance is estimated by the two samples, the correction factor $n/n - 1$ has to be used,[4] although this does not affect the result for the example under consideration

$$F = \frac{61 \cdot 0 \times 10/9}{20 \cdot 7 \times 10/9}$$

The value of F has to be compared with the appropriate *number of degrees of freedom* for each variance estimate in tables of the variance

---

[1] A further illustration of the use of this method for two samples of an hundred cities in the United States is provided by G. W. Hartman and J. C. Hook, op. cit. (1956).

[2] An arbitrary figure of 20 is usually adopted as the limiting size.

[3] I.e. to test that the variances, in addition to the means, refer to the assumed common parent population.

[4] Bessel's Correction, which is also used for the best estimate of the standard deviation (p. 487).

ratio.[1] The number of degrees of freedom is equal to the number of frequency values which it is possible to assign arbitrarily. Thus, if a sample of n observations is grouped into x classes and the frequencies of $(x - 1)$ classes are given, the frequency of the final xth class is determined by the total, Σf or n. In a similar manner, the mean and standard deviation of a series may represent additional constants ('parameters'), which are determined from the sample and subsequently used as bases for the statistical assessment of the items in the sample. Such parameters, which limit the number of degrees of freedom, are referred to as constraints or restrictions.

In the F test the only restriction upon each sample is the number of observations and so there are $(n - 1)$ degrees of freedom for both estimates of the variance. The value of F for nine degrees of freedom with each estimate is 3·2 at the 5 per cent probability level (i.e. a 5 per cent probability that the value is due to chance).[2] The calculated value of F is only 2·9, indicating a greater probability that the difference between the variances is the result of chance. The null hypothesis that the variances are not significantly different is therefore upheld and it is possible to proceed with the t test itself.

The value of t for the present purpose[3] is given by

$$t = \frac{Distance\ between\ means}{Standard\ error\ of\ the\ difference}$$

$$= \frac{|\bar{x}_a - \bar{x}_b|}{\sqrt{\left\{\left(\dfrac{(n_a - 1)s_a^2 + (n_b - 1)s_b^2}{(n_a - 1) + (n_b - 1)}\right)\left(\dfrac{1}{n_a} + \dfrac{1}{n_b}\right)\right\}}}$$

There are two alternative expressions for t. For the one above it is assumed that the two samples have the same variances. The samples are 'pooled' to obtain the best estimate of the population variance. If one considers that the sample variances are different, then the expression for t is

$$t = \frac{|\bar{x}_a - \bar{x}_b|}{\sqrt{\left(\dfrac{s_a^2}{n_a - 1} + \dfrac{s_b^2}{n_b - 1}\right)}}$$

[1] For example, D. V. Lindley and G. C. P. Miller, op. cit. (1953).

[2] The adoption of the 5 per cent level as the critical significance level is simply a convention.

[3] The standard error of the difference is $\sqrt{\left(\dfrac{\sigma_a^2}{n_a} + \dfrac{\sigma_b^2}{n_b}\right)}$ or $\sigma\sqrt{\left(\dfrac{1}{n_a} + \dfrac{1}{n_b}\right)}$ A best estimate of σ is necessary and this is derived by modifying s_a^2 and s_b^2 according to their respective number of degrees of freedom.

The latter gives a smaller value of t and provides a more stringent test. For the example of Table 4, the first expression gives

$$t = \left(\frac{8 \cdot 8}{\sqrt{\left\{ \left(\frac{9 \times 61 \cdot 0 + 9 \times 20 \cdot 7}{18} \right) \cdot \left(\frac{1}{10} + \frac{1}{10} \right) \right\}}} \right)$$

$$= \frac{8 \cdot 8}{\sqrt{8 \cdot 2}}$$

$$= 3 \cdot 08$$

This value is read against the appropriate number of degrees of freedom at different probability levels in a table of the t distribution. There are $(n - 1)$ degrees of freedom for each sample, giving $(n_a - 1) + (n_b - 1)$ degrees of freedom in all. In the example there are 18 degrees of freedom and the tabulated value of t at the 0·3 per cent probability level is 3·1. The calculated value of $t = 3·08$ is therefore significant at this probability level, which implies that there is only a 0·3 per cent (three in one thousand) probability of the difference between the mean number of live births in the two cantons resulting from chance. The more stringent test of the second expression for t would still give a very significant difference at the 1 per cent level. The example underlines the fact that a more firm conclusion may be obtained with the t test when the samples are small.

Non-Parametric Alternatives

Two non-parametric alternatives to the t test deserve mention. They are the sign test on matched pairs and the Kolmogorov-Smirnov two sample test.

The *sign test* which is used for nominal data, is based on the *binomial distribution*. This has a mean of $n/2$ and a standard deviation of $\sqrt{n}/2$ for two series. The test statistic at the 95% confidence level is $n/2 \pm \sqrt{n}$ for $n > 25$. It is best illustrated by a simple example. The sale of cars in two towns of the same size was as follows over a 37 year period:

Sales in A $>$ sales in B in 27 years
Sales in A $<$ sales in B in 9 years
Sales in A $=$ sales in B in 1 year.

Is the difference statistically significant? In order to reject the null hypothesis that no difference exists, the value of 27 (or 9) must lie outside the limits of $\frac{n}{2} \pm \sqrt{n}$. (For a two-tailed test the direction of the difference is not specified.) Ties are excluded, and the value of n is reduced accordingly, so that the limits are 18 ± 6. The test shows that we can reject the null hypothesis at the 95% level.

The *Kolmogorov-Smirnov test* for ordinal data is sensitive to any difference in two frequency distributions. The two data series must be measured in the same class intervals and the values determined as cumulative frequencies expressed as proportions, not percentages. The test statistic relates to the maximum difference between the two series in any class interval. For example, given the data in Table 5, are the farm sizes similar in the two areas?

TABLE 5

Size of Farm Area (acres)	<5	5–10	10–15	15–20	Total No. of Farms
Number of farms					
Area A	10	25	10	5	50
Area B	2	6	40	12	60
Cumulative frequency					
Area A	10	35	45	50	50
Area B	2	8	48	60	60
Proportional cumulative frequency					
Area A	0·10	0·70	0·90	1·00	
Area B	0·03	0·14	0·80	1·00	
Difference	0·07	0·56	0·10	0·0	

The maximum difference is compared with the following:

$1 \cdot 36$ K for the 5% significance level
$1 \cdot 63$ K for the 1% significance level
$1 \cdot 95$ K for the 0·1% significance level

where $K = \sqrt{\dfrac{n_1 + n_2}{n_1 n_2}}$; n_1 and n_2 must each exceed 40. Tables are available for less than 40 cases provided that $n_1 = n_2$.

In the example of Table 5, $K = 0 \cdot 191$ so that the maximum difference of 0·56 is significant at the 0·1% level. The distributions of farm size in the two areas are significantly different from one another.

Analysis of Variance

It is often necessary to extend consideration of the significance of differences between mean values to a large number of samples. In such cases, repetition of t tests would be very laborious and the technique of

Analysis of Variance provides a convenient and powerful statistical tool.[1] The method is to compare the amount of variance *between* the samples with that *within* the samples. The variance of a sample is the square of the standard deviation, but a best estimate of the population variance is obtained by dividing the sum of the squares of the deviations by $(n - 1)$, which is the number of degrees of freedom. The practical procedure is to calculate the sum of squares for the total number of cases and to obtain the within sample sum of squares by subtraction of the between sample sum of squares from the total sum of squares. The respective variances are then determined and if the between sample variance significantly exceeds the within sample variance, one may be sure that the samples do not belong to a common population. The calculations are illustrated by a worked example.

Example: Four occupation categories (A, B, C, D) are distinguished in a city survey. The number of workers in each occupation category is determined for five census tracts and it is necessary to investigate whether there is a significant relationship between the occupations and the numbers employed in them. In this case the occupation categories form the sample groups.

The first step is to take an estimated average in order to simplify the calculations. The value of 340 will be selected and all the numbers in Table 6(a) are subtracted from 340, as shown in Table 6(b). The values of the latter table are then squared and retabulated (Table 6(c)).

TABLE 6(a)
Numbers Employed in each Occupation Category

	A	B	C	D
Census tract 1	100	150	350	350
„ „ 2	250	400	400	300
„ „ 3	250	400	650	500
„ „ 4	300	450	550	550
„ „ 5	200	300	400	500
Totals	1,100	1,700	2,350	2,200
Averages	220	340	470	440

[1] For an example which employs it to compare the population of the labour force in West Virginia with other groups of states, see L. Zobler, 'Decision Making in Regional Construction', *Annals of the Association of American Geographers*, vol. 48, pp. 140–8 (Lawrence, Kansas, 1958). See also P. Haggett, 'Regional and Local Components in the Distribution of Forested Areas in Southeast Brazil,' *Geographical Journal*. vol. 130, pp. 365–80 (London, 1964).

TABLE 6(b) Retabulation					TABLE 6(c) Squares			
A	B	C	D		A	B	C	D
−240	−190	10	10		57,600	36,100	100	100
−90	60	60	−40		8,100	3,600	3,600	1,600
−90	60	310	160		8,100	3,600	96,100	25,600
−40	110	210	210		1,600	12,100	44,100	44,100
−140	−40	60	160		19,600	1,600	3,600	25,600
−600	0	650	500		95,000	57,000	147,500	97,000

With the results of the tables the requisite calculations may be performed.

It is first necessary to obtain the total sum of the squares, making an adjustment for the 'estimated average' of 340. This adjustment or correction factor is the square of the sum of the sample totals of Table 6(b) divided by the total number of items.

$$Correction\ factor = \frac{(-600 + 0 + 650 + 500)^2}{20}$$

$$= \frac{550^2}{20}$$

$$= 15,125$$

The total sum of the squares of Table 6(c) is given by the sum of the sample totals of the table less the correction factor.

$$Total\ sum\ of\ squares = 95,000 + 57,000 + 147,000 + 97,000 - 15,125$$
$$= 380,875$$

The 'between sample' sum of the squares is calculated from the squares of each total in Table 6(b) divided by the number of items in each sample, which is five, and adjusted for the estimated average.

$$Between\ sample\ sum\ of\ squares = \frac{1}{5}(600^2 + 0^2 + 650^2 + 500^2) - 15,125$$
$$= 191,375$$

The 'within sample' sum of the squares is the difference between the total sum of the squares and the 'between sample' sum of the squares.

$$Within\ sample\ sum\ of\ squares = 380,875 - 191,375$$
$$= 189,500$$

It remains to calculate the number of degrees of freedom associated with the variances between and within the samples. There are $(n - 1)$ degrees of freedom for the total sum of the squares and also for the between sample sum of the squares, which gives nineteen and three respectively. The degrees of freedom for the within sample sum of the squares is obtained from the difference between the other two, $(19 - 3)$ $= 16$. The results may be tabulated as in Table 6(d) and the estimates

TABLE 6(d)

	Sums of squares	Degrees of freedom	Variance estimate
Between sample	191,375	3	63,792
Within sample	189,500	16	11,844

$$F = \frac{Greater\ variance\ estimate}{Lesser\ variance\ estimate}$$
$$= \frac{63,792}{11,844}$$
$$= 5\cdot39$$

of the variance between and within samples are calculated by dividing the sums of squares by the appropriate number of degrees of freedom. The ratio of the greater estimate of the variance to the lesser gives Snedecor's F ratio.

References to tables of F shows that for three degrees of freedom on the greater variance estimate and for sixteen on the lesser estimate the value is significant at the 2·5 per cent probability level. It is unlikely that the variances between and within the samples differ by chance and the occupation categories, therefore, are significantly different in terms of the numbers employed in each of them.

Chi-Square (χ^2) Test

The χ^2 test provides a measure of *association*, i.e. the degree to which two or more groups of data are related to one another. It is an extremely valuable test, since the variables need not be quantified beyond their expression in a number of categories; i.e., it can be used with ordinal and even nominal data. The method is again to test the null hypothesis that the observed results do not differ significantly from those which are to be expected by chance.

Example: The size of farm areas, graded into classes, is examined for two districts. The problem is to decide whether there is a significant difference between the size of farms in the two districts (Table 7).

TABLE 7

Size of farm area (acres)	Less than 25	25–50	50–100	100–200	200–500	Total no. of farms
Number of farms in district A	10	44	57	30	9	150
Number of farms in district B	40	73	70	24	7	214
	50	117	127	54	16	364

Chi square is obtained from

$$\chi^2 = \sum_{i=1}^{n} \frac{(O_i - E_i)^2}{E_i}$$

where O is the observed frequency and E is the expected or theoretical frequency.

The value of E should not be less than six and grouping of classes may be necessary to ensure this.

The expected values are calculated for each corresponding observed frequency as the product of the total of the column and the total of the row, divided by the grand total of items. Table 8 will clarify the derivation of these values.

TABLE 8

District A	E_1	E_3	E_5	E_7	E_9	150
District B	E_2	E_4	E_6	E_8	E_{10}	214
	50	117	127	45	16	364

where $E_1 = \dfrac{50 \times 150}{364}$, $E_2 = \dfrac{50 \times 214}{364}$, $E_3 = \dfrac{117 \times 150}{364}$ and so on.

The expected frequencies are shown in Table 9.

TABLE 9

District A	20·6	48·2	52·3	22·3	6·6	150
District B	29·4	68·8	74·7	31·7	9·4	214
	50	117	127	54	16	364

$$\chi^2 = \frac{(O_1 - E_1)^2}{E_1} + \frac{(O_2 - E_2)^2}{E_2} + \cdots + \frac{(O_{10} - E_{10})^2}{E_{10}}$$

$$= \frac{10 \cdot 6^2}{20 \cdot 6} + \frac{4 \cdot 2^2}{48 \cdot 2} + \frac{4 \cdot 7^2}{52 \cdot 3} + \frac{7 \cdot 7^2}{22 \cdot 3} + \frac{2 \cdot 4^2}{6 \cdot 6} +$$

$$\frac{10 \cdot 6^2}{29 \cdot 4} + \frac{4 \cdot 2^2}{68 \cdot 8} + \frac{4 \cdot 7^2}{74 \cdot 7} + \frac{7 \cdot 7^2}{31 \cdot 7} + \frac{2 \cdot 4^2}{9 \cdot 4}$$

$$= 16 \cdot 6$$

The number of degrees of freedom is (number of rows − 1) × (number of columns − 1), which is 4 × 1 = 4. Reference to a table for the χ^2 distribution shows that for $\chi^2 = 16 \cdot 6$ with 4 degrees of freedom, the difference between the districts is significant at the 0·5 per cent probability level.

An interesting study of the significance of regional boundaries is made by Zobler.[1] He employs the χ^2 test to examine the relationships between soil categories, between types of land-use, and between population groups in a number of regions in Salem County, New Jersey, which are selected on the basis of physiography and geology.

Correlation

The consideration of association between variables may now be extended to cases with quantitative data. The relationship between two quantitative variables is readily displayed graphically by a scatter diagram,[2] but a more precise measure of the correlation or concomitant variation may be required.

The most common measure of the relationship between two variables is K. Pearson's *Product Moment Correlation Coefficient*. The value of the coefficient ranges from − 1 to + 1. A correlation of + 1·0 implies that one variable increases or decreases exactly as the other increases or decreases. A correlation of − 1·0 denotes an exactly inverse relationship, while 0·0 signifies no relationship between the variables. In practice, even independent variables usually show a small chance correlation. The computation of the coefficient is demonstrated by the example in

[1] L. Zobler, 'Statistical Testing of Regional Boundaries', *Annals of the Association of American Geographers*, vol. 47, pp. 83–95 (Lawrence, Kansas, 1957); and also 'The Distinction between Relative and Absolute Frequencies in using Chi-Square for Regional Analysis', *Annals of the Association of American Geographers*, vol. 48, pp. 456–7 (Lawrence, Kansas, 1958).

[2] For an example with population data, see p. 383.

Table 10, examining the production of iron ore and crude steel.

TABLE 10
Production of Iron Ore and Crude Steel in France

	Iron ore (million tons)	$(x - \bar{x})$	Crude steel (million tons)	$(y - \bar{y})$	$(x - \bar{x}) \times (y - \bar{y})$
1919	9·4	−22·8	2·2	−3·3	75·2
1920	13·9	−18·3	2·7	−3·8	69·5
1921	14·2	−18·0	3·1	−3·4	61·2
1922	21·1	−11·1	4·5	−2·0	22·2
1923	23·3	−8·9	5·2	−1·3	11·6
1924	29·0	−3·2	6·7	0·2	−0·6
1925	35·6	3·4	7·5	1·0	3·4
1926	39·3	7·1	8·3	2·1	14·9
1927	45·5	13·3	8·3	1·8	23·9
1928	49·2	17·0	9·5	3·0	51·0
1929	50·7	18·5	9·7	3·2	59·2
1930	48·6	16·4	9·4	2·9	47·6
1931	38·6	6·4	7·8	1·3	8·3
1932	27·6	−4·6	5·6	−0·9	4·1
1933	30·2	−2·0	6·6	0·1	−0·2
1934	32·0	−0·2	6·2	−0·3	0·1
1935	32·0	−0·2	6·3	−0·2	0·0
1936	33·3	1·1	6·7	0·2	0·2
1937	37·8	5·6	7·9	1·4	7·8
1938	33·1	0·9	6·1	−0·4	−0·4
					459·0

$$\bar{x} = 32 \cdot 2 \qquad\qquad \bar{y} = 6 \cdot 5$$
$$\sigma_x = 11 \cdot 46 \qquad\qquad \sigma_y = 2 \cdot 04$$

Source: *Annuaire Statistique de la France*, pp. 135–9 (Paris, 1961).
The Product Moment Correlation Coefficient, r

$$= \frac{Covariance\ of\ x\ and\ y}{\sigma_x \cdot \sigma_y}$$

$$= \frac{\frac{1}{n}\Sigma(x - \bar{x})(y - \bar{y})}{\sigma_x \cdot \sigma_y}$$

$$= \frac{459 \cdot 0}{20 \times 11 \cdot 46 \times 2 \cdot 04}$$

$$= 0 \cdot 98$$

There is clearly a close positive relationship between the output of iron ore and crude steel in each year. Care must always be taken with time series to ensure that a general trend does not obscure short-period variations of opposite sign in the two variables.

The statistical significance of a correlation coefficient may be determined by calculation of its standard error, if the sample is large and the correlation is not high. In these circumstances the standard error, σ_r, is

$$\sigma_r = \frac{1 - r^2}{\sqrt{n}}$$

Otherwise the t test may be used with $(n - 2)$ degrees of freedom and

$$t = \sqrt{\left(\frac{r^2 (n - 2)}{1 - r^2}\right)}$$

where n refers to the number of pairs of data.[1] In the above example the degrees of freedom should strictly be modified to allow for the year-to-year interdependence in the data.

Linear Regression

The correlation coefficient provides a numerical index of the degree of relationship, but it may be preferable to draw the line which most closely approximates to the general trend of the points plotted on a scatter diagram. This is a regression line, or the line of 'best fit'. There are in fact two regression lines, for y on x and for x on y. In the first instance, it is then possible to 'predict' the approximate value of y, given a figure for x and vice versa for the regression of x on y.

The equation for y (unknown) on x (known) is

$$(y - \bar{y}) = r\frac{\sigma_y}{\sigma_x}(x - \bar{x})$$

For x (unknown) and y (known) the expression is

$$(x - \bar{x}) = r\frac{\sigma_x}{\sigma_y}(y - \bar{y})$$

The regression equation of y on x for the data referring to iron ore and crude steel in France becomes

$$y - 6{\cdot}5 = 0{\cdot}98\frac{2{\cdot}04}{11{\cdot}46}(x - 32{\cdot}2)$$

Thus, $y = 0{\cdot}17x + 0{\cdot}9$

[1] It should be remembered that while the correlation coefficient itself does not depend upon a normal distribution, the standard error and t tests do.

Taking arbitrary values,

$$if\ x = 20,\ y = 4 \cdot 3\ and$$
$$if\ x = 40,\ y = 7 \cdot 7$$

These values may be used to draw the regression line on the scatter diagram (Fig. 236), and in the same manner the regression line for y on

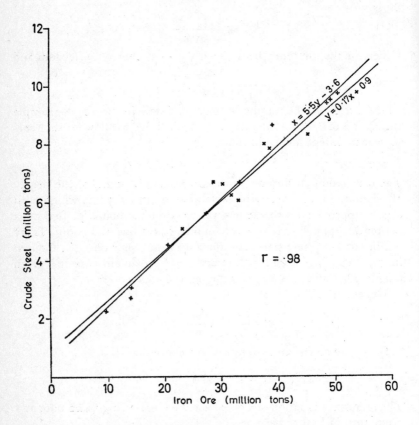

Figure 236. REGRESSION LINES FOR THE PRODUCTION OF IRON ORE AND CRUDE STEEL IN FRANCE, 1919–38
Source: *Annuaire Statistique de la France, 1961*

x is obtained. The equation of the latter is

$$x = 5 \cdot 5\,y - 3 \cdot 6$$

Increasing use is being made of these techniques in studies of urban geography (see also p. 460).[1]

The relationship between two variables may be in some cases non-linear and the best fit for the scatter diagram will then be provided by some form of curve. The precise calculation of such curves is difficult, although an approximate one may be sketched through the points on the graph. Alternatively, the data may be transformed (p. 491) in order to ensure linearity.

Measures of association and correlation can be used for areal distributions[2] and a full account of the application of a variety of techniques in a study of industrial geography is provided by McCarty et al.[3] Nevertheless, caution must be employed in areal studies in view of the need to make adjustment for the size of the areas for which correlation coefficients are being calculated.[4]

Various attempts have been made to devise coefficients of association for geographical purposes. In one of the earliest considerations of statistical methods in the analysis of distributions, Wright proposes a number of coefficients to measure evenness and association, which are derived mainly by graphical techniques.[5]

Florence suggests further coefficients of localization and association (or linkage) and demonstrates their use in a study of industry in Britain.[6] The coefficient of localization is obtained by summing either the positive or the negative deviations of the percentage of all workers in industry X, in regions A–F, from the total occupied population in regions A–F as a percentage of all workers. Table 11 demonstrates the procedure.

[1] See, for example, E. N. Thomas, 'Toward a Central Place Model', *Geographical Review*, vol. 51, pp. 400–11 (New York, 1961). The basic assumptions of regression analysis and related problems are fully reviewed by M. A. Poole and P. N. O'Farrell, 'The Assumptions of the Linear Regression Model', *Institute of British Geographers, Transactions*, No. 52, pp. 145–58 (London, 1971).

[2] Elementary examples of the application of the correlation coefficient are given by J. G. Chadwick, 'Correlation between Geographical Distributions. A Statistical Technique', *Geography*, vol. 46, pp. 25–30 (London, 1961).

[3] H. H. McCarty, J. C. Hook, D. S. Knos and G. R. Davies, *The Measurement of Association in Industrial Geography*, State University of Iowa (Iowa City, 1965).

[4] A. H. Robinson, 'The Necessity of Weighting Values in Correlation Analysis of Areal Data', *Annals of the Association of American Geographers*, vol. 46, pp. 223–6 (Lawrence, Kansas, 1956).

[5] Reference must be made to the original paper for details: J. K. Wright, 'Some Measures of Distributions', *Annals of the Association of American Geographers*, vol. 27, pp. 177–211 (Lancaster, Pa., 1937).

[6] P. S. Florence, 'The Selection of Industries Suitable for Dispersion into Rural Areas', *Journal of the Royal Statistical Society*, A, vol. 107, pp. 93–116 (London, 1944).

TABLE II

Region	Occupied population as a percentage of total national employment	Percentage of national employment in industry X	Positive deviations
A	20	25	5
B	30	45	15
C	15	10	
D	10	5	
E	5	5	
F	20	10	
			20

Coefficient of Localization = 20/100 = 0·2

A coefficient of 0·0 represent complete coincidence of the selected industry with all occupations and 1·0 represents extreme differentiation. A recent investigation of the distribution of manufacturing industry in New Zealand[1] illustrates the application of Florence's coefficient and of other indices.

The coefficient of association between two industries is obtained by subtracting the positive deviations of industry A over industry B, divided by 100, from one another. A coefficient of 1·0 implies absolute coincidence of the industries in each region and 0·0 extreme differentiation.[2]

These coefficients may be used with a wide variety of percentage figures; for example, factory size and population groups according to age or colour group.[3] There is, however, some doubt as to the validity of the technique. H. H. McCarty et al.[4] carry out detailed evaluations of the relative merits of the simple correlation coefficient, r, and the coefficient of association, and show that, although the latter appears to

[1] G. J. R. Linge, 'Some Measures of the Distribution of Manufacturing applied to New Zealand', *New Zealand Geographer*, vol. 17, pp. 195–208 (Auckland, 1961).

[2] Naturally, the actual degree of 'coincidence' will depend upon the scale of region which is selected. H. H. McCarty et al. (op. cit) point out that for counties in the United States, assuming square areas, 'identical locations' implies within 16 miles of each other, on the average.

[3] J. W. Alexander, 'Location of Manufacturing: Methods of Measurement', *Annals of the Association of American Geographers*, vol. 48, pp. 20–6 (Lawrence, Kansas, 1958).

[4] H. H. McCarty et al. (op. cit.) (1956), pp. 30–44.

operate satisfactorily in practice, its theoretical basis is not very sound and that it gives poor results with hypothetical data. A more fundamental criticism of the use of employed labour (labour 'input') as a criterion of the concentration of manufacturing is that it overlooks the different regional levels of labour productivity (manufacturing 'output'). This concept of 'input-output', developed by American economists, is now being applied to regional production in several countries.[1]

Further Correlation Methods

The correlation techniques discussed above can be extended to more complex problems. For example, it may be essential to consider the mutual variation of three variables. The simple correlation coefficient is inadequate for this purpose and *partial correlation* coefficients must be calculated. The basic formula for the partial correlation coefficient $r_{xy} \cdot z$, between the two variables x and y with the effect of z, the third variable removed is

$$r_{xy} \cdot z = \frac{r_{xy} - (r_{xz} \times r_{yz})}{\sqrt{\{(1 - r_{xz}^2)(1 - r_{yz}^2)\}}}$$

It involves, therefore, the caluculation of the three simple correlation coefficients between the variables taken in pairs (r_{xy}, r_{xz} and r_{yz}). The method can be developed to handle more than three variables.[2]

The regression equations for the simple case may also be developed to treat partial regression with several variables. This involves fitting a linear surface to measurements of a given dependent variable. A typical problem for which these methods are invaluable is the relationship between crop yield and weather, where it is known that rainfall, evapotranspiration, sunshine, temperature and humidity are contributory and inter-related weather factors. Regression techniques also find application in population studies; Thomas, for example, assesses the effect of nine independent variables on population growth in the

---

[1] See, for example, F. A. Leeming, 'Problems in the Evaluation of Local Outputs in the United Kingdom', *Transactions and Papers, 1962, Institute of British Geographers*, no. 30, pp. 45–58 (London, 1962); and J. H. Cumberland, 'Inter-regional and Regional Input-Output Techniques, in W. Isard, *Methods of Regional Analysis: an Introduction to Regional Science* (New York, 1960).

[2] An example of the application of simple and more advanced correlation methods is given by A. H. Robinson, J. B. Lindberg and L. W. Brinkman, 'A Correlation and Regression Analysis applied to Rural Farm Population Densities in the Great Plains', *Annals of the Association of American Geographers*, vol. 42, pp. 211–21 (Lawrence, Kansas. 1961).

Chicago area.[1] Modern computer programmes for multiple regression commonly use a 'stepwise' procedure which incorporates the independent variables one by one and computes a new regression equation at each step.

The coefficient of correlation between values predicted by the partial regression equations and the observed values is termed the *multiple correlation coefficient*. It indicates the maximum correlation which can be obtained between one variable (the observed values) and the combination of several independent factors. The square of this coefficient shows what proportion of the total variance is 'explained' by the characteristics selected for investigation and thereby provides a check on the validity of the hypothesis upon which the selection of those characteristics is based.[2]

Often it is useful to examine the residuals from the regression. In spatial studies a map of these residuals may suggest further variables to be included in the regression.[3]

Trend surface analysis is a special form of multiple regression. Any isopleth map may be described by a best-fitting polynomial equation of the form

$$y = a + bx + cx^2 + dx^3 + \cdots$$

The surface may be linear, quadratic, cubic, quartic or higher order but, in general, the aim is for maximum simplicity combined with a reasonably high explained variance. This approach is being widely explored in geomorphology and climatology although there are certain problems of methodology and interpretation.[4]

[1] E. N. Thomas, 'Areal Associations between Population Growth and Selected Factors in the Chicago Urbanized Area', *Economic Geography*, vol. 36, pp. 158–70 (Worcester, Mass., 1960). See also L. J. King, 'A Multivariate Analysis of the Spacing of Urban Settlements in the United States', *Annals of the Association of American Geographers*, vol. 42, pp. 222–33 (Lawrence, Kansas, 1961).

[2] The square of the correlation coefficient (r^2) is called the *coefficient of determination* and the principle applies for the simple and multiple correlation coefficients.

[3] E. N. Thomas, 'Maps of Residuals from Regression: Their Characteristics and Uses in Geographic Research', *Department of Geography, State University of Iowa, Report 2*, 60 pp. (Ames, Iowa 1960); S. Gregory, 'The Orographic Component in Rainfall Distribution Patterns', in J. A. Sporck (ed.), *Mélanges de Géographie Offerts à M. Omer Tulipe*, vol. 1, pp. 234–254 (Gembloux, Belgium, 1969).

[4] R. J. Chorley and P. Haggett, 'Trend-Surface Mapping in Geographical Research', *Institute of British Geographers, Transactions* , no. 37, pp. 47–67 (London, 1965); G. B. Norcliffe, 'On the Use and Limitations of Trend Surface Models', *Canadian Geographer*, vol. 13, pp. 338–348 (Toronto, 1969). G. Robinson, 'Some Comments on Trend-Surface Analysis,' *Area*, vol. 2(3), pp. 31–36 (London, 1970); K. Bassett and R. J. Chorley, 'An Experiment in Terrain Filtering', *Area*, vol. 3(2), pp. 78–91 (London, 1971).

A rather different procedure involves the mathematical combination of orthogonal polynomial equations to provide a description of an isopleth map.

The method may be used to state what proportion of a distribution pattern is accounted for by various elementary surfaces or to assess the correspondence of two distributions. The former usage is exemplified by investigations of pressure maps,[1] and the latter by a study which compares the distribution of population density and of precipitation in Nebraska.[2] Such comparisons necessitate initial computations to make the vertical scale of the topographies comparable.

If the isopleths are regarded as topographic contours, it can be visualized that a number of simple surface configurations may be combined to produce the 'relief'. The individual surfaces consist of rectilinear, convex and concave slopes, and symmetrical and asymmetrical combinations of these, each of which can be specified mathematically.[3] The orthogonality of the equations implies that the surfaces are uncorrelated. The combination of the surfaces specified by these equations allows a complex equation describing the actual 'topography' to be built up until the description is sufficiently precise.[4] Computer programmes for this purpose are available.

Finally, we must briefly outline *principal components analysis*, to which reference has already been made (p. 465, p. 469).

This is an extension of the principles of multiple correlation and regression. A covariance or correlation matrix is first set up between the measurements on each pair of objects (Q mode) or each pair of attributes (R mode).

The basic aim of this technique is to express the observed variance in the data set by the minimum possible number of new functions (components or eigenvectors) and to concentrate as much of the variance of

[1] F. K. Hare, 'The Dynamic Aspects of Climatology', *Geografiska Annaler*, vol. 39, pp. 87–104 (Stockholm, 1957).

[2] A. H. Robinson and R. A. Bryson, 'A Method for Describing Quantitatively the Correspondence of Certain Geographical Distributions', *Annals of the Association of American Geographers*, vol. 47, pp. 379–91 (Lawrence, Kansas, 1957). This paper provides a not too technical account of the necessary computations for the technique.

[3] Schematic surfaces are illustrated by T. F. Malone, 'Applications of Synoptic Climatology to Weather Prediction', p. 242, Chapter 28, of S. Petterssen, *Weather Analysis and Forecasting*, vol. 2 (New York, 1956).

[4] Orthogonal polynomial equations provide a transformation of the constants and powers of x to allow the successive addition of further constants without necessitating the recalculation of the previous 'constants' at each addition. See F. E. Croxton and D. J. Cowden, *Applied General Statistics*, pp. 433–5 (New York, 1939).

the correlation matrix as possible in the first component.[1] The components are mutually uncorrelated (or orthogonal). The reduction in the number of variables is feasible because in general there is considerable correlation between the variables. The new components have to be interpreted by the investigator and it is important to realize that the successful application of the method depends how well this interpretation is performed. The relationship between each of the new components and the original variables is expressed by a weighting coefficient ('factor loading') and these coefficients provide the basis for the interpretation of the components. Mapping the spatial distribution of the components may also be helpful. In a climatological study of the United States, Steiner[2] reduced 16 climatic parameters to four principal components which accounted for 88% of the original variance. They were interpreted as representing measures of 'humidity,' 'atmospheric turbidity,' 'continentality' and 'thermality.' Principal components analysis has been widely applied to such problems as regional delimitation,[3] the classification of towns[4] and of agricultural systems[5]. However, in order to derive satisfactory results it is important to have considerable prior understanding of the problem under investigation and to select, and if necessary to scale, the variables with care.

Details of the procedures may be found in the sources footnoted but mention must be made of some points which sometimes apparently cause confusion. First, the components, which can be regarded as axes in multi-dimensional space, are sometimes 'rotated' in order to maximize the fit to the data. A frequent computer solution is known as 'Varimax'. Second, factor analysis is akin to principal components

[1] A graphical illustration of the principles is given by P. R. Gould, 'The Geographical Interpretation of Eigenvalues', *Institute of British Geographers. Transactions*, no. 42, pp. 53–86 (London, 1967). Another useful discussion is R. B. Cattell, 'Factor Analysis: an Introduction to Essentials, Parts I and II', *Biometrics*, vol. 21, pp. 190–215, 405–35, (London, 1965).

[2] D. Steiner, 'A Multivariate Statistical Approach to Climatic Regionalization', *Tijdschrift van het Koninklij Nederlandsch Aardrijkskundig Genootschap*, vol. 82, pp. 329–347 (Amsterdam, 1965).

[3] M. J. Hagood, N. Danilevsky and C. O. Beum, 'An Examination of the Use of Factor Analysis in the Problem of Sub-regional Delineation', *Rural Sociology*, vol. 6, pp. 216–33, (Raleigh, N. C., 1941); B. J. L. Berry, 'A Method for Deriving Multifactor Uniform Regions,' *Przeglad Geograficzny*, vol. 33, pp. 263–82 (Warsaw, 1961); R. J. Johnston, 'Grouping and Regionalization: some Methodological and Technical Observations', *Economic Geography*, vol. 46, pp. 293–305 (Worcester, Mass., 1970).

[4] C. A. Moser and W. Scott, *British Towns: a Statistical Study of their Social and Economic Differences*, (London, 1961).

[5] J. D. Henshall and L. J. King, 'Some Structural Characteristics of Peasant Agriculture in Barbados', *Economic Geography*, vol. 42, pp. 74–84 (Worcester, Mass., 1966).

although it differs in technical procedure and in fundamental concept. The technical difference involves the estimation of the correlation of a variable with itself in the principal diagonal of the correlation matrix. In principal components the value is assumed to be unity implying that the variables included in the analysis perfectly account for the variance of the underlying factors. In factor analysis an estimate, referred to as a *communality*, is substituted where random and other disturbances unique to the data sample are eliminated and allowance is made for unrepresented variables. Conceptually, principal components involves an internal structuring and simplification of the data set as an aid to developing hypotheses. In factor analysis a basic model is assumed and the data are examined in the light of this.

Classification and Clustering

The organization of observational data into groups is fundamental to much of geographical research although the conceptual aspects need not concern us here.[1] Groups may be formed by division (classification) of the data into subsets or by agglomeration (clustering) of individuals into groups. The procedures involve the identification of appropriate characteristics (or attributes) describing the individuals to be classified and the selection of a suitable measure of similarity or distance between the individuals and groups. One distance measure which has already been referred to (p. 489) is based on Euclidean distance but others may be encountered in the literature.[2] Similarity measures include coefficients of association (such as χ^2) and correlation.

Modern classification procedures are based primarily on techniques related to principal component analysis and there have been extensive developments along these lines in ecology[3] and in urban and economic geography.[4] There are now many computer programmes available to undertake such analysis.

Clustering procedures are still in an experimental phase. Simple

[1] See D. Harvey, *Explanation in Geography*, pp. 326–338 (London, 1969); D. B. Grigg, 'The Logic of Regional Systems', *Annals of the Association of American Geographers*, vol. 55, pp. 465–491 (Lawrence, Kansas, 1965).

[2] The literature of numerical taxonomy deals extensively with this topic: R. R. Sokal and P. H. A. Sneath, *Principles of Numerical Taxonomy*, 359 pp. (San Francisco, 1963); N. A. Spence and P. J. Taylor, 'Quantitative Methods in Regional Taxonomy', *Progress in Geography*, vol. 2, pp. 1–64 (London, 1970).

[3] See P. Greig-Smith, *Quantitative Plant Ecology*, 256 pp. (London, 1964).

[4] B. J. L. Berry, 'Grouping and Regionalization' in W. L. Garrison and D. F. Marble (editors), *Quantitative Geography*, Part I, pp. 219–251 (Evanston, 1967). See also references on p. 514 above.

linkage methods based on correlation coefficients between pairs of individuals, such as nearest neighbour analysis, can be applied without computational aids,[1] but the more sophisticated procedures based on the relationships between all members of a group or the relationship of the individuals to the group mean, or some other reference item, require computer programmes.[2] It is important that in using such techniques careful thought be given to the one most suitable for the problem under consideration. There is a risk, otherwise, that the selection of the procedure used will be determined by the availability of particular programmes.

It will be appreciated that only an outline of the vast field of statistical methods can be given here. These methods are finding wide application in geographical studies, although it must be emphasized that the correct use of the methods calls for an understanding of the principles underlying them. In particular, refined methods of analysis will not produce precise results if the initial data are themselves crude and inaccurate; in other words, one must not use 'high power methods on lower power data'. Statistical techniques should be used primarily in the support of reasoned arguments, but when they are handled with care the techniques provide the geographer with an invaluable tool.

Note: Symbols used in the Appendix

c cell interval (interval between class boundaries)
d deviation or departure
E expected frequency
F the variance ratio (Snedecor)
f frequency
n number of observations
O observed frequency
Q_1 upper quartile
Q_2 lower quartile
r correlation coefficient
$r_{xy \cdot z}$ partial correlation coefficient between x and y with the effect of z removed
s standard deviation of a sample
t a frequency distribution used in the significance test of 'Student'

[1] A review of linkage methods is provided by R. J. Johnston, 'Choice in Classification: the Subjectivity of Objective Methods', *Annals of the Association of American Geographers*, vol. 58, pp. 575–589 (Lawrence, Kansas, 1968).

[2] See G. N. Lance and W. T. Williams, 'A General Theory of Classificatory Sorting Strategies', *Computer Journal*, vol. 10, pp. 373–380 and pp. 271–277 (London, 1967, 1968).

x any variable

x_n the nth value of x

$\bar{x}$ the arithmetic mean of a series of x's

x_0 the assumed mean

Σ the summation sign

$\sum_{i=1}^{n} x_i$ the sum of the values of the variable x for $i = 1, 2 \ldots n$

σ standard deviation of the population

$\sigma_{\bar{x}}$ standard error of a sample mean.

$\hat{\sigma}$ best estimate of the standard deviation of the population

χ^2 Chi-square a frequency distribution used to test association (K. Pearson)

Index

Cumulative graphs, climatic, 233; population, 391–2
Cumulative residuals, 493 (fig.)
Curve fitting, 387–8
Curve-parallels, 232 (fig.)
Curve-plotter, 81
Curves, smoothed, 231 (fig.), 233
Cycle diagrams, 37 (fig.), 38

DASYMETRIC TECHNIQUE, 337–9
Data, climatic, 191–202; economic, 259–67; population, 312–31; settlements, 396–401
Date isopleths, 206–7 (and fig.), 278
Datum, O.S., 88
Day-degrees, 212–14 (and fig.)
Daylight, daily variation of, 220 (fig.), 221
'Dead-ground', 165–7 (and fig.)
Death-rate, crude, 323, 379 (fig.)
Degrees of freedom, 498, 505
Demangeon's coefficient of dispersion, 429
Demographic coefficients, 321
Dendrograms, 456–7
Density, of population, 321–2, 333–9 (and fig.); comparative, 385–6; of housing, 432 (and fig.)
Depressions, tracks of, 254–6 (and fig.)
Derived elements of climate, 191
Design of maps, 13
Desk calculators, 476–7
De Smet's curve, 156–7
Determination of slope, 129–58 (and figs.)
Deviation, mean, 486; standard, 210, 391, 486–8
Deviational graphs of population, 388, 389 (fig.), 390
Diagonal-scales, 20–1 (and fig.)
Dines Tube Anemometer, 197
Dip of strata, 189
Direct contact negative, 70
Directional diagrams, 376
Direction of wind, 197–8
Directories, 399
Discs, proportional, see Circles, proportional
Dispersion diagrams, 37–8; rainfall, 243–6 (fig.)
Dispersion of settlements, 429–31
Distance-isopleths, 280–1
Divided circles, see Circles, divided
Divided rectangles, see Rectangles, divided

Divided slope histograms, 143
Divided strips, see Strips, divided
d-Mac Pencil Follower, 82
Documentary sources, for settlements, 397–8
Dots, 25, 27–9; agricultural, 288–91 (and fig.); population, 334 (fig.), 335 (fig.); of slope analysis, 137 (fig.), 140–1
Dot planimeter, 75–7 (and figs.)
Drawing instruments, 1–7
Drawing-table, 7
Drycolour, 9
Dry-transfer lettering, 62
Duplication, map, 68–71
Duration isopleths, 206
Duration of rainfall, 196
Durational graphs, 225
Dyeline process, 69–70
Dynamic maps, 307–11 (and figs.)
Dynamic scale models, 84
Dyrite, 11

ECOLOGICAL MODELS, 453–5
Econographs, 306–7
Economic data, 259–67
Economic regions, delimitation of, 281–2
Edge-lines (of relief), 93
Education, data, 316
Effectiveness of rainfall, 196
Eigenvalues, 468
Eigenvectors, 514
Electrofax map-copier, 71
Electronic computers, 81–2, 150, 283, 478
Elements, climatic, 191–9
Elements of settlement pattern, 410–15
Elliot's method of slope analysis, 147
Enclosure maps, 397
Engraving, map, 71–2
Enlargement of maps, 64–8
Enumeration units, areas of, 314
Equicorrelatives, 211
Equipluves, 209
Equivariables, 209–11
Ergographs, 296, 298 (fig.), 299
Estate maps, 397
Ethnic structure, 317–18
Ethnographic distributions, 330 (fig.), 331–2, 371 (and figs.)
Evaporation, 196
Evapo-transpiration, 196, 215–16
Expectation of life, by isopleths, 358–9
Experimental models, 84